# Essentials of Physics

HERMAN GEWIRTZ

Chairman, Physical Science Department
Bronx High School of Science, New York City

BARRON'S EDUCATIONAL SERIES, INC.
WOODBURY, NEW YORK

All inquiries should be addressed to:
Barron's Educational Series, Inc.
113 Crossways Park Drive
Woodbury, New York 11797

*Library of Congress Catalog Card No. 67-30941*
International Standard Book No. 0-8120-0278-4

PRINTED IN THE UNITED STATES OF AMERICA

6 7 8 9 10 11 M 3 2 1 0 9 8

# CONTENTS

# INTRODUCTION

This book is intended primarily to help the students to master the essential and basic concepts, principles, laws, and facts of physics which a modern first course in the subject stresses. The basic guide has been the most recent New York State syllabus which was developed in the light of PSSC modernization. However, the design of the book is broader than this. Other material is included which should help students of other physics courses, including those which use some of the PSSC contents and approach. In addition, stress has been placed on historical development, on the process of growth of physical theories, and on the tentative nature of physical laws and theories.

Material not required for the New York State syllabus is clearly marked with a double asterisk (**). Items marked with a single asterisk (*) are included as optional material in the New York syllabus. However, each student is expected to know the optional material from two of the following four groups:

1) Chapters 1, 4–7  2) Chapters 8–11  3) Chapters 12–15  4) Chapters 17, 18.

Many illustrative EXAMPLES are used in the text to clarify quantitative relationships. In addition, each chapter is provided with a wide variety of Questions and Problems which may be used for daily home study. Also, each chapter has a set of Test Questions consisting of objective questions of different types to help the student prepare for tests and final examinations. Answers are given for these questions. The teacher may also find it desirable to use some or all of these Test Questions for homework assignments or chapter summary. Furthermore, Topic Questions have been provided for each chapter to focus attention on some of the important facts and ideas. In each chapter these questions are numbered to refer to the topic with the corresponding numbers. For example, questions 1.1 and 1.2 refer to topic 1; questions 2.1 and 2.2 refer to topic 2.

As a final review, the book also includes three of the most recent New York State Regents Examinations in Physics. These are the actual exams given to New York State residents at the end of a basic course of study in physics. Whether a resident of the state or not, the student will find that these year-end examinations provide an overview of the subject and a general test of knowledge.

This volume is dedicated to my wife, Matilda, and to Paul, Jonathan, and George.

# 1 Measurement and Units

## 1 The Nature of Physics

Underlying the physicist's work, like that of other scientists, is the belief that nature is not whimsical. For example, we observe day and night following each other over and over again. We expect this to continue. We observe the change of seasons over and over again, and no matter how bitter the winter we expect it to be followed by spring and summer.

The physicist is concerned with finding such regularities in nature, and when he makes a generalization about how nature behaves, we say we have a *law* of nature. Frequently, in order to be able to make such generalizations, the physicist has to make careful measurements. All measurements, as we shall see, have a certain unavoidable inaccuracy. It should therefore be no surprise that sometimes the laws of physics, generalizations based on such measurements, have to be revised or, occasionally, discarded. For example, the apparent motion of the sum and its planets suggested a circular motion of the earth and other planets around the sun. More careful measurements indicated that the path of each planet is an ellipse.

Frequently, questions come up which can not be answered satisfactorily on the basis of existing laws or theories. To find answers for these questions, to explore new hypotheses, the physicist often designs experiments which will permit measurements. These measurements may help to formulate a new law or to express an old law in more precise form. Sometimes experiments lead to unexpected results, but as Pasteur said, "Chance favors the mind that is prepared."

2 **Why do we measure things?** Careful measurements are important to the businessman as well as to the scientist. In business, the customer wants to get at least what he pays for. An honest salesman wants the customer to get the correct quantity of merchandise, and, to be on the safe side, may even throw in a little bit extra, but the more he gives away without charge, the smaller his profit. It pays him to have good equipment for measuring his merchandise carefully.

The physicist wants to explain the phenomena he observes in nature. Before he can even attempt an explanation, he must be sure of his observations. Careful observation usually requires careful measurement. As Lord Kelvin (1824–1907) said: "When you can measure what you are speaking about and express it in numbers, you know something about it; and when you cannot measure it, when you cannot express it in numbers, your knowledge is of a meager and unsatisfactory kind; it may be the beginning of knowledge, but you have scarcely in your thoughts advanced to the stage of a science."

**3 How do we measure things?** It may not be obvious, but when we measure something we are comparing it with a standard. When you measure the height of a room, you may use a 12-inch ruler and see how many times the ruler fits from floor to ceiling. In that case the ruler provides you with a standard of length. By international agreement we use something other than the length of your ruler as the official standard for length: the meter.

For a long time English-speaking countries used the yard as the official standard of length. Now we also use the meter as the official standard, although in many non-scientific situations we still speak about the yard, the foot, the inch, etc. Note that these are all *units* of length. However, the meter is now the *standard* unit on which the others are based. Until recently the meter was defined by scientists as the distance between two scratches on a special bar kept under controlled conditions in France. Now the meter is defined in terms of a special property of light. We shall give this definition when we discuss light.

1 meter = 39.37 inches

**4 Systems of Measurement** In the next few chapters we shall be concerned with quantities or characteristics that can be expressed in terms of length, mass, and time. We have noted above that many different units exist for reporting the result of the same measurement: 2 yards, 6 feet, 72 inches, etc. There are three systems of units commonly used, one English system and two based on the metric system: the *MKS* system and the *CGS* system. In the *CGS* system the basic units of length, mass, and time are, respectively, the centimeter (cm), the gram (gm), and second (sec). In the *MKS* system the corresponding units are meter (m), kilogram (kg), the second. The Kelvin degree is often added as a basic unit for temperature measurement. In the *English system* the three basic units are the foot (ft), pound (lb), and second.

**5 Powers of Ten** In writing or using large or small numbers it is often convenient to use powers of ten:

$10^0 = 1$ | $5 \times 10^4 = 50{,}000$
$10^1 = 10$ | $10^{-1} = 0.1$
$10^2 = 100$ | $10^{-2} = 0.01$
$10^3 = 1000$ | $10^{-3} = 0.001$
$10^6 = 1{,}000{,}000$ | $5 \times 10^{-4} = 0.0005$

**6 Prefixes** Another method frequently used to express large or small numbers is to use a prefix with the appropriate unit:

Tera-(T)-$10^{12}$ | centi-(c)-$10^{-2}$
Giga-(G)-$10^9$ | milli-(m)-$10^{-3}$
Mega-(M)-$10^6$ | micro-($\mu$)-$10^{-6}$
kilo-(K)-$10^3$ | nano-(n)-$10^{-9}$
 | pico-(p)-$10^{-12}$

Here are some examples of the use of the more common prefixes:

1 kilometer (Km) = $10^3$ meters = 1000 meters

1 meter = 100 centimeters = 1000 millimeters

1 centimeter (cm) = $10^{-2}$ meter = 0.01 meter
1 millimeter (mm) = $10^{-3}$ meter = 0.001 meter
1 Megaton = $10^6$ tons = 1,000,000 tons

7 **Units of Length** To measure length we measure the distance between two points. We may use instruments such as a ruler, yardstick, meterstick, and tape measure. Some units which we may use to express length are the meter, yard, foot, inch, kilometer, mile, and centimeter.

1 meter = 39.37 inches
1 inch = 2.54 cm
1 Km = 0.62 mile

8 **Units of Area** Area is a measure of the amount of surface. The area of a rectangle is equal to the product of its length and its width. Some units of area are the square centimeter ($cm^2$), square meter ($m^2$), square foot ($ft^2$), and square inch ($in^2$).

9 **Units of Volume** The volume of an object is a measure of the amount of space included by the surfaces of the object. The volume of a rectangular solid is equal to the product of its length, width, and height. Some units of volume are the cubic centimeter ($cm^3$), cubic meter ($m^3$), cubic foot ($ft^3$), cubic inch ($in^3$), quart and liter. The volume occupied by one kilogram of water at a certain temperature (4° Celsius) is one liter.

1 liter = 1000 $cm^3$
1 milliliter = 1 $cm^3$
1 quart $\doteq$ 946 $cm^3$
1 cubic meter = $10^6$ $cm^3$

**Note:** The equal sign with a dot over it ($\doteq$) stands for: is approximately equal to. Another symbol used is $\approx$.

The volume of a liquid can be measured by using a graduated cylinder. The volume of non-porous, insoluble solids can be measured by submerging them in the liquid in a graduated cylinder. The liquid rises to a higher level. The change in level gives the volume of the submerged solid.

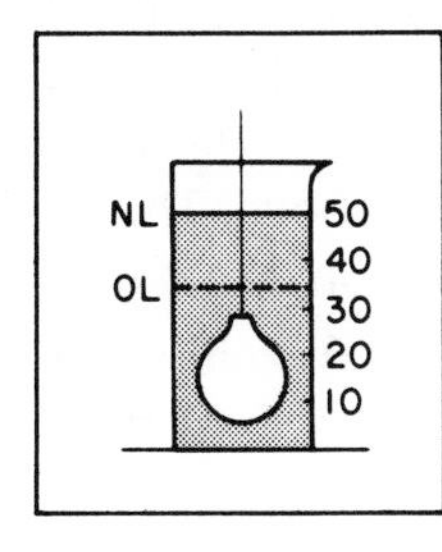


**Measuring the volume of a liquid using graduated cylinder. OL indicates original level, NL new level. Therefore volume is 15cm³.**

10 **Units of Mass** The concept of mass is not simple. The mass of a substance is a measure of the amount of matter in the substance. It is also a measure of the inertia of the substance. The standard of mass is the mass of a block of metal kept in France. It is the standard kilogram.

1 kilogram (Kg) = 1000 grams (gm)
the weight of 1 Kg $\doteq$ 2.2 pounds (lb)
the weight of 454 gm $\doteq$ 1 lb

You are familiar with the use of the *pound* as a unit of money in England, and also as a unit of weight. In addition, it has been used as a unit of mass. Mass and weight are related, but not the same thing. We shall discuss mass, weight, and the pound when we get to Newton's Second Law. The mass of 1 liter of water = 1 Kg. The mass of 1 $cm^3$ of water = 1 gm.

11
** **Density** The density of a substance is mass per unit volume. Density = mass/volume. For example, the density of water is 1 kg/liter, or 1 $gm/cm^3$, or 1,000 $kg/m^3$.

## 12 Significant Figures

We can count exactly but we cannot measure exactly. If we ask ten students to count the number of chairs in the room, we expect the same answer from each student. However, if we ask the same students to measure very carefully the width of one of the chairs, we can expect different answers such as 52.3 cm, 52.4 cm, and 52.5 cm. Each of these measurements has three digits or, we sometimes say, three figures. The third digit is an estimate, but it tells us that the student is sure that the width is less than 52.6 cm. The third digit, therefore, has some significance, but there is some doubt as to exactly what it ought to be. We say that this digit is doubtful. We say that the number 52.4 cm has three significant figures or three significant digits. If, as far as we can tell, the last digit should be zero, we write the zero down and thus indicate the accuracy with which we tried to measure. In 52.0 cm, the zero is significant and we have three significant figures. If we express this measurement as 0.000520 Km, we still have only three significant figures. The first four zeros are used merely to indicate the location of the decimal point, the last zero being used to show how closely we tried to measure. The number 40.70 has four significant figures. In scientific work, when we make a measurement we estimate the last digit and write it down. It is the doubtful digit, but it is included in counting the number of significant figures.

In describing the results of measurements, the numbers we use will contain only significant figures. The last digit, and only the last one, will be doubtful; but it is still significant. It may be a zero, as in 52.0 cm in the above paragraph. Notice that zeros between nonzero digits are always significant, as in 40.70 above. The zero between the 4 and the 7 is significant. Zeros preceding the first nonzero digit are not significant; in 0.000520 Km shown above, the first 4 zeros are not significant.

One advantage of the standard notation discussed below, is that it uses only significant figures in the first factor.

13 **Standard Notation** This is also called scientific notation. This notation is often useful in dealing with very large and very small numbers, and in indicating significant figures. When using the standard notation we write a number as two factors; the first factor has one digit (but not zero) in front of the decimal point, and the second factor is a power of 10.

**EXAMPLES:** $6.05 \times 10^3$ (= 6,050); $2.70 \times 10^5$ (= 270,000); $5.2 \times 10^{-4}$ (= 0.00052).

Notice two additional things. First, the exponent of 10 is a whole number; secondly, the first factor gives us the number of significant figures. In $2.70 \times 10^5$ we can tell that there are three significant figures. When we look at 270,000, we can't tell how many zeros are significant.

**14 Order of Magnitude** The order of magnitude of a number is the power of 10 closest to the number. This rounding off can usually be done readily by first writing the number in standard notation.

**EXAMPLE:** 1. $186{,}000 = 1.86 \times 10^5 \doteq 10^5$.

The order of magnitude of 186,000 is $10^5$.

Or, look at it this way. 186,000 is between 100,000 (or $10^5$) and 1,000,000 (or $10^6$). It is closer to 100,000.

Therefore, its order of magnitude is $10^5$.

**EXAMPLE:** 2. 973 is between 1000 (or $10^3$) and 100 (or $10^2$). It is closer to 1000.

Therefore, the order of magnitude of 973 is $10^3$.

The other way of looking at it is: $973 = 9.73 \times 10^2 \doteq 10 \times 10^2 = 10^3$.

**EXAMPLE:** 3. $0.0027 = 2.7 \times 10^{-3} \doteq 10^{-3}$.

The order of magnitude of 0.0027 is $10^{-3}$.

**15 More on Significant Figures** (This may be postponed until forces have been studied.) As we indicated, all measurements involve an estimate. Usually we *report a measurement in which only the last digit is in doubt*. Frequently we have to perform calculations with these numbers. We should write the result of these calculations with only one doubtful digit. Before you study and memorize the rules, note the basic principles. If you add a doubtful digit to a certain or a doubtful digit, the sum must be doubtful. For example, if you don't know how much money there is in a coin box, you still won't know its contents after you put a dime into it. A similar idea applies to subtraction.

If you multiply a doubtful digit by a certain or a doubtful digit, the product is doubtful. The corresponding idea applies to division.

Unless told otherwise by your teacher, for work involving significant figures, the final answer should contain one and only one doubtful digit, the last one. All other doubtful digits are dropped.

**16 Rules for Addition and Subtraction** Write the numbers under each other as in the usual way for addition or subtraction. Disregard all columns containing doubtful figures except the first one (starting the count from the left).

**EXAMPLE:** Add 103.2, 6.14, and 25,324.

$$
\begin{array}{r}
103.\bar{2} \\
+\ 6.1\bar{4} \\
+25.32\bar{4} \\
\hline
134.\bar{6}
\end{array}
$$

Remember that the last digit in each number is doubtful. To make it easier to recognize this we put a bar over the last digit. The tenths column is the first column which has a doubtful figure in it: the 2 in 103.2. Therefore we don't bother adding the hundredths or thousandths columns. The sum is 134.6; it has four significant figures, with the 6 a doubtful digit. The 6 is doubtful because we obtained it by adding a doubtful 2 to the 1 and 3.

17 **Rules for Multiplication and Division** When multiplying or dividing two numbers, the number of significant figures of the answer is usually the same as that of the less accurate of the two numbers.

**EXAMPLE:** Multiply 2.2341 by 3.2.
There are five significant figures in the first number and two in the second. This makes the second number, 3.2, the less accurate one. It has two significant figures. The answer will then be rounded off to two significant figures: 7.1.

18 **Using Standard Notation** When multiplying two numbers together which are written in standard form, there is no special problem: multiply the first two factors and the second two factors.

**EXAMPLE:** $(6.0 \times 10^3)(4.0 \times 10^2) = 24.0 \times 10^5$, or $2.4 \times 10^6$.

When adding or subtracting two numbers written in standard form, there is no special problem if both numbers have the same power of 10.

$$6.0 \times 10^3 - 4.0 \times 10^3 = 2.0 \times 10^3.$$

If the two numbers do not have the same power of 10, we must first write the two numbers with the same power of 10, or, if we prefer, change both to the conventional form which does not use powers of 10.

**EXAMPLE:** $6.0 \times 10^3 - 4.0 \times 10^2$.

METHOD 1. $(6.0 \times 10^3) - (0.40 \times 10^3) = 5.6 \times 10^3$.

METHOD 2. $(6000) - (400) = 5{,}600$ or $5.6 \times 10^3$.

# Questions and Problems

1. Convert to meters: *a.* 5 Km *b.* 5 cm *c.* 5 mm.
2. Convert to centimeters: *a.* 5 Km *b.* 45 m *c.* 45 mm.
3. Convert to millimeters: *a.* 37 Km *b.* 37 m *c.* 37 cm *d.* 37 mm.
4. Convert to inches: *a.* 5.2 m *b.* 5.2 cm.
5. Convert to centimeters: *a.* 37 inches *b.* 37 feet *c.* 5.5 inches.
6. Give three units of area in *a.* the English system *b.* the metric system.
7. Give three units of volume in *a.* the English system *b.* the metric system.
8. State the number of significant figures in each of the following: *a.* 3.24 *b.* 0.0273 *c.* 20.03 *d.* 30.20 *e.* 5.0 *f.* 7.020.

9. Add and express the sum to the proper number of significant figures: *a.* $3.24 + 0.0273$ *b.* $5.0 + 7.063$ *c.* $30.27 + 8.0$.

10. Subtract (and express the difference to the proper number of significant figures): *a.* $3.24 - 0.0273$ *b.* $5.0 - 7.063$ *c.* $30.27 - 8.0$.

11. Express each of the following in standard notation: *a.* 30,275 *b.* 0.00273 *c.* 0.0723 *d.* 4,789.

12. Express in standard notation to three significant figures: *a.* 4567 *b.* 0.00245 *c.* 45 *d.* 5003.

13. Express without the use of powers of 10: *a.* $4.0 \times 10^{3}$ *b.* $4 \times 10^{-4}$ *c.* $52 \times 10^{-3}$ *d.* $4.7 \times 10^{-4}$.

14. Multiply and express to the proper number of significant figures: *a.* $2.24 \times 2.54$ *b.* $4.5 \times 39.37$ *c.* $2.03 \times 2520$.

15. The dimensions of a rectangle are $7.5 \times 10^{2}$ cm by $1.5 \times 10^{3}$ cm. *a.* express the dimensions in meters *b.* calculate the area of the rectangle *c.* calculate the perimeter of the rectangle in centimeters.

16. A tank 40 cm $\times$ 20 cm is filled with water to a depth of 10 cm. Calculate *a.* the area of the bottom of the tank *b.* the volume of water in the tank *c.* the mass of water in the tank.

17. Add the pairs of numbers shown, expressing the answer to the proper number of significant figures. *a.* $6.23 \times 10^{4}$ and $2.70 \times 10^{3}$ *b.* $4.19 \times 10^{3}$ and $2.3 \times 10^{3}$ *c.* $2.52 \times 10^{-3}$ and $3.47 \times 10^{-2}$.

18. For each of the pairs of numbers shown in question 17, subtract the second from the first.

19. For each of the pairs of numbers shown in question 17, multiply the second by the first.

20. For each of the pairs of numbers shown in question 17, divide the first by the second.

21. Give the order of magnitude of each of the following: *a.* $3.13 \times 10^{4}$ *b.* $6.023 \times 10^{23}$ *c.* $1.60 \times 10^{-19}$ *d.* $2.998 \times 10^{10}$ cm.

## Test Questions

1. The amount of space an object occupies is known as its ............
2. One kilogram is equal to ........... gram(s).
3. One liter is equal to ........... cubic centimeter(s).
4. Ten inches equal ........... centimeter(s).
5. One cubic inch equals ........... cubic centimeter(s).
6. 325 centimeters = ........... meter(s).
7. 42.5 millimeters = ........... centimeter(s).
8. One square inch = ........... square centimeter(s).
9. Five feet = ........... meter(s).
10. A reported measurement is 269.12 cm. The number of significant figures in this measurement is ............
11. A reported measurement is 3.025 cm. The number of significant figures in this measurement is ............
12. The measurement which has the same number of significant figures as 50.00 cm is 1. 0.0234 cm 2. 0.2 cm 3. 0.2341 cm 4. 0.023 cm.

13. The number 32,700 may be expressed in standard notation as ............
14. Without using powers of ten, the number $4.5 \times 10^3$ may be written as ............
15. Without using powers of ten, the number $4.5 \times 10^{-3}$ may be written as ............
16. State the number of significant figures in each of the following: *a.* 4.23 *b.* 0.00819.
17. Add (and express to the proper number of significant figures): $4.23 + 10.02 + 0.00234$.
18. The sum of 3.00000000 kilometers, 60.0000 meters, and 7 millimeters is ............ cm.
19. Multiply (and express to the proper number of significant digits): *a.* $3.03 \times 3.03$ *b.* $4.202 \times 10.1$ (express in standard notation).

## Topic Questions

(In each chapter these questions are numbered to refer to the topic with the corresponding numbers. For example, questions 1.1 and 1.2 refer to topic 1; questions 2.1 and 2.2 refer to topic 2.)

1.1 What is meant by the expression *A Law of Nature*?

1.2 Why is it sometimes necessary to revise a law of physics?

2.1 Why is careful weighing important to an honest salesman?

2.2 Why is careful measurement important to a physicist?

3.1 What is the official standard of length?

3.2 How many inches are there in a meter?

4.1 What are the basic units of length, mass, and time in the MKS system of measurement?

5.1 How can the following numbers be written as powers of 10? *a.* 100 *b.* 0.01.

6.1 What does the prefix *kilo-* stand for?

6.2 How many centimeters are there in one meter?

7.1 How does the kilometer compare with the mile?

8.1 What are two units of area in the metric system?

9.1 What are three units of volume in the metric system?

9.2 For what measurement is the graduated cylinder intended?

10.1 What is one definition of mass?

10.2 Approximately how much does one kilogram weigh?

11.1 What is the density of water?

12.1 The length of a board is reported as 47.3 cm. Assume that the digit 3 is doubtful. *a.* How many significant figures does the number 47.3 have? *b.* If the 3 is doubtful, why don't we just drop it and report the length as 47 cm?

13.1 What is another name for *standard notation*?

13.2 How many significant figures are there in the number $2.70 \times 10^3$?

14.1 What is meant by the order of magnitude of a number?

15.1 In a calculation involving significant figures, how many doubtful digits should the answer usually contain?

16.1 When dealing with significant figures, what is the rule for addition and subtraction?

17.1 When dealing with significant figures, what is the rule for multiplication and division?

# 2 Forces and Vector Quantities

## 1 Introduction

Imagine the following situation. You start for a walk from in front of your house. You walk four miles the first hour and three miles the second hour. At the end of the second hour, how far are you from your house?

If you said seven miles, you made the mistake which is made by many people. Actually, with the information given, all we know is that you can't be more than seven miles away. But if you changed direction during your walk, if, for example, you started walking back towards your house any time before the end of the two hours, you would be less than seven miles away from home. In fact, you might be right back where you started from. This example shows that in some situations we must know not only *how far* or *how fast*, but also *in what direction*.

## 2 Vector and Scalar Quantities

A *vector quantity* is a quantity which can only be specified completely by giving its magnitude and direction. By magnitude we mean size; it is usually expressed with a number and a unit; for example, 5 ft and 60 miles/hour. If we say, *five feet to the north*, the *five feet* gives the magnitude, and *to the north* gives the direction. Examples of vector quantities which we shall consider soon are force, displacement, and velocity.

A *scalar quantity* is one that is fully specified by giving its magnitude. Examples are time, distance, area, volume, mass, speed, energy, and amount of money.

3 **Vectors** A vector quantity may be represented graphically by a vector. A *vector* is an arrow whose length is proportional to the magnitude of the vector quantity and whose direction is parallel to the direction of the vector quantity.

For example, suppose we want to draw a vector to represent *4 miles eastward*. If we face north, east is to our right. On paper we usually select the top edge of the paper to represent north; then the right edge of the paper represents east. The vector will be an arrow drawn with the arrowhead pointing to the right edge. How long shall we draw the arrow?

We use a convenient scale. In this case, a suitable scale is, 1/4 inch equals 1 mile. On the same diagram we must use the same scale for all the different vectors.

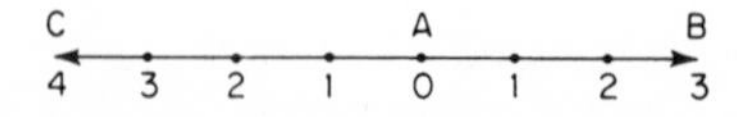

*AB* represents 3 miles eastward; *AC* represents 4 miles westward. Sometimes, to emphasize the fact that *AB* and *AC* are vectors, we write them with arrows above them, like this $\overrightarrow{AB}$ and $\overrightarrow{AC}$.

## 4 Composition of Vectors

**Composition of Vectors** Suppose we take a trip. We go 4 miles eastward and then continue another 3 miles eastward. This is represented by vectors *EF* and *FG*.

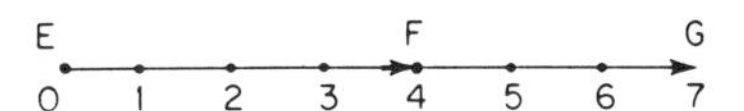


In this simple case it is obvious that at the end of this trip we are 7 miles farther eastward. In physics we call a change of position a *displacement*. Notice that displacement is more than a distance; it is a vector quantity. In addition to the 7 miles we specify a direction. We can now say more briefly, that the combined effect of displacement of 4 miles eastward and 3 miles eastward is 7 miles eastward.

Suppose a displacement of 4 miles eastward is followed by a displacement of 3 miles westward? What is the combined effect of these two displacements? This is represented by vectors $\overrightarrow{HI}$ and $\overrightarrow{IK}$. We can see that we end up at *K*, one mile

H K I

east of our starting point, *H*. The resultant displacement is one mile eastward, and is represented by $\overrightarrow{HK}$; insert the arrowhead.

What is the combined effect of a displacement of 4 miles eastward followed by a displacement of 3 miles northward? This is represented by vectors *LM* and *MN*. The distance from our starting point is now given by the length of *LN*, and our net displacement has been from *L* to *N*. Vector *LN* therefore gives the combined effect of the two displacements. To get the magnitude of *LN* we use the same scale. We measure off on *LN* the unit we selected (1/2 inch to represent a mile). We notice that the length of *LN* is 2 1/2 inches. Therefore it represents a distance of 5 miles in the direction shown.

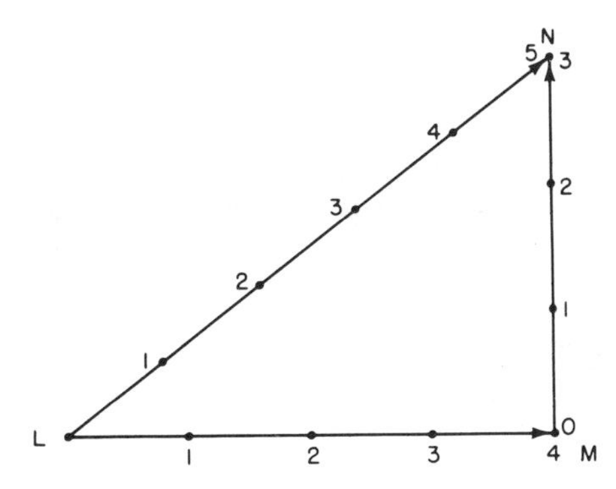


*Measuring the Magnitude of displacement*

We can also get the length represented by *LN* if we use the *Pythagorean Theorem*: In a right triangle, the square of the hypotenuse is equal to the sum of the squares of the other two sides. $(LN)^2 = (LM)^2 + (MN)^2$.

$$(LN)^2 = 4^2 + 3^2. \quad (LN)^2 = 25. \text{ Therefore, } LN = 5.$$

The combined effect of two or more vectors is often called the resultant vector or just the *resultant*. The process of finding the resultant of two or more vectors is called composition of vectors, or combination of vectors, or addition of vectors. Notice that in this kind of addition, three plus four does not have to equal seven.

1. The resultant of two vectors in the same direction is the sum of the two vectors.
2. The resultant of two vectors in opposite directions is their difference; the direction of the resultant is the direction of the larger given vector.
3. The resultant of two vectors at right angles to each other may be found graphically or by the Pythagorean Theorem as shown above. The resultant of two vectors at any angle with each other will be discussed later when dealing with composition of forces.

We can now give a more complete definition: A *vector quantity* is a quantity which has magnitude and direction and is added the way displacements are added.

# 5 Concurrent Forces

A *force* is a push or pull. It produces or prevents motion, or has a tendency to do so. It is a vector quantity. For example, hold this book in the palm of your hand. If you quickly move your hand aside, the book falls. It starts to move because the earth pulls on it. This pull of the earth is known as gravity; it is a force towards the center of the earth. The earth's pull on an object on earth is the *weight* of the object on earth. The weight of a 5-pound bag of sugar is about five pounds (5 lb). Why didn't the book fall while you held it in your hand? The earth pulled down on it, but you exerted an upward force on the book to cancel the pull of gravity; you exerted an upward force just equal to the weight of the book.

Some units of force are: newton, pound, dyne. You are familiar with the pound: if you are holding a 5-pound bag of sugar in your hand, your hand is exerting a force of 5 pounds (5 lb). If you are holding 1 kilogram in your hand, your hand is exerting a force of about 9.8 newtons. If you are holding one gram, you are exerting a force of 980 dynes. One pound = 4.5 newtons. More exact definitions of these units of force will be given with the discussion of Newton's second law of motion. A spring balance (also called spring scale) is commonly used to measure forces.

Since force is a vector quantity it can be represented by a vector. An arrow is drawn whose length is proportional to the magnitude of the force. The direction of the arrow gives the direction of the force. The arrow representing a weight is usually drawn to point towards the bottom of the page. A 5-pound weight is repre-

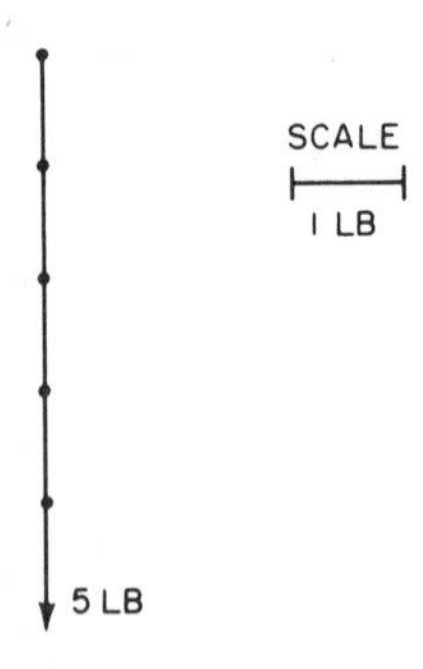


sented by the diagram. A scale is selected arbitrarily to suit the conditions and indicated in the drawings. A certain length (e.g. 1/4 inch) is used to represent one pound, and five of these lengths are laid off on a vertical line to represent five pounds. The arrowhead indicates the direction.

*Concurrent forces* are two or more forces acting simultaneously on the same point of an object. We are often concerned with getting the combined effect of these forces. The *resultant* of two or more forces is a single force which can produce the same effect as these forces (and can therefore be used to replace them). For example, if a 4-newton force and a 3-newton force act in the same direction, their resultant is a 7-newton force in the same direction; if they act in opposite directions (at an angle of 180°), their resultant is a 1-newton force in the direction of the 4-newton force.

## 6 Parallelogram of Forces

When the angle between two forces is other than 0° or 180°, we use a method similar to that described above under Composition of Vectors. Let us consider first the *graphical method*, or scale drawing method.

**EXAMPLE:** Two forces, one of 3.0 newtons and the other of 4.0 newtons, act on an object. The angle between the forces is 60°. What is their resultant?

**SOLUTION:** The series of diagrams below indicates the method. First draw two lines to represent the direction of the forces, in this case 60°.

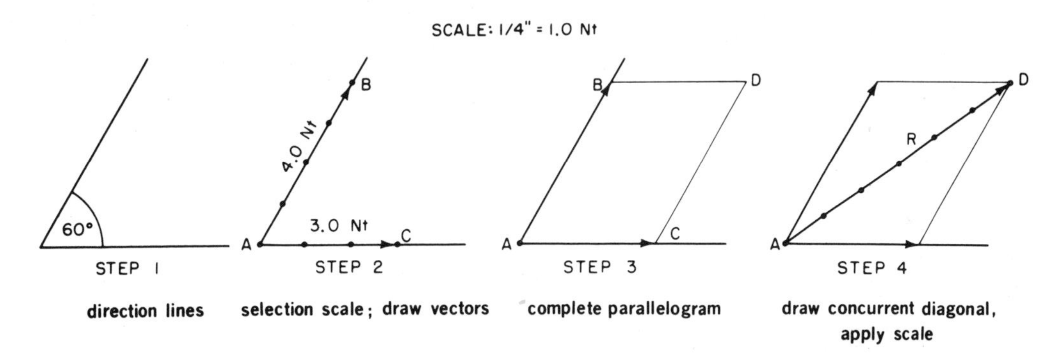


Then select a suitable scale to represent the known forces as vectors; in this case, the scale selected is 1/4 inch = 1.0 newton. Notice that in the case of concurrent forces we draw the vectors from the same starting point, *A*.

With the two known vectors *AB* and *AC* as sides, complete the parallelogram *ABDC*.

In Step 4, draw the concurrent diagonal: the diagonal which starts from the common point *A*. This diagonal *AD* represents the resultant of the two given forces. To get the magnitude of this resultant we apply the scale: on *AD* lay off the unit, in this case 1/4 inch. We get just a little more than six of these units. Since each unit represents 1.0 newton, the resultant is a little more than 6.0 newtons. We estimate 6.1 newtons. It makes an angle of about 36° with the 3-newton force.

### 6a Two Forces at Right Angles

**Two Forces at Right Angles** Let us consider another common *example*: The two forces are perpendicular to each other.

**SOLUTION:** Even if we end up using an algebraic solution, it is a good idea to make a careful sketch using the four steps outlined above. This is indicated in the diagram at the right. The resultant is *EH*. The scale length fits into this diagonal five times. One scale length represents 1.0 newton; therefore the resultant is approx. 5 newtons. We notice that *EGH* is a right triangle. The two arms represent 3.0 and 4.0 newtons, respectively. The hypotenuse represents the resultant. According to the *Pythagorean Theorem*, the hypotenuse squared is equal to the sum of the squares of the two arms:

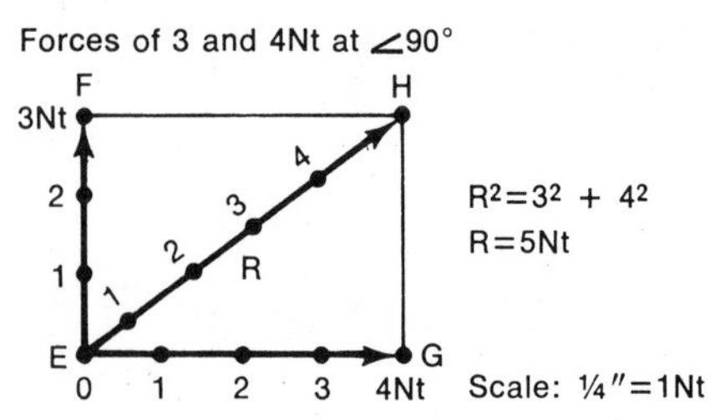


$$R^2 = 3^2 + 4^2 = 9 + 16.$$

Therefore,

$$R = 5.0 \text{ newtons.}$$

You should also recognize the 3-4-5 right triangle: If two arms of a right triangle are 3 and 4, respectively, the hypotenuse is 5. But remember this is for a right triangle.

**6b Triangle of Forces** You may have noticed that when we drew *forces*, all the vectors started at the same point. When we wanted the resultant of two *displacements*, the second displacement started at the end of the first vector (see paragraph 4); the resultant of the two was the vector drawn from the beginning of the first (point *L* above) to the end of the second (point *N*). Actually we could use the same method with forces. Remember that opposite sides of a parallelogram are equal. In the above parallelogram, *ABDC*, *CD* = *AB*. Instead of drawing vector *AB*, we could first have drawn *CD*. However, it will probably be less confusing if you use the parallelogram method outlined above for concurrent forces: draw all force vectors from the same point, including the resultant.

If you use the triangle method, be careful about *the angle between the two forces*. It is the angle between the two forces when they are drawn from the same point. For the above two examples the diagrams would be:

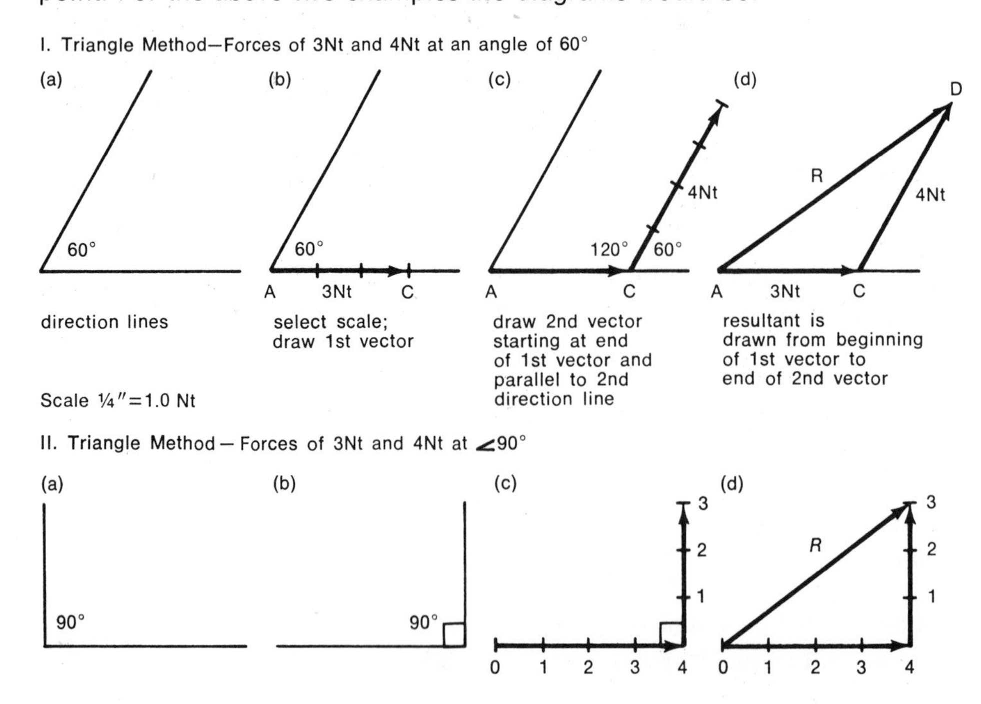

**Three Concurrent Forces** To find the resultant of three concurrent forces we apply a similar procedure: first replace any two forces by their resultant, and then combine this resultant with the third force to get the final resultant.

**7 Subtraction of Vectors** We have already mentioned that combining vectors is also referred to as adding vectors or getting the sum of vectors. Occasionally we need to get the difference between vectors, that is, to subtract one vector from another. This is of only minor importance in an elementary course; it may be used when studying circular motion.

Subtraction of vectors is easy if you remember that, when drawing a graph, $+x$ is in the opposite direction from $-x$. This also applies to vectors. If you have $+\overrightarrow{AB}$, then you know that $-\overrightarrow{AB}$ is a vector of the same magnitude in the opposite direction. Once you have drawn $-\overrightarrow{AB}$, you combine it with a second vector by following the rules for addition of vectors. The diagram on the left shows how we get the

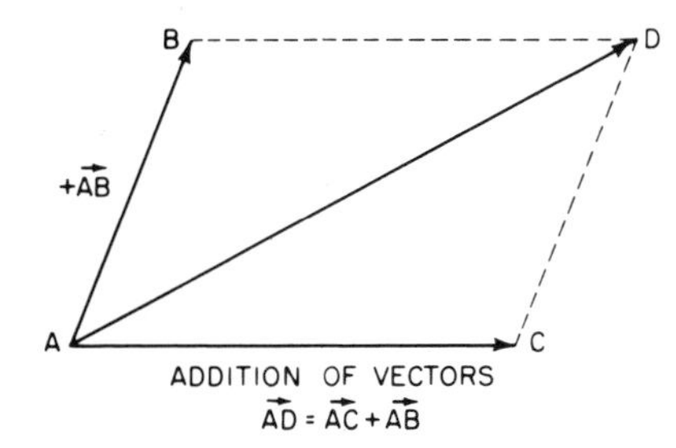

ADDITION OF VECTORS
$\overrightarrow{AD} = \overrightarrow{AC} + \overrightarrow{AB}$

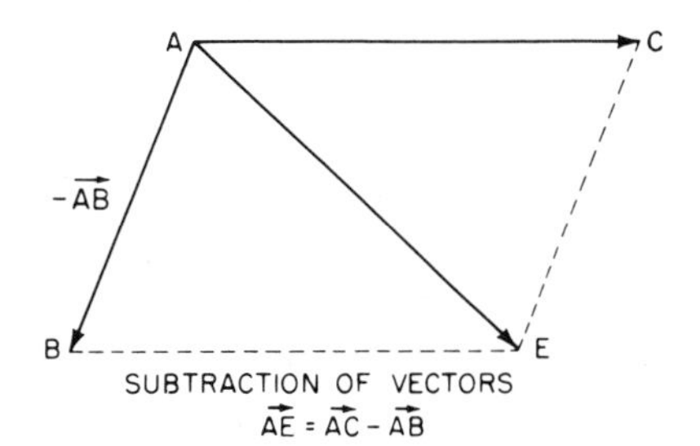

SUBTRACTION OF VECTORS
$\overrightarrow{AE} = \overrightarrow{AC} - \overrightarrow{AB}$

resultant of vectors $AC$ and $AB$. The diagonal is the resultant, and therefore, $\overrightarrow{AD} = \overrightarrow{AC} + \overrightarrow{AB}$; remember, we are speaking about *vector* addition. The diagram on the right shows how we get the resultant of $\overrightarrow{AC} - \overrightarrow{AB}$. First we draw $-\overrightarrow{AB}$ in the direction opposite to that of $\overrightarrow{AB}$, and of the same length. The diagonal of the second parallelogram is the (vector) sum of $\overrightarrow{AC}$ and $-\overrightarrow{AB}$, or the (vector) difference between $\overrightarrow{AC}$ and $\overrightarrow{AB}$.

**8 Equilibrium** A force applied to an object tends to move the object. In order to prevent this motion we apply another force, equal and opposite to the first one. We say this force produces equilibrium and call it the *equilibrant*.

The equilibrant force is equal and opposite to the resultant force. In the example (¶ 6a) we found that if a 3-newton force acts at right angles to a 4-newton force,

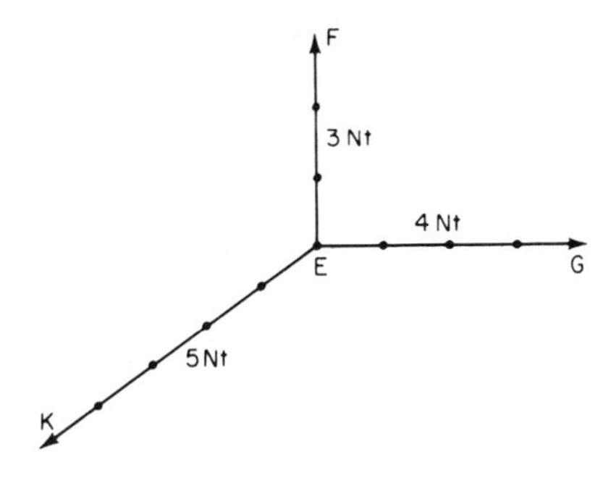


the resultant is a 5-newton force (given by $EH$ in the diagram of ¶ 6a). The equilibrant of the 3- and 4-newton forces, therefore, is a 5-newton force in the direction opposite to $\overrightarrow{EH}$. This is shown in the diagram; $\overrightarrow{EK}$ is the equilibrant of $\overrightarrow{EF}$ and $\overrightarrow{EG}$. The combined effect, or vector sum, or resultant, of these three forces is zero.

The equilibrant cancels the moving effect of the other forces. When an object is in equilibrium, the vector sum of all forces acting on it is zero.

If an object is in equilibrium under the action of several forces, any force may be considered the equilibrant of the other forces.

**FOR EXAMPLE:** In the above diagram, three forces of 3, 4, and 5 newtons are in equilibrium with each other. The equilibrant of the 3-newton and 5-newton forces is the 4-newton force to the right.

9 **Bearing** The direction of a vector may be given by using words such as north or south. A second method is to give the angle made by the vector with some other vector or direction. A third method is to give its bearing. The *bearing* of a vector is the angle it makes with the north-south line, measured clockwise from the northerly direction. In the diagram, the bearing of *OA* is 030°, of *OB* is 110°, and of *OC* is 225°.

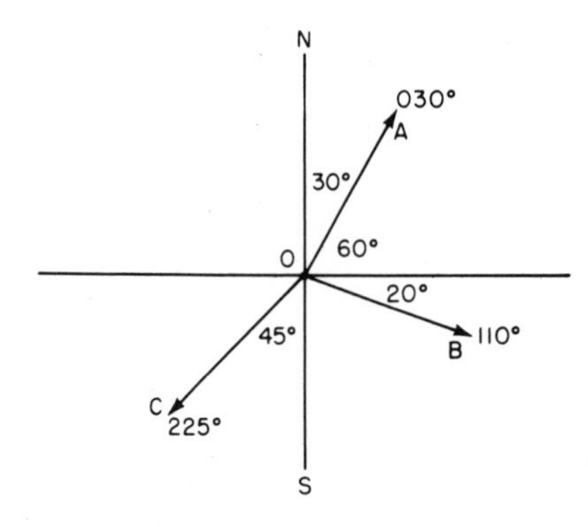


# 10 Resolution of Forces

We have seen that two or more forces can be replaced by a single force, their resultant. The reverse process is also possible. A single force may be replaced by two or more forces—its *components*. Usually we use two components at right angles; these are sometimes called rectangular components. For example, if we pull a sled by means of a rope, the pull or force we exert is in the direction of the rope. This direction is usually at some acute angle with the ground, but the sled moves along the ground. In order to find out how effective our pull is, we find its component parallel to the ground.

## HOW TO FIND THE COMPONENTS OF A FORCE

**EXAMPLE 1** A box is pulled along the horizontal ground by means of a rope making an angle of 30° with the horizontal. The rope is pulled with a force of 10 newtons. Find the horizontal and vertical component of this force.

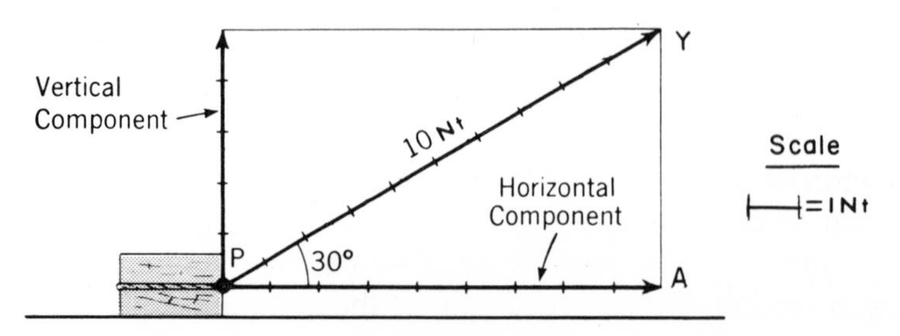

## GRAPHICAL METHOD

1. First we draw the three direction lines starting from a common point P on the object: horizontal, vertical and at 30° to the horizontal.
2. Using an arbitrary scale, we measure off the known force: 10 newtons at 30° with the horizontal; vector *PY*.
3. From the end of this known vector, Y, we draw horizontal and vertical lines to complete the rectangle.
4. The intersections of these two lines with the original vertical and horizontal lines mark the ends of the two desired components; we put arrowheads at these ends.
5. On each component vector we mark off the scale unit to see how many times the unit fits into the length. In this example, the scale unit fits into the horizontal component almost nine times—estimate 8.7 times. Since each unit represents one newton, the horizontal component is 8.7 newtons. The vertical component is 5.0 newtons.

**EXAMPLE 2** A 90-lb object is suspended from the middle of a rope. The ends of the rope are attached to the ceiling. What is the tension in the rope if the two halves of the rope make an angle of 90° with each other?

**Note:** The *tension* in the rope is the pull exerted by the rope.

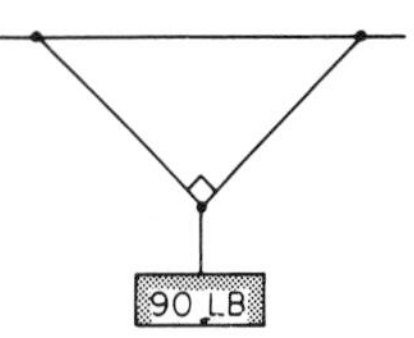


After making a sketch of the situation, we make a vector diagram, following the steps outlined under Example 1. We note in addition that there is equilibrium, and that the 90-lb weight can be thought of as the equilibrant of the pulls exerted by the two halves of the rope. We select a suitable scale to draw the known 90-lb weight, *OE*. Since the equilibrant is opposite and equal to the resultant, we draw the arrow *R* opposite and equal to *OE*. From the end of the known vector *R*, we draw lines parallel to the known direction of the ropes. This gives us the rectangle, of which *R* is the resultant, and the sides shown (*OC* and *OA*) are the components. To get the magnitude of these components, we mark off the scale unit to see how many times the unit fits into the length. In this case the length is about 3.2 scale

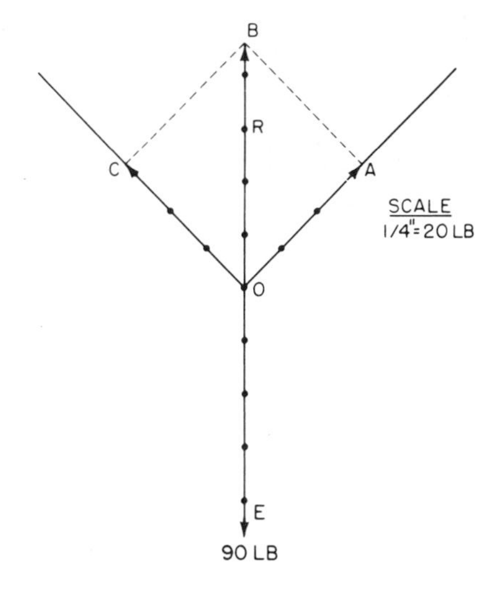

units. Since each unit represents 20 lb, the tension in each half of the rope is about 64 lb.

**EXAMPLE 3** *The Inclined Plane.* Sometimes we want to find components which are not vertical and horizontal. This is illustrated in the following problem. Let *AE* represent an inclined plane making an angle of 30° with the horizontal. An object weighing 100 newtons is placed on it. Assume friction is negligible. The object tends to slide down the plane. What force must be applied in the opposite direction to keep the object from sliding down? The weight of the object always acts vertically downward. This is our known force of 100 newtons, and we represent it by vector *PF*, using a scale of 1/4 inch = 20 newtons. (*P* is the center of the object.

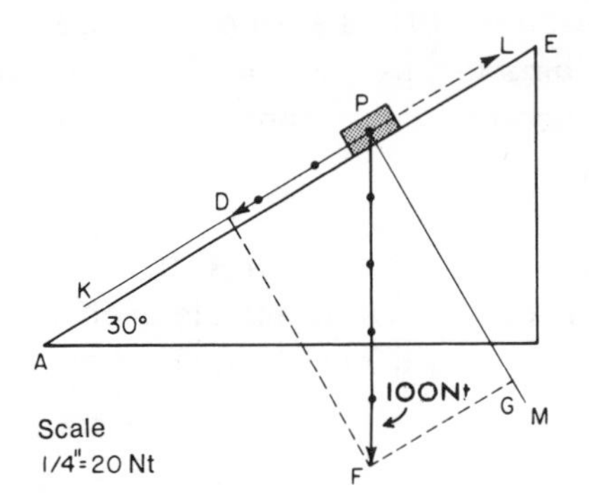


This is the center of gravity of the object and will be discussed later.) Since the object tends to slide down the plane, we draw direction line *PK* parallel to the plane to find the component of the weight in that direction. If we want the component perpendicular to the plane we also draw *PM* perpendicular to the plane. From *F*, the end of the known vector, we draw lines parallel to the direction lines *PK* and *PM*. *PK* is intersected at *D*. Therefore *PD* is the component (of the weight) parallel to the plane. In order to overcome its tendency to pull the object down the plane, we exert a force equal and opposite to it; this force is represented by *PL*. On *PD* we lay off our scale unit. The 2 1/2 units represent 50 newtons, since each unit represents 20 newtons. Therefore the required force *PL* has a magnitude of 50 newtons.

## 11 ** Using Simple Trigonometry

You must understand the graphical method for finding resultants and components of vectors. However, you can often save time by using a few simple facts about right triangles. In a right triangle: the sine of an acute angle is the side opposite the angle divided by the hypotenuse; the cosine of the angle is the side adjacent to the angle divided by the hypotenuse; the tangent of the angle is the side opposite the angle divided by the side adjacent.

$\sin A = a/c$; $\sin B = b/c$
$\cos A = b/c$; $\cos B = a/c$
$\tan A = a/b$; $\tan B = b/a$

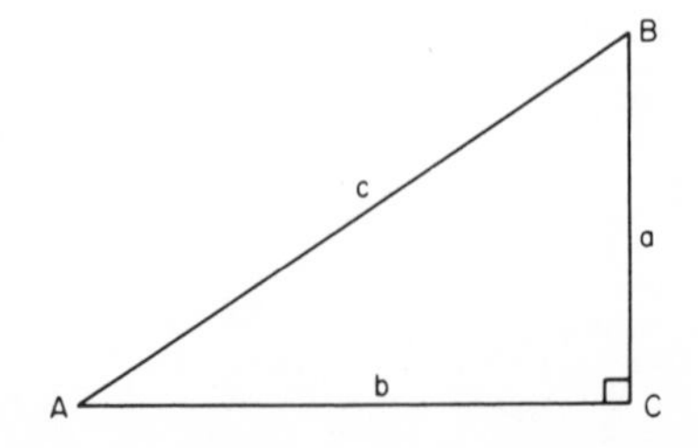

**FOR EXAMPLE:** In Figure 1, *PY* is given as 10 newtons. The horizontal component, *PA*, can be found as follows:

$$\cos 30° = PA/PY.$$
$$0.866 = PA/10\ Nt;$$
$$PA = 8.66\ Nt.$$

The graphical method gave 8.7 newtons.

## OTHER RIGHT TRIANGLE FACTS

1. In a 30°-60°-90° right triangle, the side opposite the 30° is equal to one-half of the hypotenuse; the side opposite the 60° is equal to one-half the hypotenuse times $\sqrt{3}$.
2. In an isosceles right triangle, the length of the hypotenuse is equal to the length of a side times $\sqrt{2}$.

## OTHER TRIANGLE FACTS

1. *Law of Sines.* In any triangle, the sides are proportional to the sines of the opposite angles.

$$\frac{a}{\sin A} = \frac{b}{\sin B} = \frac{c}{\sin C}$$

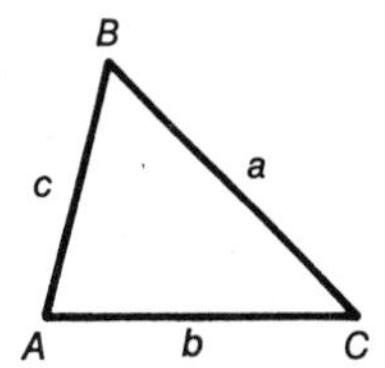


2. In any triangle, the square of any side equals the sum of the squares of the other 2 sides, minus twice the product *of these two* sides and the cosine of their *included* angle.

# Questions and Problems

**Note:** Most problems in this book use numbers with an indicated number of significant figures. Where this is not done, assume a precision allowing two significant figures.

1. A sled is pulled by means of a rope 2 meters long. The rope makes an angle of 30° with the ground. If the force pulling the rope is 80. *Nt*, what is the horizontal component of this force? Use a graphical solution.
2. Two forces act at an angle of 120° with each other. One is 40. *Nt* and the other is 60. *Nt*. Find the resultant by means of a graphical method. Label *a.* the resultant and *b.* the equilibrant.
3. Two forces act at an angle of 60° with each other. One is 50. *Nt* and the other is 70. *Nt*. Find, by means of a graphical method, *a.* the resultant *b.* the equilibrant.
4. From a given point, an object is first displaced 10.0 m in an easterly direction and then 24.0 m in a southerly direction. Determine the mag-

nitude and direction of the resulting displacement graphically and by computation.

5. Two forces act on a body. One is 90. *Nt* and the other is 50. *Nt*. The two forces act at right angles to each other. *a.* Using a graphical method find the resultant force. *b.* Draw and label the equilibrant.

6. A force of 12 *Nt* 090° and a force of 20. *Nt* 270° act on a point. Calculate the magnitude and bearing of the resultant force. This means that the bearing of the forces are 90° and 270°, respectively.

7. A force of 40. lb 090° and a force of 50. lb 180° act on a point. Calculate the magnitude and bearing of the resultant force.

8. A boy weighing 120. lb sits on a swing. The swing is pulled aside with a horizontal force of 90. lb. *a.* What is the force exerted by the two swing ropes? *b.* What is the angle made by the ropes with the vertical?

9. A man pushes on the handle of a lawn mower with a force of 120. lb. The angle between the handle and the ground is 30°. Calculate the magnitude of the horizontal and vertical components of this force.

10. A force of 20. *Nt* 030° acts on a point. Calculate the magnitude of the northerly and easterly components of this force.

11. A boy pulls a sled with a force of 30. lb. If the rope makes an angle of 45° with the ground, what are the magnitudes of the horizontal and vertical components?

12. A force of 50. *Nt* acts at an angle of 20° with the horizontal. Calculate the magnitudes of the horizontal and vertical components.

13. A rectangular picture weighing 12 *Nt* is supported by a wire attached to its two top corners. The wire passes over a nail in the wall where the two portions of the wire make an angle of 90° with each other. Calculate the pull exerted by each portion of the wire.

14. A man suspends a 100-lb object by means of two ropes attached to the same point on the object. The two ropes make an angle of 90° with each other. Calculate the pull in each rope when one of the ropes makes an angle with the vertical of *a.* 45° *b.* 30°

15. A weight of 200 newtons is supported by a wire making an angle of 30° with a boom as shown. Calculate *a.* the tension in the wire *b.* the outward thrust of the boom.

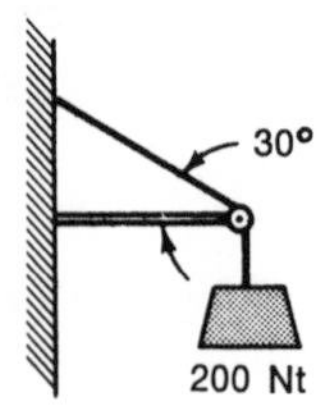


**Note:** For problems involving combination of velocities, see the end of Chapter 4.

## Test Questions

1. The force required to balance the earth's pull on a 1-kilogram object is 1. 1 *Nt* 2. 9.8 *Nt* 3. 980 *Nt* 4. 1 lb

2. The force required to balance the earth's pull on a 10-kilogram object is ............ newtons.

3. The minimum resultant possible for displacements of 8 ft and 3 ft is ............ ft.
4. The minimum resultant of two displacements occurs when their direction is ............
5. The maximum resultant of two forces of 80 lb and 70 lb acting on the same point is ............ lb.
6. If two forces act on the same particle, the resultant will be a maximum if the direction between the two forces is ............ degrees.
7. The single force that has the same effect as two forces which it replaces is known as their 1. replica 2. equilibrant 3. resultant 4. vector
8. Two forces act together on an object. The magnitude of their resultant is least when the angle between the forces is 1. 0° 2. 90° 3. 180° 4. 45°
9. Two forces act together on a particle. As the angle between the two forces increases from 0° to 180°, the magnitude of the resultant ............ (increases, decreases, remains the same).
10. The resultant of a 12-*Nt* force and a 5-*Nt* force acting simultaneously on an object in the same direction is, in newtons, 1. zero 2. 5 3. 7 4. 17
11. The resultant of a 3-*Nt* force and a 4-*Nt* force acting simultaneously on an object at right angles to each other is, in newtons, 1. 1 2. 5 3. 7 4. 12
12. Three forces act on an object: 5 lb, 8 lb, and 12 lb. The object remains stationary and no other forces act on it. The resultant of all three forces is 1. zero lb 2. 25 lb 3. 21 lb 4. cannot be determined since no direction is given.
13. Three forces act from a single point. One force is 300 newtons due south, a second force is 500 newtons due west, and the third force is 100 newtons due east. The magnitude of the resultant force, in newtons, is 1. 100 2. 300 3. 500 4. 583
14. A 100-lb object is supported by two wires making an angle of 45° with the vertical and an angle of 90° with each other. The pull exerted by each wire is ............ lb.
15. Two forces act on an object. One force is 30 *Nt* acting west and the other is 30 *Nt* acting south. The resultant force is 1. $30\sqrt{2}$ *Nt* at 225° 2. $60\sqrt{2}$ *Nt* at 225° 3. $30\sqrt{2}$ *Nt* at 45° 4. 60 *Nt* at 45°
16. The resultant, in newtons, of a 5-*Nt* and a 12-*Nt* force, both pulling west on the same object at the same time, is ............
17. Three forces act simultaneously on an object. One is 10 lb to the north, the second is 6 lb to the south, and the third is 3 lb to the east. The magnitude of the resultant force is ............ lb.
18. A force of 20.0 *Nt* (bearing 090.0°) acts upon an object at point 0. Another force of 15.0 *Nt* (bearing 180.0°) acts on the body at the same point. The resultant force is approximately ............
19. A constant force is applied to the handle of a lawnmower. As the handle is lowered, the horizontal component of the force applied to the handle ............ (increases, decreases, remains the same).
20. As the angle between a force and its vertical component decreases from 80°, the difference in magnitude between the force and its vertical component ............ (decreases, increases, remains the same).

21. As the resultant of a group of concurrent forces increases, the equilibrant........... (increases, decreases, remains the same).
22. When three forces are in equilibrium, any one of these is the ........... of the other two.
23. A vector quantity always has magnitude and ............
24. A window pole is used to close a window. The angle between the pole and the window which would require the least force is, in degrees, 1. zero 2. 30 3. 45 4. 90
25. As the resultant of two concurrent forces increases, their equilibrant 1. increases 2. decreases 3. remains the same.
26. Two forces of 100 newtons each act at an angle of 120° with each other. Their resultant, in newtons, is 1. 100 2. 141 3. 173 4. 200

## Topic Questions

1.1 Why is it sometimes important to give direction as well as distance?

2.1 What is meant by a vector quantity? Give an example.

2.2 What is meant by a scalar quantity? Give an example.

3.1 What is a vector?

4.1 What is meant by *displacement*?

4.2 What is meant by *resultant vector*?

4.3 How do we get the resultant of two vectors at right angles to each other?

5.1 What is meant by the weight of an object on earth?

5.2 What instrument may be used to measure forces?

5.3 What is the term which is used for the single force which can be used to replace two other forces?

6.1 What are the important steps for getting the resultant of two forces graphically?

7.1 If vector $+K$ is directed to the north, what is the direction of vector $-K$?

8.1 If two forces act on a small object, how does their resultant compare with their equilibrant?

8.2 What is the vector sum of two forces and their equilibrant?

9.1 If a bearing is given, from what direction is the angle measured?

10.1 What are the important steps in finding the horizontal and vertical components of a force making an angle of 40° with the horizontal? Discuss the graphical method.

10.2 What is meant by the *tension* in a rope?

# 3 **Parallel Forces and Moment of a Force

## 1 Introduction

Have you ever tried to push a door open absent-mindedly and found it difficult to do because you were pushing close to the hinges? Doorknobs are put near the edge of the door away from the hinges because experience has shown that this location makes it easier to open and close doors.

We mentioned earlier that a force has a tendency to produce or prevent motion. This motion may be along a straight or a curved line.

The *torque* or *moment of a force* is the effectiveness of a force in producing or tending to produce rotation. The axis around which the rotation takes place, or tends to take place, is called the *fulcrum* or *pivot*. In the case of a door, the hinges form a convenient fulcrum. We saw from the above illustration that the moment of a force depends not only on the force but also on how far from the fulcrum the force is applied. The moment of a force is equal to the force multiplied by its perpendicular distance from the fulcrum. This perpendicular distance from the fulcrum to the line of action of the force is known as the length of the *moment arm*.

$$\text{moment of a force} = \text{force} \times \text{length of moment arm}$$

In the diagram are shown four forces acting on a slender rigid rod, *XY*. Such a rod is known as a *lever*. The fulcrum is represented as a little triangle, *F*. Force *B* is 3.0 newtons and is applied perpendicularly to the lever at a distance of 0.40 meter from the fulcrum. Therefore its moment is $3.0\ Nt \times 0.40\ \text{m} = 1.2$ meter-newton. The length of *A*'s moment arm is 0.60 meter. Therefore the moment of force *A* is $10\ Nt \times 0.60\ \text{m} = 6.0$ meter-newton. Force *C* is not perpendicular to the lever. To get its moment arm we must drop a perpendicular from the fulcrum to the direction of force *C*. This is shown in the diagram. The length of the moment arm is given as 0.30 m. Therefore the moment of force *C* is $2.0\ Nt \times 0.30\ \text{m}$ or 0.60 meter-newton.

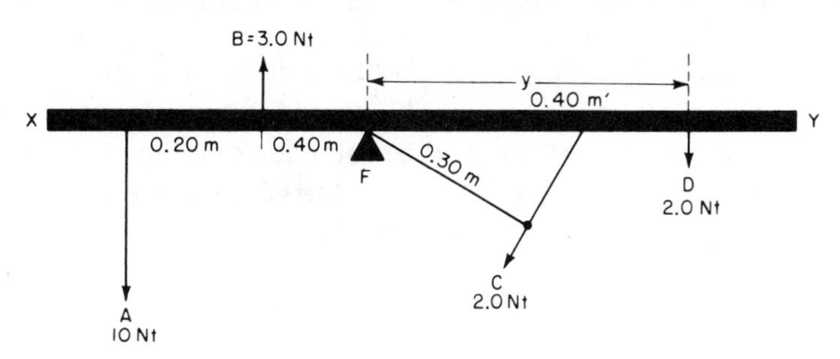


**2 Equilibrium of Lever** If the lever is to be in equilibrium, how far from the fulcrum must force *D* be applied? So far we discussed equilibrium only when all the forces are applied to the same point of the object. Under that condition, if the object is in equilibrium, one force (the equilibrant) is equal and opposite to the resultant of the other forces. However, in the above diagram we show forces applied to different points of the object. Under these conditions something additional has to be true for *equilibrium*:

The sum of the Clockwise Moments = The sum of the Counterclockwise Moments.*

* **Note:** Technically it is not the moments that are clockwise or counter-clockwise; the tendency for rotation is.

Forces *B*, *C*, and *D* have clockwise moments about fulcrum *F* because they tend to produce clockwise rotation about the fulcrum. Force *A* tends to make the lever rotate counterclockwise; therefore it has a counterclockwise moment. If we let the distance from the fulcrum to force *D* be *y*, we can write, for equilibrium:

$$3.0 \times 0.40 + 2.0 \times 0.30 + 2.0y = 10. \times 0.6$$
$$2.0y = 6.0 - 1.2 - 0.6$$
$$y = 2.1 \text{ m}$$

The fulcrum has to exert an upward force; this is not shown in the diagram.

**3 Other Units** A unit for moment of a force is a force unit multiplied by a distance unit. Three common units are: in the *MKS* system, the meter-newton; in the *CGS* system, the centimeter-dyne; in the English system, the pound-foot.

**4 Center of Gravity** We know that the earth pulls on objects. The earth's attraction for objects is known as *gravity*. The weight of an object on earth is the earth's pull on it. Actually, of course, the earth pulls on every part of the object, and the weight of the object is the resultant of all these practically parallel forces. This resultant goes through a point known as the center of gravity. If the fulcrum is placed directly below the center of gravity, there will be no tendency for the object to rotate. For example, balance a pencil in a horizontal position using a finger tip as the fulcrum. The center of gravity of the pencil will then be directly above your finger tip. This will not necessarily be the geometric center or middle of the pencil, especially if the pencil has an eraser at the end. If the object is uniform, the center of gravity is at the geometric center of the object. For example, the center of gravity of a meter stick is at the 50-centimeter mark. When talking about moments of forces we can think of all the weight of an object as being concentrated at the center of gravity.

**EXAMPLE:** A meter stick weighing 1 newton is supported by a pivot at the 40-*cm* mark. Where should a 2-newton weight be placed in order to maintain equilibrium? A quick sketch shows the weight of the meter stick acting at the 50-*cm* mark, the center of gravity of the stick. The fulcrum, at the 40-*cm* mark, is 10 cm away. Therefore the 1-*Nt* weight exerts a clockwise moment. In order to produce equilibrium,

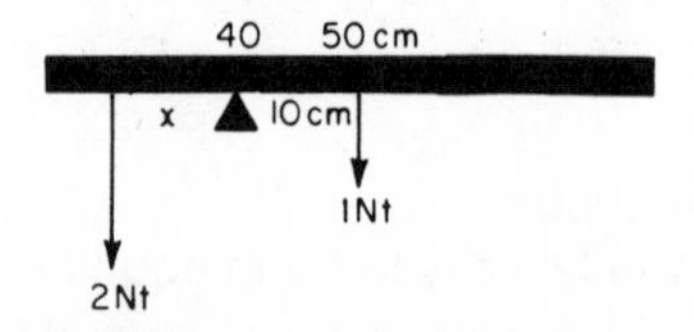

the 2-newton weight is hung on the other side of the fulcrum, $x$ centimeters away. This will then give us a counterclockwise moment. For equilibrium,

$$\text{counterclockwise moments} = \text{clockwise moments}$$
$$2x = 1\ Nt \times 10\ \text{cm}$$
$$x = 5\ cm$$

The 2-newton weight must be hung 5 cm to the left of the fulcrum, that is, on the 35-*cm* mark. In order to keep the total weight of 3.0 newtons from falling, the pivot must exert an upward force of 3.0 newtons.

5 **Lever Supported At Two Points** If an object is supported at two points, we can apply the above methods by using one of the points as the fulcrum; at the other point we use the force exerted by the support. The method will be explained with the aid of a problem.

**EXAMPLE:** A uniform stick weighing 20. pounds is carried in a horizontal position by two boys holding it, one at each end. A load of 80. pounds is suspended 3.0 ft from one end. If the stick is 10. ft long, how much force is exerted by each boy?

Make a sketch of the situation. We'll use end *A* as the fulcrum, and let *F* stand for the force exerted by the boy at the other end, *B*.

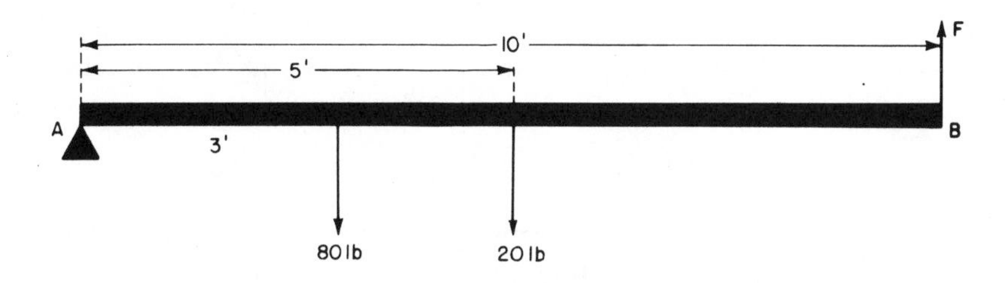


$$\text{counterclockwise moment} = \text{sum of clockwise moments}$$
$$10\,F = (80 \times 3) + (20 \times 5)$$
$$10\,F = 240 + 100$$
$$F = 34\ \text{lb.}$$

The boy at *B* has to exert a force of 34 lb upward. The other boy has to supply a force of 66 lb to make up for the total downward force of 100 lb. This can be checked by using end *B* as the fulcrum. Don't make the mistake of switching pivots while setting up the equation.

How can the *center of gravity of an irregular object* be determined? Consider the object shown below. First the object is suspended so that it swings freely about point *A*, as in Position 1. When the object comes to rest, use a plumb line (a weight suspended from a string) to draw a vertical line through point *A*. Repeat this process at least once by suspending the object from another point such as *B* (Position 2). Where the vertical lines intersect is the center of gravity, *C*.

Note that if the object is in some position such as Position 3, (suspended from *A*) the weight of the object acting at *C* produces a clockwise moment about the pivot at *A* (equal to $W \times S$) which causes the object to swing back towards Position 1. *When the object comes to rest, the center of gravity is on a vertical line going through the fulcrum.* The moment due to the weight is then zero because its moment arm equals zero.

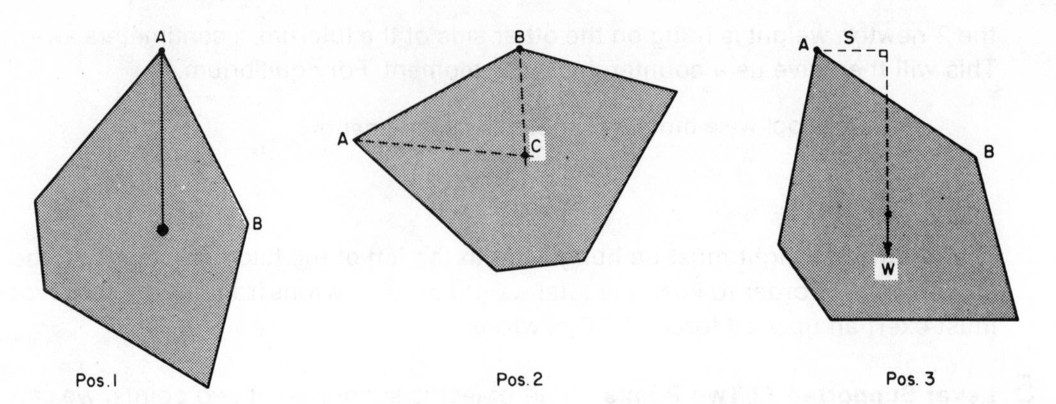


6 **Stability** The above indicates that the stability of objects can be increased by building them with centers of gravity as low as possible. The stability of objects is also increased by providing them with as big a base as possible. A stable object is not toppled readily. Any attempt to topple the object will fail as long as a vertical line through the center of gravity falls within the base. As shown in the figure *b.*, when the object has been tipped slightly, the object's weight acting along the vertical direction from the center of gravity provides a moment of a force which tends to turn the object back to its original position. The broader the base *AB* is

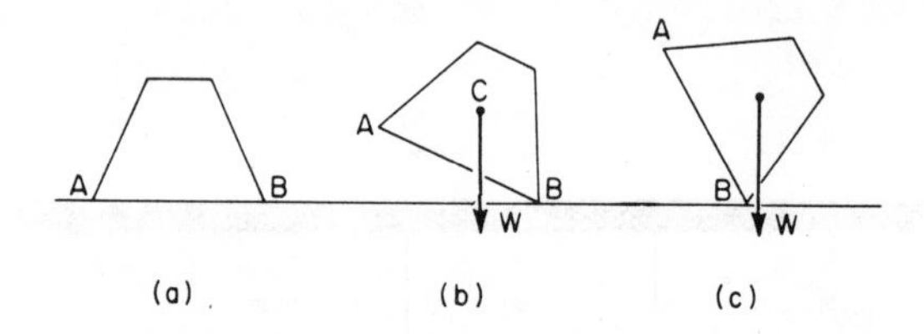


made, the more the object can be tipped without toppling it. In *c.* the object has been tipped too far. Now the vertical line from the center of gravity goes outside the base, and the weight produces a clockwise moment to topple the object.

# Questions and Problems

1. Two boys balance each other on a see-saw which is 12 feet long. The boy sitting at one end weighs 60 pounds. Where does the other boy have to sit if he weighs 80 pounds?
2. A yardstick weighing 8 ounces is balanced at the center. Two weights are suspended from it. One weight is suspended six inches from the left end and the other eight inches from the right end. The one at the left weighs 12 ounces. *a.* What is the torque due to this weight? *b.* Where must the second weight be suspended so that the stick will remain in the horizontal position?
3. A yardstick weighing 8 ounces is balanced in the horizontal position by pivoting it at the 10 inch mark and suspending an unknown weight from the 3-inch mark. The yardstick is uniform. How heavy is the unknown weight?
4. A uniform plank weighing 40 pounds is carried in the horizontal position by two men holding it, one at each end. A boy weighing 60 pounds sits on top of the plank, 4.0 feet from one end. The plank is 10. feet long. What is the upward force exerted by each man?

5. A uniform plank weighing 60 pounds lies on level ground. What force is required to just pick up one end?
6. A non-uniform plank 10. feet long and weighing 50. pounds lies on level ground. Its center of gravity is 4.0 feet from one end. How much force is required to just lift *a.* the end closer to the center of gravity? *b.* the other end?

## Test Questions

1. The center of gravity of a uniform meter stick is nearest to the line marked ............
2. If two parallel forces of 20 *Nt* and 10 *Nt*, respectively, are in equilibrium with a force of 30 *Nt*, the resultant of the three forces is ............
3. A uniform plank weighing 100 lb lies on level ground. To lift one end off the ground requires a force of ............
4. A tapered pole is 8 ft long and weighs 56 lb. If the center of gravity is one foot from the thick end, the force required to lift the thick end is ............
5. A uniform meter stick is balanced on the 50-*cm* mark. A 100-*gm* object is suspended from the 70-*cm* mark and another 100-*gm* object from the 90-*cm* mark. Where must a 200-*gm* object be suspended to maintain equilibrium?
6. The center of gravity of a baseball bat is *a.* in the middle *b.* near the thin end *c.* near the thick end.
7. A force of 20. *Nt* 30 *cm* from the fulcrum has the same moment as a force of ............ *Nt.* 15 *cm* from the fulcrum.
8. Jonathan and George are balanced on a see-saw. They have equal 1. distances from the fulcrum 2. weights 3. masses 4. torques
9. Three moments act on a balanced rod. One is 10 meter-newtons clockwise, the second is 15 meter-newtons counterclockwise. The third moment has a magnitude, in meter-newtons, of 1. 5 2. 10 3. 15 4. 25

## Topic Questions

1.1 What is meant by *a.* the moment of a force *b.* the fulcrum *c.* the moment arm?

1.2 How can we calculate the moment of a force?

2.1 If a lever is in equilibrium, what must be true of the moments of the forces acting on it?

3.1 In the English system of measurement, what is a unit for the moment of a force?

4.1 Where is the center of gravity of a uniform meter stick?

5.1 When a lever is supported at two points, *a.* where is the fulcrum *b.* what is one of the upward forces?

5.2 How can the center of gravity of an irregular object be found experimentally?

6.1 When building an object, what are two things that can be done to give it stability?

# 4 Motion

## 1 Introduction

You probably think that you are not moving around much as you are looking at this page. Actually, of course, you are moving at a very high speed. If you are at the equator you are moving around the earth's axis at a speed of about 1000 miles per hour because of the earth's rotation; if you are in New York City it is about 800 miles per hour. We are moving around the sun at an even greater speed.

## 2 Speed and Velocity

When we speak about motion or rest, the surface of the earth is our reference point, unless we specifically state otherwise. (We sometimes say the earth is our *frame of reference.*) This is what you would have in mind if you say that you are now at rest; you are not moving with respect to the earth. If you are moving (with respect to the earth), you have a certain speed.

*Speed* is distance covered per unit time. Your speed may be 60 miles per hour, 60 ft per hour, 60 meters per hour, etc. When we use the term speed we are not concerned with the direction of motion. Speed is a scalar quantity.

*Velocity* is a vector quantity. The velocity of an object is its speed in a certain direction; it has magnitude and direction. The magnitude of the velocity is the object's speed. The velocity changes if either the speed or the direction of motion, or both, changes. If an object is moving at constant speed around a circle, its velocity is changing because its direction is constantly changing. At any instant its direction is the tangent to the circle.

3 **Combination of Velocities** The motion of an object can sometimes be thought of as the result of a combination of velocities.

**FOR EXAMPLE:** If you walk at 3 miles per hour from the last car of a train towards the front while the train is moving at 50 miles per hour, your velocity with respect to the earth is 53 miles per hour in the direction of the train's motion. (Note that it was implied that the 3 mi/hr was with respect to the train.) If the two given velocities are not in the same direction, we can get the combined or resultant velocity by using combination of vectors as we did for forces and displacements.

**EXAMPLE:** An airplane is moving east in still air at 200 Km/hr. A steady south wind starts blowing at 100 Km/hr. What will be the plane's velocity? The graphical solution is shown in the diagram. Recall that a north wind is a cold wind—it comes from the north; a south wind comes from the south.

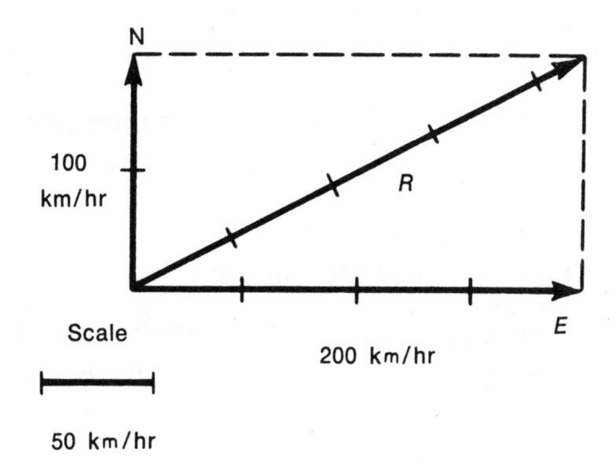


The plane's velocity will be approximately 220 Km/hr in the direction shown by $R$. We can also get the magnitude by using the *Pythagorean Theorem*:

$$R = \sqrt{(100)^2 + (200)^2} = 220 \text{ Km/hr.}$$

## 4 Speed and Average Speed

**Speed and Average Speed** In elementary school you probably learned that distance equals rate times time. In physics we say speed instead of rate.

$$\text{Distance } (s) = \text{speed } (v) \times \text{time } (t)$$
$$s = vt$$

This formula applies when the speed of an object remains constant.

**FOR EXAMPLE:** If a train goes at a constant speed of 60 mi/hr, the distance traveled in 3 hours = 60 mi/hr × 3 hr = 180 mi. Usually it is not possible to maintain a constant speed. We define *average speed* ($\bar{v}$) as distance traveled divided by time of travel:

$$\bar{v} = s/t$$

If we clear fractions we get:

$$s = \bar{v}t,$$

the same formula we used to calculate the distance traveled when the speed is constant except that $\bar{v}$ (read: $v$-bar) is the average speed. If the speed remains constant the average speed is the constant speed.

**EXAMPLE:** A car moves north with a constant velocity of 50 kilometers per hour. How far will it have traveled in 4 hours?

$$s = vt;$$
$$s = 50 \text{ km/hr} \times 4 \text{ hr} = 200 \text{ km}$$

The return trip takes 8 hours. What was the average speed for the round trip?

$$\bar{v} = s/t;$$
$$\bar{v} = 400 \text{ km}/12 \text{ hr} = 33.3 \text{ km/hr}$$

Motion in which the velocity remains constant is known as *Uniform Motion*.

# 5 Accelerated Motion

Motion in which the velocity changes is known as accelerated motion. *Acceleration* (a) is the rate of change of velocity. It is a vector quantity.

$$\text{acceleration} = \frac{\text{change in velocity}}{\text{time required for change}}$$

In order to save space the physicist uses the Greek letter delta ($\Delta$) when he wants to say *change in*; the formula for acceleration then becomes:

$$a = \frac{\Delta v}{t}$$

read this as: $a$ equals delta $v$ over $t$.

Remember delta $v$ is the change in velocity; this means the velocity we end up with, minus the velocity we started with. Some books write this as:

$$a = \frac{v_f - v_i}{t}; \qquad v_f = \text{final velocity}, v_i = \text{initial velocity}.$$

**EXAMPLES: 1.** A train moving along a straight track required 5.0 seconds to change its velocity from 15 miles per hour to 60 miles per hour. What was its acceleration?

$$a = \Delta v/t = (60 \text{ mi/hr} - 15 \text{ mi/hr})/5.0 \text{ sec}$$
$$= 9.0 \text{ miles per hour per second.}$$

This means that each second the velocity increased 9.0 miles per hour. Notice that the unit of acceleration requires two time intervals; in this case hour and second. Sometimes we prefer to have the two time intervals be the same: 9.0 mi/hr is approximately equal to 13 ft/sec; therefore 9.0 mi per hr per sec = 13 ft per sec per sec. This is usually abbreviated as 13 ft/sec$^2$. It is best to read this as "13 feet per second every second."

**2.** A car starting from rest has a velocity of 20 m/sec at the end of the first second, 40 m/sec at the end of the second second, and 60 m/sec at the end of the third second. What was its acceleration?

During each second the velocity increased 20 m/sec. Therefore, acceleration = ($\Delta v/t$) = (20 meters per second/1 sec) = 20 m/sec$^2$.

**Note:** In order to stress the fact that a vector is being described, sometimes an arrow is placed above the letter, $\vec{a}$ and $\vec{v}$, to refer to acceleration and velocity.

## 6 Deceleration

**Deceleration** When an object slows down its velocity is decreasing. By definition we are still dealing with accelerated motion. We say the object is decelerated. Sometimes such an object is said to have negative acceleration. This should be interpreted to mean that the direction of the acceleration is opposite to the direction of the initial velocity.

## 7 Uniformly accelerated motion

**Uniformly accelerated motion** is motion with constant acceleration. Let us first consider such *motion starting from rest*: the initial velocity is zero. Then whatever velocity the object has at any instant is also the change in velocity. The expression for acceleration then becomes:

$$a = v/t,$$

where $v$ is the velocity after time $t$ (or instantaneous velocity) and $t$ is the time since the motion started.

**EXAMPLE:** An object starting from rest increases its velocity steadily to 48 m/sec in 4 seconds. Its acceleration is $\frac{48 \text{ m/sec}}{4 \text{ sec}}$, or 12 m/sec$^2$.

The above formula can be written as:

$$v = at.$$

During the motion the velocity has increased steadily from zero to $v$. The average velocity ($\bar{v}$) during this motion is $\frac{1}{2}v$, or $\frac{1}{2}at$. Since distance ($s$) equals average speed × time,

$$s = \tfrac{1}{2}at^2.$$

**EXAMPLE:** How far did the above object travel in 4 seconds?

$$s = \tfrac{1}{2}at^2$$

$s = \frac{1}{2} \times 12 \text{ m/sec}^2 \times (4 \text{ sec})^2 = 96$ meters. We could have solved this problem by using the other formula for distance:

$$s = \bar{v}t$$

For motion with constant acceleration starting from rest,

$$\bar{v} = \tfrac{1}{2}v = \tfrac{1}{2} \times 48 \text{ m/sec} = 24 \text{ m/sec}$$

and

$$s = (24 \text{ m/sec}) \times 4 \text{ sec} = 96 \text{ meters}$$

It is possible to derive another formula for this kind of motion: $v^2 = 2as$. In this example we know that $v = 48$ m/sec, and that $a = 12$ m/sec$^2$. Therefore, if we wish, we may use the new formula to solve for the distance $s$:

$$(48 \text{ m/sec})^2 = 2(12 \text{ m/sec}^2)s; \text{ and } s = 96 \text{ meters.}$$

# 8 Falling Objects

A special case of motion with constant acceleration is the motion of objects falling freely near the surface of the earth. Galileo (1564–1642) showed that this acceleration is the same for heavy objects as for light objects, if air resistance is negligible. Its value is somewhat different for different locations on the earth, but for most calculations we may use the same numerical value. This is:

9.8 m/sec$^2$, 980 cm/sec$^2$, or 32 ft/sec$^2$.

The letter $g$ is often used in formulas instead of $a$ when speaking about the acceleration of such falling objects since it is produced by the earth's gravity.

*Leaning Tower of Pisa*
*("Emit" Italian Governments Travel Office)*

**EXAMPLE:** A stone is dropped from the George Washington Bridge. What is its speed at the end of the third second?

$v = gt$
$v = (9.8\ \text{m/sec}^2)(3\ \text{sec}) = 29.4\ \text{m/sec}.$

## 9 Summary of Formulas for Uniformly Accelerated Motion (*starting from rest*)

1. $v = at$
2. $s = \frac{1}{2}at^2$ where $v$ is the velocity or speed after time $t$
3. $v^2 = 2as$ — $t$ is the time since the motion started
4. $\bar{v} = \frac{1}{2}at$ — $s$ is the distance traveled, or displacement of the object, in time $t$
5. $v = gt$
6. $s = \frac{1}{2}gt^2$ — $a$ is the object's acceleration
7. $v^2 = 2gs$ — $g$ is the acceleration due to gravity
8. $\bar{v} = \frac{1}{2}gt$ — $\bar{v}$ is the average velocity or speed during the motion
9. $s = \bar{v}t$

10 **Which formula to use?** Formulas 5–8 are really not needed. They are usually presented merely as a reminder that for falling objects we know the numerical value of the acceleration. On some examinations its value will be given as: $g = 9.8\ \text{m/sec}^2$. Therefore memorize formulas 1–4, especially 1–3; but learn at

the same time what the letters stand for and that the formulas are for motion with constant acceleration. Formula 9 is sometimes useful; you became familiar with it a few pages earlier. Should you use formula 1, 2, 3, or 9? That depends on what information you are given and what you have to find. Each of the formulas has three letters. You must know the numerical value of two of these in order to calculate the value of the third.

**FOR EXAMPLE:** In the above problem with the George Washington Bridge, we were asked to find speed. The letter $v$ appears in formula 1 and 3; it stands for speed. Which of the two formulas shall we use? Both formulas use acceleration $a$; we know that for falling objects this equals 9.8 m/sec$^2$. Formula 3 requires knowledge of $s$, the distance traveled. This information is not given. On the other hand, formula 1 requires knowledge of $t$, the time traveled. This is given in the problem as 3 seconds. Therefore we use formula 1 (or its equivalent, formula 5).

**CAUTION** In using the above formulas use consistent units.

**FOR EXAMPLE:** In the *MKS* system express velocity ($v$) in m/sec, acceleration ($a$) in m/sec$^2$, time ($t$) in sec, and displacement or distance ($s$) in meters.

## 11 Graphs for Motion

Sometimes graphs provide a convenient summary for information. As an illustration let us consider a problem for which we have already done some calculations. An object starts from rest and moves with a constant acceleration of 12 m/sec$^2$. Draw three graphs for this motion: $a$ vs. $t$; $v$ vs. $t$; and $s$ vs. $t$.

First let us make a table of values covering a suitable time interval. Below is such a table for a five-second interval. Column I gives the time ($t$) in seconds. Column II gives the acceleration ($a$) in meters/sec$^2$; the acceleration is constant and given as 12 m/sec$^2$; therefore it has this value throughout the motion. We use columns I and II to make the graph $a$ vs. $t$. Note that the graph is a horizontal straight line.

| I | II | III | IV |
|---|---|---|---|
| $t$ (sec) | $a$ (m/sec$^2$) | $v$ (m/sec) | $s$ (m) |
| 0 | 12 | 0 | 0 |
| 1 | 12 | 12 | 6 |
| 2 | 12 | 24 | 24 |
| 3 | 12 | 36 | 54 |
| 4 | 12 | 48 | 96 |
| 5 | 12 | 60 | 150 |

Column III gives the velocity ($v$) at the start and at the end of 1 second, 2 seconds, etc. The value can be calculated by using the formula

$$v = at.$$

We use columns I and III to plot graph A. Notice that the graph of velocity vs. time for uniformly accelerated motion is an oblique straight line. It goes through the origin (the intersection of the axes) because the motion starts from rest. The steeper the graph, the greater the acceleration. The motion described by graph B has an acceleration greater than that of A.

Since for motion starting from rest, acceleration is equal to velocity divided by time, the acceleration for graphs A or B may be obtained by taking any point on the graphs and dividing the velocity it represents by the corresponding value of the time. (You may recall the term *slope* used to describe the steepness of a straight line. For graphs A and B it is the value of $v$ divided by the corresponding value of $t$; for example, for graph A, at $t = 3$ seconds, $v = 36$ m/sec. Therefore the slope is 36 m/sec divided by 3 seconds. This is 12 m/sec$^2$, which is the acceleration. The slope of the graph of velocity vs. time is the acceleration of the motion described.)

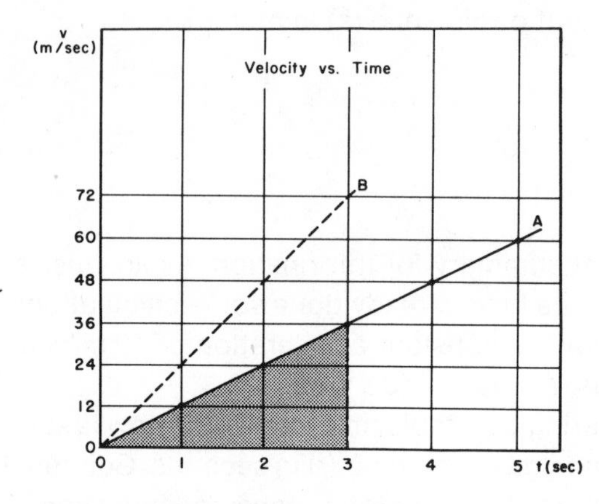


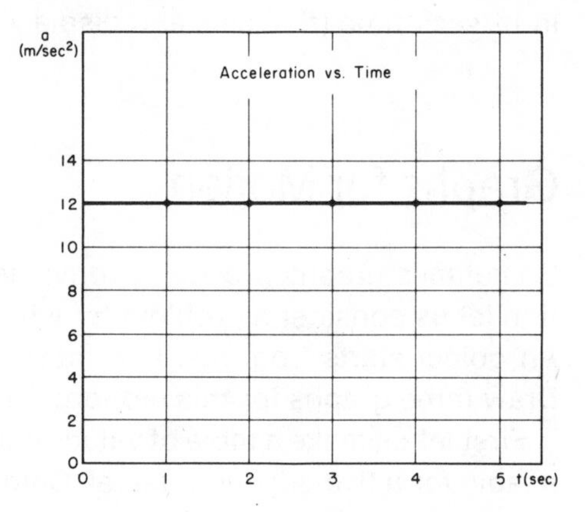


Column IV gives the displacement or distance ($s$) traveled by the object at the end of 1 second, 2 seconds, etc. We use columns I and IV to plot the graph of $s$ vs. $t$. Notice that the smooth curve through the plotted points is not a straight line; it is a *parabola* curving upwards. You may have expected this, since $s$ is proportional to $t^2$. Similarly, perhaps you expected the straight-line graph above for velocity vs. time (graph A), since the velocity is proportional to time.

**Note on Proportion:** What do we mean when we say that *velocity is proportional to time*? We mean that when time changes, the velocity (of the object which is described) must also change in such a way that the ratio of the velocity to the time remains constant:

$$\frac{\text{velocity}}{\text{time}} = \text{constant}.$$

If we use the letters which we used before, $v$ for velocity and $t$ for time, this can be written as:

$$\frac{v}{t} = \text{constant}.$$

Let us look at the above table of values. When $t = 2$ sec, $v = 24$ m/sec. For this ratio of $v/t$ the constant is 24/2, or 12 (m/sec$^2$). A little later when $t = 4$ sec, $v = 48$ m/sec. For this ratio of $v/t$ the constant is 48/4, or again 12 (m/sec$^2$). From this example we can see that two other things are true when we say that velocity is proportional to time:

FIRST: $\frac{v_1}{t_1} = \frac{v_2}{t_2}$.

SECOND: When the time is doubled, the velocity also doubles. When the time is tripled, the velocity also triples, etc.

We also note from the example that the constant is the acceleration of the object:

$$\frac{v}{t} = a.$$

If we clear fractions we get

$$v = at,$$

or formula 1 in paragraph 9.

Finally, instead of saying that velocity is proportional to time, your mathematics teacher might say *velocity varies directly as time*.

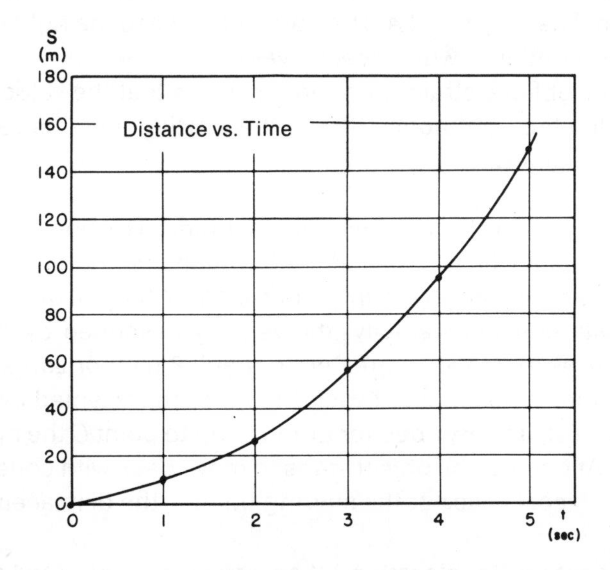


The distance traveled by the object may be obtained from the graph of velocity vs. time.

**FOR EXAMPLE:** The area of the shaded triangle under graph A gives the distance traveled by the object during the first three seconds. Let us check this. You recall that the area of a triangle equals one-half the base times the altitude. The base of the shaded triangle is 3 seconds; the altitude is 36 m/sec.

The area = (1/2)(3 sec)(36 m/sec) = 54 m.

This checks with the distance shown in the table for $t = 3$ sec. In general, *the area under the velocity vs. time graph gives the distance or displacement.* (This can be proved by noting that the altitude is $v$; the average speed is $v/2$; and the base is $t$.

Therefore $\frac{1}{2} \times \text{altitude} \times \text{base} = \frac{1}{2}vt = \bar{v}t$. Compare this with formula 9 above: $s = vt$.)

**More on Graphs** The above graphs described the motion of objects starting from rest and moving with constant acceleration. Such objects move along a straight line. Now look at the following graphs and see if you can tell what kind of motion is described by each of them, again assuming that in each case the object moves along a straight line.

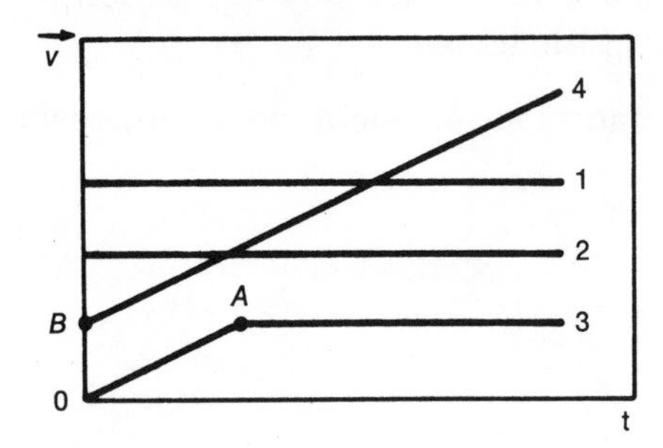


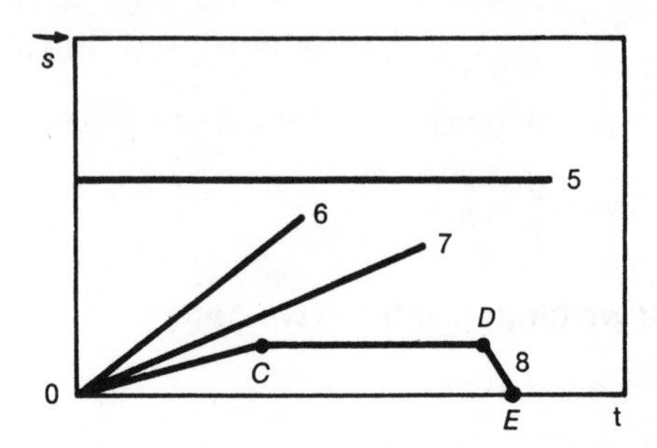


**GRAPHS** 1–4 represent the velocity against time. In graphs 1 and 2 the velocity remains constant, but the velocity in 2 is less than in 1. In graph 3 the motion starts from rest. The velocity increases with a constant acceleration as shown by the oblique straight line until point *A*. The horizontal line to the right of *A* indicates that the motion then continues with constant velocity.

In graph 4 the oblique straight line again shows that the velocity increases with constant acceleration; but notice that at the starting time the velocity is not zero, as indicated by point *B*.

**GRAPHS** 5–8 represent displacement against time. The horizontal line of graph 5 represents an object whose displacement from the reference point is constant. In other words, the object is at rest. In graphs 6 and 7 the oblique straight lines indicate motion with constant velocity; the velocity described by graph 6 is greater since its steepness (or slope) is greater. In graph 8 the horizontal portion between points *C* and *D* indicates that between the times represented by these two points the object is at rest, as described for graph 5. Up to point *C* the motion is with constant velocity. At point *D* the object starts to move back with constant velocity, and at point *E* the object is back at the starting point: the displacement is zero.

## 12 Motion with Constant Deceleration

If an object is decelerated at a constant rate, the same formulas may be used as given above for motion with constant acceleration *provided the object comes to rest* at the end of time *t*. The letters in the formulas have the same meaning except that *a* stands for the numerical value of the constant deceleration and *v* stands for the initial speed.

**FOR EXAMPLE:** An automobile traveling at a speed of 20 m/sec is brought to a stop in 5.0 sec by a steady application of the brakes. Calculate the deceleration and distance traveled during this interval.

1. $v = at;$

$$20\ \text{m/sec} = a(5\ \text{sec})$$

$$a = 4\ \text{m/sec}^2.$$

2. $s = \frac{1}{2}at^2$

$s = \frac{1}{2}(4\ \text{m/sec}^2)(5\ \text{sec})^2$

$= 50\ \text{m}.$

## 13 Accelerated Motion due to Gravity

We have already described the motion of freely falling objects. The acceleration ($g$) of such objects near the surface of the earth is practically constant. If an object is tossed vertically upwards, the motion is decelerated, the value of the deceleration again being $g = 9.8\ \text{m/sec}^2$. At the highest point of its motion the value of the velocity is momentarily zero. Therefore formulas 5–8 may be used for the upward motion as well as for the downward motion. It also turns out that the time required for the object to go up is equal to the time required for the object to come down.

**EXAMPLE:** A stone is thrown vertically upward. It takes 5.0 seconds for it to reach the highest point. With what velocity was it thrown upward, and how high did it go?

1. $v = gt$

$v = (9.8\ \text{m/sec}^2)(5\ \text{sec})$

$= 49\ \text{m/sec}.$

2. $s = \frac{1}{2}gt^2$

$s = \frac{1}{2}(9.8\ \text{m/sec}^2)(5\ \text{sec})^2$

$= 123\ \text{m}.$

** If an object is projected horizontally into the air, the path it describes is a parabola. The motion can be thought of as a combination of two separate motions: a horizontal motion with constant velocity, the velocity being the one with which it was projected; and a vertical motion due to gravity. The vertical component of the velocity is the same as though the object were falling vertically.

** **EXAMPLE:** A stone is thrown from a bridge. It is thrown horizontally with a velocity of 20 ft/sec and hits the water in 5 seconds. How high is the bridge?

Since the vertical motion is the same as though the object were falling vertically, the information about the horizontal velocity is not needed in this part of the problem. We use the formula for motion of freely falling objects:

$s = \frac{1}{2}gt^2$

$s = \frac{1}{2}(32\ \text{ft/sec}^2)(5\ \text{sec})^2$

$= 400\ \text{ft}.$

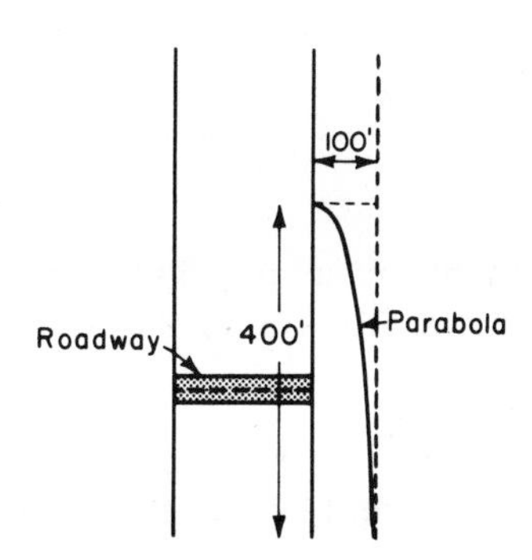

**Note:** The roadway of the George Washington Bridge is 240 ft above the water. However, its towers are 600 ft high.

The horizontal component of the velocity is not affected by the vertical motion. Therefore the velocity in the horizontal direction remains at 20 ft/sec if air resistance is negligible. To calculate the distance traveled by the stone in the horizontal direction use the formula for motion with constant velocity:

$$s = vt$$

$$s = (20 \text{ ft/sec})(5 \text{ sec}) = 100 \text{ ft}.$$

14 ** **General Formulas for Uniformly Accelerated Motion** The above formulas for uniformly accelerated motion are correct when the initial or final velocity is zero. When neither is zero, the following more general formulas apply:

1. $v_f = v_i + at$
2. $s = v_i t + \frac{1}{2}at^2$
3. $v_f^2 = v_i^2 + 2as$
4. $s = \bar{v}t$

where $v_f$ is the velocity or speed after time $t$
$v_i$ is the initial velocity or speed
$t$ is the time for the motion being considered
$a$ is the object's constant acceleration; if the acceleration is due to gravity, $g$ may be used
$s$ is the distance traveled, or displacement of the object, in time $t$
$\bar{v}$ is the average velocity or speed.

EXAMPLE: A car is moving at 15 mi/hr when the driver "steps on the gas," thus accelerating the car to a speed of 45 mi/hr in 10 sec. Calculate the acceleration and the distance traveled during these 10 seconds.

First change the speed to feet per second by multiplying the speed in miles per hour by 22/15.

$$v_i = 15 \text{ mi/hr} = 22 \text{ ft/sec}$$

$$v_f = 45 \text{ mi/hr} = 66 \text{ ft/sec}$$

$$v_f = v_i + at$$

$$66 \text{ ft/sec} = 22 \text{ ft/sec} + (a)(10 \text{ sec})$$

$$a = 4.4 \text{ ft/sec}^2$$

$$s = v_i t + \tfrac{1}{2}at^2$$

$$s = (22 \text{ ft/sec})(10 \text{ sec}) + \tfrac{1}{2}(4.4 \text{ ft/sec}^2)(10 \text{ sec})^2$$

$$= 220 \text{ ft} + 220 \text{ ft}$$

$$= 440 \text{ ft}.$$

# Questions and Problems

1. The speed of an automobile traveling along a straight highway is changed from 50 mi/hr to 80 mi/hr in 5 seconds. Find its average acceleration.

2. A ball rolls down an inclined plane. It has a velocity of 6 ft/sec at the end of the first second, 12 ft/sec at the end of the second second, 18 ft/sec

at the end of the third second. Calculate *a.* its acceleration *b.* the distance it travels during the 3 seconds *c.* the distance it travels during the third second.

3. An object starts from rest and travels with uniformly accelerated motion. If it travels 10 ft during the first second, what is its acceleration?

4. An object is projected upward with a velocity of 150 m/sec. Calculate *a.* the time required for it to come back to the ground *b.* the greatest height it reaches.

5. An object falls freely from rest. What distance does it travel during the first second? Express your answer in meters and in feet.

6. What distance is traveled by an object during the first 4 seconds of free fall?

7. Of the following, which graph best shows the relation between total distance traveled and time during free fall?

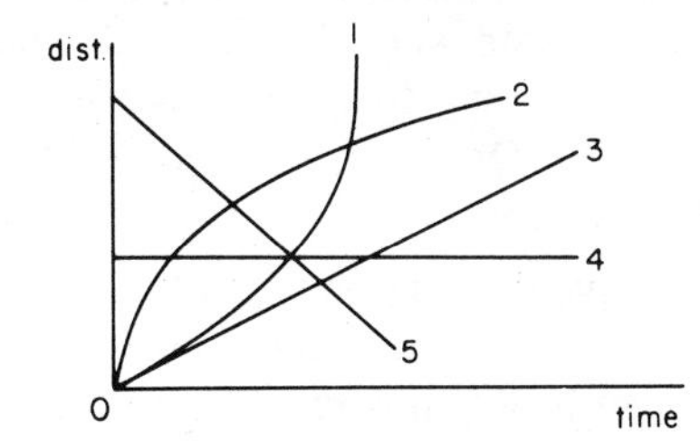


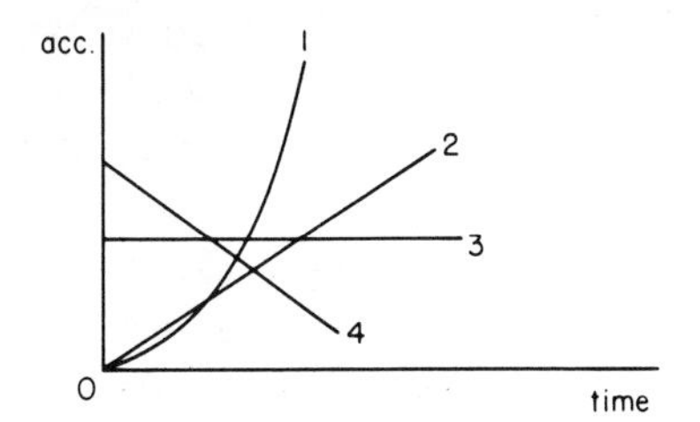


8. Which of the above graphs best shows the relation between acceleration and time during free fall?

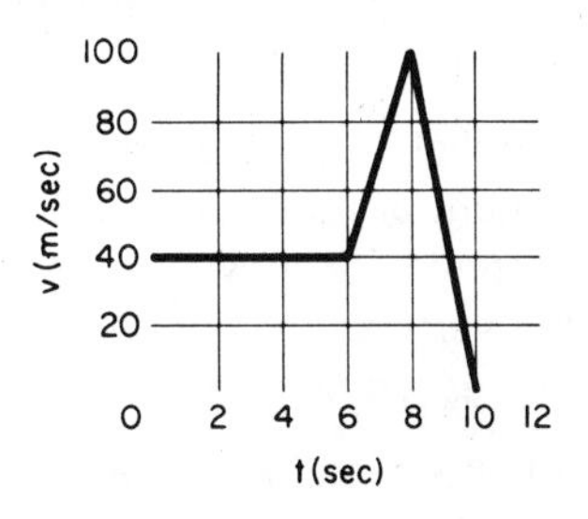


9. The above graph of velocity vs. time represents the motion of a certain object. *a.* What is the velocity of the object when $t = 2$ sec? *b.* What is the acceleration of the object when $t = 2$ sec? *c.* What is the distance traveled by the object during the first five seconds? *d.* What is the velocity of the object when $t = 7$ sec? *e.* What is the acceleration of the object when $t = 7$ sec? *f.* What is the distance traveled by the object during the interval $t = 6$ sec and $t = 8$ sec? *g.* What is the velocity of the object when $t = 9$ sec? *h.* What is the acceleration of the object when $t = 9$ sec? *i.* What is the distance traveled by the object during the interval $t = 8$ sec and $t = 10$ sec? *j.* What is the average speed for the whole trip?

10. An object has a constant acceleration of 6 m/sec$^2$. Assuming it starts from rest, *a.* calculate its velocity at the end of each of the first five seconds *b.* calculate the distance it has traveled at the end of each of the first five seconds *c.* make a graph for its motion of velocity vs. time *d.* make a graph for its motion of distance traveled vs. time.

11. An automobile is traveling along a straight highway at a speed of 60 km/hr when the brakes are applied. It takes 6 seconds for the car to

stop. *a.* Calculate its acceleration. *b.* Calculate the distance it travels in coming to a stop. *c.* Make a graph of velocity vs. time to represent its motion after the brakes are applied. *d.* Make a graph of distance vs. time to represent its motion after the brakes are applied.

12. A boat can travel with a speed of 15 km/hr in still water. If it is used in a river which flows with a speed of 4 km/hr, what will be the boat's velocity when it moves *a.* upstream *b.* downstream *c.* at right angles to the river flow?

13. An airplane is traveling north in still air with a velocity of 200 km/hr. When it is pushed by a westerly wind having a velocity of 100 km/hr, what will be the plane's velocity?

14. ** *a.* What velocity is attained by an object which is accelerated at 0.25 m/sec² for 10 seconds if its initial velocity is 5.0 m/sec? *b.* What distance does the object travel during this time?

15. When Lunar Orbiter I was approximately 550 miles from the moon in August 1966, its retrorockets were fired for 9.5 minutes. This cut its speed from 6000 miles per hour to 2000 miles per hour. Calculate *a.* Orbiter's acceleration *b.* Orbiter's average speed during the 9.5 minutes.

16. A boat moves across a river with a velocity of 12 km per hour at an angle of 30° with the current. *a.* What is the speed at right angles to the current? *b.* If the river is 400 meters wide, how long does it take for the boat to cross it?

17. ** A bullet is fired horizontally from a height of 200 meters and hits the level ground 6000 meters away. With approximately what velocity does the bullet leave the gun?

18. ** A bullet is shot up obliquely from the ground. The horizontal component of its velocity is 40 m/sec. If the bullet remains in the air for 4 seconds, *a.* how far from the gun does it hit the ground? *b.* what is the vertical component of its velocity? *c.* What is the altitude of the highest point in its path?

19. ** A ball is pushed down an inclined plane. It has a velocity of 6 ft/sec at the end of the first second, 10 ft/sec at the end of the second second, 14 ft/sec at the end of the third second. Calculate *a.* its constant acceleration *b.* its initial velocity *c.* the distance it travels during the three seconds *d.* the distance it travels during the third second.

## Test Questions

1. An object moves with a constant velocity of 9.8 m/sec. Its acceleration is, in m/sec², 1. zero 2. 4.9 3. 9.8 4. 32

2. An object moves with a constant acceleration of 15 cm/sec². If it starts from rest, its speed at the end of the 4th second is, in cm/sec, 1. zero 2. 15 3. 60 4. 39.2

3. The distance traveled by the object in question 2 during the first four seconds is, in cm, 1. 15 2. 60 3. 120 4. 240

4. During the third second, the acceleration of the object in question 2 is, in cm/sec², 1. 5 2. 10 3. 15 4. 20

5. A stone is dropped from a bridge which is 256 ft above the water. The time required for the stone to reach the water is, in seconds, 1. 10 2. 8 3. 6 4. 4

6. A rock is tossed upward with a vertical velocity of 49 meters per second. The time required for it to reach its greatest height, in seconds, is 1. 3 2. 5 3. 10 4. 15

7. The maximum height reached by the rock in question 6 is, in meters, approx., 1. 50 2. 100 3. 120 4. 240

8. A speed of 90 mi/hr, expressed in feet/sec, is 1. 90 2. less than one 3. 130 4. greater than 180

9. An airplane moves in level flight with a speed of 90 mi/hr at an altitude of 2.0 miles. If a bag of flour drops out of this plane and air resistance is negligible, the vertical distance the bag drops in 3 seconds is, approx., 1. 2.70 miles 2. 96 ft 3. 48 ft 4. 144 ft

10. The horizontal distance moved by the bag in 3 seconds is, approx., 1. 390 ft 2. 270 miles 3. 600 miles 4. 5 miles

Graphs of velocity vs. time for questions 11–16.

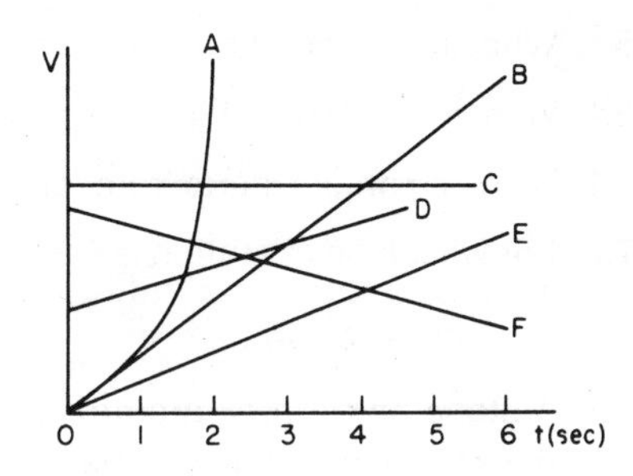


11. For the object in question 1, the graph which best shows how its speed changes with time is 1. A 2. B 3. C 4. E

12. For the object in question 2, the graph which best shows how its speed changes with time is 1. A 2. B 3. C 4. D

13. Of the following graphs, the one which represents motion with the greatest acceleration is 1. B 2. C 3. D 4. E

14. The graph which represents motion with the greatest speed at $t = 1$ sec is 1. A 2. B 3. C 4. D

15. The graph which best represents decelerated motion is 1. C 2. D 3. E 4. F

16. Of the following, the graph which best represents the motion of a stone dropped from a bridge is 1. A 2. B 3. C 4. D

17. A car travels a distance of 400. kilometers in 6 hours. Its average speed for the trip, in Km/hr, is, approx., 1. 67 2. 35 3. 140 4. 2400

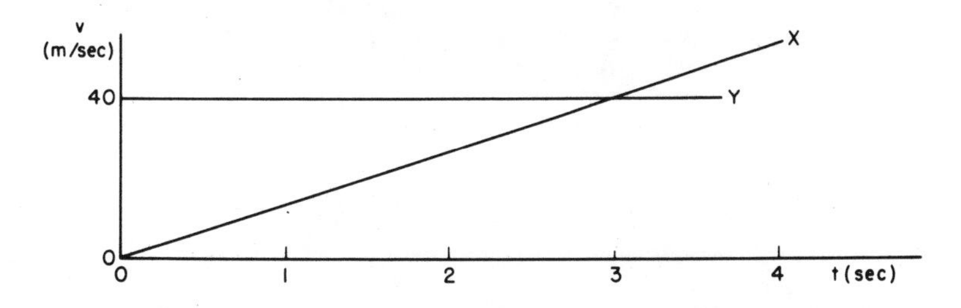

18. The distance traveled in 3 seconds by the object represented by graph Y is, approx., 1. 40 m 2. 120 m 3. 60 m 4. 160 m

19. The acceleration of the object represented by graph X is, in m/sec$^2$, approx., 1. 13 2. 40 3. 120 4. 67

20. The distance traveled in 3 seconds by the object represented by graph X is, approx., 1. 40 m 2. 60 m 3. 80 m 4. 120 m

# Topic Questions

1.1 How fast is the Empire State Building in New York City moving around the earth's axis?

2.1 What is meant by *a.* speed *b.* velocity?

3.1 How does the method for getting the resultant of two velocities compare with the method for getting the resultant of two forces?

4.1 The formula *distance* = *speed* × *time* is correct for two kinds of speed. What are they?

5.1 What is the definition of acceleration?

6.1 Why does the definition of acceleration apply to decelerated motion?

7.1 When an object starts from rest, what is its *initial velocity*?

7.2 For what type of motion is each of these two formulas correct:

$$s = vt; s = \tfrac{1}{2}vt?$$

8.1 A stone and a rock are dropped at the same time from a bridge. How does the acceleration of the stone compare with that of the rock?

11.1 For motion with constant acceleration, what is the graph of acceleration vs. time?

11.2 On a plot of velocity against time, what does the steepness of the graph tell us?

12.1 When dealing with decelerated motion, what do the letters stand for, if we use the formula:

$v = at$?

13.1 When an object is tossed vertically upwards, *a.* what is its speed at the highest point? *b.* how does the initial speed at the bottom compare with the object's speed when it comes down again to the same point?

# 5 Forces and Motion — Newton's Laws

## 1 Introduction

Everyday experience tells us that if we want to set an object into motion we have to push or pull it. In other words, a force is needed to accelerate an object. On the other hand, we can also be misled by everyday experience to believe that a force is needed to keep an object from slowing down. In fact, the ancient Greek philosophers believed that if an object is moving with constant velocity, a force must be acting on it to keep it moving. We no longer believe this is true for all types of motion. We shall discuss this further after we have discussed some types of motion in detail.

*Isaac Newton*
*Niel Bohr Library A.I.P.*

Galileo recognized that the problem is basically the same for accelerated and decelerated motion. It was Isaac Newton (1642–1727) who first clearly stated and developed the *Laws of Motion*.

# 2 Newton's First Law of Motion

One way of stating this law is as follows: *If the net force acting on an object is zero, the velocity of the object does not change.* That is, if the resultant force acting on the object is zero, an object at rest remains at rest, and an object in motion keeps moving with constant speed in a straight line. In other words, if an object is at rest and remains at rest, the net force acting on it is zero; the object is in equilibirum. However, the *First Law* also tells us, that if an object keeps moving with constant speed in a straight line, it too is in equilibrium since the resultant force acting on it is zero. People as well as inanimate matter are subject to the *First Law*. When a bus starts, the passengers tend to *lurch backwards*; when the bus stops suddenly, the passengers *lurch forward.* Actually, of course, the lurching is with respect to the bus. When the bus starts, the passenger tends to remain in his original position with respect to the earth and tends to be left behind in his original position with respect to the earth and tends to be left behind as the bus picks up speed. Seatbelts in automobiles are intended to keep this lurching from becoming too violent when the car decelerates.

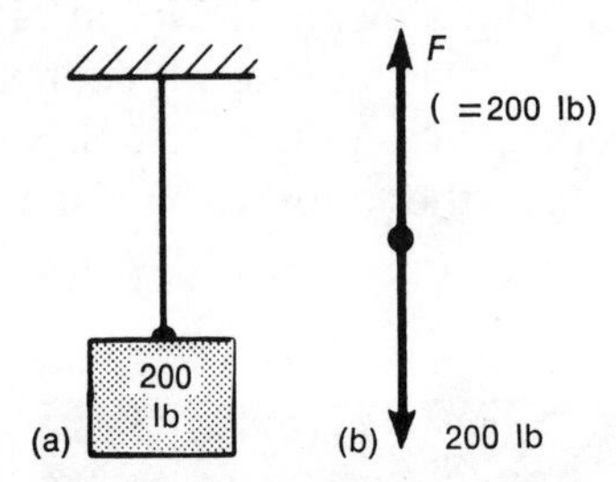


In the figure, a 200-pound object is suspended from a ceiling by means of a rope. The object is at rest; what are the forces acting on it? Since the object is at rest and stays at rest, it is in equilibrium. Therefore *Newton's First Law* applies and the resultant of all forces acting on it is zero. The earth supplies a 200-lb force acting downward (gravity). An equal and opposite force must be applied to the object. This is done by the rope. This pull of the rope is referred to as the tension in the rope. The tension in this rope must be 200 lb. This is shown by the vector diagram where the object is represented by the dot and the rope's upward pull by the vector *F*. Note that the tension of the rope also acts downward on the ceiling, but we are making no use of this because we are concerned with the forces acting on the *object.*

It also follows from *Newton's First Law* that if an object is moving with constant velocity, any force acting on it in one direction, must be balanced by an equal force in the opposite direction.

**EXAMPLE:** Consider this diagram, which represents an object being pulled with constant velocity along level ground by a rope making an angle of 30° with the

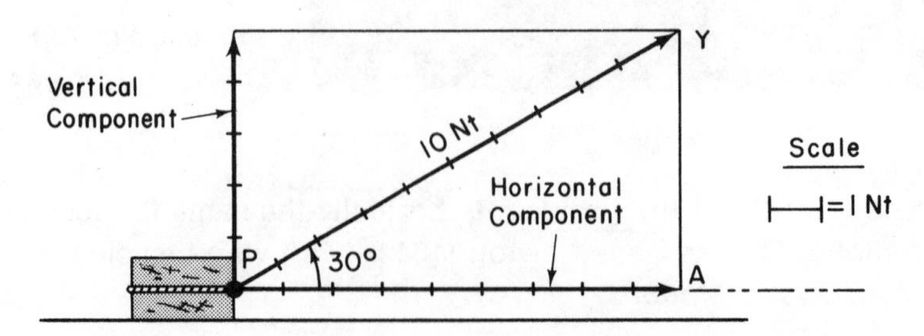

horizontal. The pull in the rope is 10 newtons. We saw in an earlier chapter that the horizontal component of this pull is 8.7 newtons to the right. (*See page 16*). According to the *First Law*, this component must be balanced by another force to the left of 8.7 newtons: exactly equal and opposite to the one towards the right. This force is provided by friction between the ground and the object.

## 3 Newton's Second Law of Motion

We have referred to common experience that we must supply a force to keep a box or sled moving with constant velocity along level ground. The *First Law* suggested that our pull is needed to overcome friction. What if there were no friction? If there is no friction, if there is no force opposing the motion, then whatever force we apply in the direction of motion is the net force, or resultant force, or unbalanced force; (all three terms are used and mean the same thing).

If a net force is applied to an object, the object is accelerated. Experiment shows that the *acceleration is proportional to the net force*; that is, if the force is doubled, the acceleration is doubled; if the force is tripled, the acceleration is tripled. If a force of 10 lb acting on an object produces an acceleration of 4 ft/sec$^2$, then a force of 20 lb acting on the same object will produce an acceleration of 8 ft/sec$^2$.

On the other hand, the same force that is enough to give a 3000-pound car an acceleration of 4 ft/sec$^2$, will produce a smaller acceleration when applied to a heavier truck. It turns out that the *acceleration is inversely proportional to the mass* of the object. If a certain force produces an acceleration of 6 m/sec$^2$ when acting on a 30-kg object, it will produce an acceleration of 12 m/sec$^2$ if applied to a 15-kg object—an acceleration twice as great on an object with one-half the mass.

These ideas are combined in *Newton's Second Law of Motion*—If the net force acting on an object is not zero, the object will be accelerated in the direction of the force; the acceleration will be proportional to the net force and inversely proportional to the mass of the object. With the proper choice of units this can be written as: $a = F/m$. or, in the usual way, as

$$F = ma$$

We shall discuss the proper units shortly.

We can think of the force as being used to overcome the inertia of the object. *Inertia* then is the property of matter by which an object resists being accelerated. Inertia has no units. We measure the inertia of an object by measuring its mass; mass has units (the kilogram, gram, etc.). *Mass* is the measure of an object's inertia. In some respects inertia is like wealth. Wealth has no units. Some wealth, at least, can be measured in terms of money. Money has units (dollars, cents, etc.). In this sense money is a measure of wealth just as mass is a measure of inertia.

The proportion between net force and acceleration can be represented as:

$$\frac{F_1}{F_2} = \frac{a_1}{a_2}$$

When the net force acting on the object is equal to the weight of the object, as in free fall, the acceleration is equal to $g$, the acceleration due to gravity. This leads to the useful proportion for a given object:

$$\frac{F}{W} = \frac{a}{g}$$

Note that the weight $W$ and the force $F$ must be in the same units, and that $g$ and the acceleration must be in the same units.

**EXAMPLE 1.** An automobile weighs 3200 lb. What unbalanced force will produce an acceleration of 4 ft/sec²?

$$\frac{F}{3200\ \text{lb}} = \frac{4\ \text{ft/sec}^2}{32\ \text{ft/sec}^2};$$
$$F = 400\ \text{lb}.$$

**EXAMPLE 2.** What is the acceleration of an object weighing 50 newtons if the net force acting on it is 100 newtons?

$$\frac{F}{W} = \frac{a}{g};$$
$$\frac{100\ Nt}{50\ Nt} = \frac{a}{9.8\ \text{m/sec}^2};$$
$$a = 19.6\ \text{m/sec}^2$$

**EXAMPLE 3.** An object has an acceleration of 4.2 m/sec² when a net force of 6 newtons acts on it. What will be the acceleration when the net force is 15 newtons?

$$\frac{F_1}{F_2} = \frac{a_1}{a_2};$$
$$\frac{15\ Nt}{6\ Nt} = \frac{a_1}{4.2\ \text{m/sec}^2};$$
$$a = 10.5\ \text{m/sec}^2.$$

As was said above, the Second Law can be expressed by the equation

$$F = ma.$$

We must be very careful to use the proper units. In the *MKS* system, we already know, the mass $m$ must be expressed in kilograms, and the acceleration $a$ in meters/sec²; the unit of force $F$ is then in newtons. This formula is actually used to define the newton; one *newton* is the net force which will give an acceleration of one meter/sec² to a mass of one kilogram. In the *CGS* system the corresponding units are gram, cm/sec², and dyne; the dyne is the unit of force in the *CGS* system.

Suppose the net force is the weight ($W$) of the object, as in free fall. The acceleration of the object then is $g$, the acceleration due to gravity. The above equation ($F = ma$) then becomes

$$W = mg.$$

This reminds us of what we already know, namely that the weight of an object is proportional to its mass.

**EXAMPLE:** By using this equation, we can find the weight of a one-kilogram object:

$$W = (1\ \text{kg})(9.8\ \text{m/sec}^2)$$
$$= 9.8\ Nt.$$

The weight of one kilogram is 9.8 newtons. In other words, the earth's pull on a one-kilogram mass on its surface is about 9.8 newtons. Since the weight of 1 kilogram also equals 2.2 lb,

$$2.2 \text{ lb} = 9.8\ Nt$$

$$\text{and } 1 \text{ lb} \doteq 4.5\ Nt$$

** The choice of units can become quite complex, especially in the English system. Is the pound a unit of mass or a unit of force? Usually we use the pound as a unit of force or weight. However, the pound was originally defined as a unit of mass: one pound is equal to approximately 454 grams. Then one pound of force was defined as the earth's pull on one pound of mass. In other words, a mass of one pound has a weight of one pound. If we have a five-pound object, we have an object whose mass is five pounds, and its weight is also five pounds. Of course, it is confusing to use the same word for both units of mass and force; and in the formula, $F = ma$, we must not express both force and mass in pounds. If possible, avoid the use of this formula, and use instead:

$\frac{F}{W} = \frac{a}{g}$, as illustrated above in Examples 1 and 2.

**4 Momentum and Impulse** The momentum of an object is defined as the product of its mass and velocity. It is a vector quantity. Mass is a scalar quantity.

$$\text{momentum} = mv$$

**EXAMPLE:** George drops a 3.0 *kg* object from the Bear Mountain Bridge. What is the object's momentum after 2 seconds? The mass is 3.0 kilogram. We need to know the velocity. For free fall,

$$\begin{aligned} v &= gt \\ &= (9.8 \text{ m/sec}^2)(2 \text{ sec}) \\ &= 19.6 \text{ m/sec.} \end{aligned}$$

$$\begin{aligned} \text{momentum} &= mv \\ &= (3.0\ kg)(19.6 \text{ m/sec}) \\ &= 59.\ kg\cdot\text{m/sec in the downward direction.} \end{aligned}$$

**Note:** That the momentum of an object is proportional to its velocity; if the velocity is tripled, the momentum is tripled.

Athletes, and others, make use of *Newton's Second Law*. To see how this applies, let us use a little algebra. The formula for *Newton's Second Law* can be written in terms of momentum by substituting for the acceleration $a$ its equivalent: $v/t$; or, if the motion doesn't start from rest, $\Delta v/t$.

$$F = ma$$

$$F = mv/t$$

$$\therefore Ft = mv.$$

This shows that the longer the time ($t$) that the net force ($F$) is applied, the greater the change in momentum ($mv$) which is produced. Ballplayers therefore *follow through*, trying to keep the bat, racquet, etc. in contact with the ball as long as possible.

The term *impulse* is sometimes used to refer to the product of net *force* × *time*, Ft. Notice that *impulse equals change in momentum*.

**EXAMPLE:** A net force of 5.0 newtons acts on a 3.0 kilogram object for 2.0 seconds. What is the change in the object's momentum?

$Ft =$ change in momentum

change in momentum $= (5.0\ Nt)(2.0\ \text{sec})$
$= 10\ Nt \cdot \text{sec}$
$= 10\ kg \cdot \text{m/sec}.$

Remember that momentum and impulse are vector quantities.

**EXAMPLE:** Suppose a 2-kilogram ball drops to the ground; it hits the ground with a speed of 3 meters per second and bounces back with the same speed. What is the ball's change in momentum?

Be careful. The speed didn't change, but the velocity did; there was a change in direction. Suppose we make the downward direction negative. Then the original velocity is $-3$ m/sec, and the velocity after rebound is $+3$ m/sec. Momentum = mass × velocity. The downward momentum is $-6\ kg \cdot$m/sec, and the upward momentum is $+6\ kg \cdot$m/sec. The change in momentum is $(+6\ kg \cdot \text{m/sec}) - (-6\ kg \cdot \text{m/sec})$. This equals $12\ kg \cdot$m/sec.

## 5 Newton's Third Law of Motion

This law may be stated as follows: *When one object exerts a force on a second object, the second object exerts an equal and opposite force on the first object.* (This is sometimes stated as: Action equals reaction.)

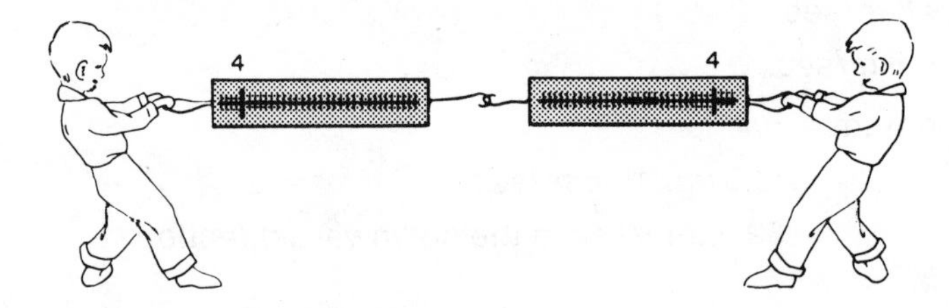


The spring balance on the left pulls on the spring balance on the right with a force of 4 newtons. The spring balance on the right must pull on the spring balance on the left with a force of 4 newtons. If the pull on the left increases, the pull on the right must increase exactly the same amount.

By combining *Newton's Second* and *Third Laws* it is possible to derive the *Law of Conservation of Momentum: When no external force acts on a system of objects, the total momentum of the system is unchanged.* (This is true for collisions, explosions, etc. The objects in the system exert forces on each other, but these are internal forces.)

**EXAMPLE:** A bullet which has a mass of 0.01 *kg* is fired from a rifle with a velocity of 300 meters/sec. If the rifle has a mass of 2.0 *kg*, what is the recoil velocity of the rifle?

The change in momentum of the rifle is equal and opposite to that of the bullet. The bullet and rifle start from rest. Their initial momentum is zero. After firing, the momentum of the rifle is numerically equal to that of the bullet.

$$m_1v_1 = m_2v_2$$
$$(2.0\ kg)v_1 = (0.01\ kg)(300\ \text{m/sec})$$
$$v_1 = 1.5\ \text{m/sec}$$

**ILLUSTRATION:** A rowboat stops at a pier and the boy in the boat steps toward the pier. The boat moves away from the pier and the boy falls into the water or barely makes the edge of the pier.

**EXPLANATION:** In order to move forward from rest the boy must have a force acting on him (*Newton's Second Law*); this force is supplied by the boat as a reaction to the boy's foot pushing against it (*Newton's Third Law*). But when the boy's foot pushes against the boat, it moves away from the pier. Looking at this from the point of view of Conservation of Momentum: The boy and the boat form a system. This system's momentum is zero when the boat stops at the pier. When the boy jumps towards the pier, the boat must simultaneously acquire a momentum equal to that of the boy's, but opposite to it. Both the boy and the boat will move with respect to the spot where the boat has stopped, but the boy will probably have misjudged the needed push.

# 6 Newton's Law of Gravitation

Another great achievement of Isaac Newton was the formulation of the *Law of Universal Gravitation: Any two particles attract each other with a force which is proportional to the product of their masses and inversely proportional to the square of the distance between them.* If the distance between two particles is doubled, the force between them becomes one-fourth as great; if the distance is tripled, the force becomes one-ninth as great.

This law can be applied to large objects also. It applies to the sun's pull on its planets. It also accounts for the weight of objects on the earth, which exerts a gravitational pull on all objects. The distance must be measured from the center of the earth to the object outside the earth. The *weight* of an object is defined as the net gravitational pull on it; for objects near the surface of the earth, the weight of the object is practically the earth's gravitational pull on it. The earth's gravitational pull is also called gravity.

**EXAMPLE:** If the earth's radius is taken as 4000 miles, an object weighing 20 lb on the surface of the earth will weigh only 5 lb at a distance 4000 miles from the earth, because the distance from the center of the earth has been doubled. Of course, the mass of the object has not changed by moving it away from the earth.

Also, the gravitational pull on a six-kilogram object is twice as great as on a three-kilogram object, since the force is proportional to the mass of the object.

Don't overlook the word *Universal*. This means that the same principle applies not only to terrestrial things, but to matter everywhere in the universe. Newton applied it to attraction between the earth and the moon, and we confidently expect it to apply between any two objects, no matter how far from us.

Perhaps you expect physical laws and principles to apply universally, but medieval man thought celestial objects were quite different from terrestrial objects. He took it for granted that the sun and planets are not made of the same matter as the earth. Our chemists and physicists tell us otherwise. Medieval scholars liked to think of the heavenly bodies as moving in circles, *perfect curves*, around the earth. *Kepler's Observations* as explained by *Newton's Laws*, showed that the earth and other planets move around the sun in ellipses.

7 **The Basis of Laws and Theories in Physics** Our knowledge of nature is based on observation and measurement. Our laws of physics are generalizations based on these measurements. Our theories are based on these laws and observations, and are only as good as these. Theories help to unify our knowledge, to explain the unknown in terms of the known, to predict. We observe that the force of attraction between two particles becomes one-fourth as great if we double the distance between them. We generalize and say that the gravitational force varies inversely as the square of the distance between any two particles. Our precision of measurement keeps increasing We may find that the generalization holds no matter how refined our measurement. On the other hand, we may discover that the law is only approximately correct. We must then be willing to change our theory or find a new one. We shall come back to this idea after we have studied more physics.

★ The *Universal Law of Gravitation* can be expressed by the following formula:

$$F = \frac{Gm_1m_2}{r^2}$$ where $F$ is the gravitational pull between the two objects

$m_1$ is the mass of object 1

$m_2$ is the mass of object 2

$r$ is the distance between the two objects

$G$ is the universal gravitational constant (in the *MKS* system the value of $G$ is $6.67 \times 10^{-11}$ $Nt \cdot m^2/kg^2$

8 **Triumph of Theory** One of the most beautiful illustrations of how laws and ★★ theories sometimes develop in physics is the explanation of planetary motion. Many centuries ago, people who watched the sky noticed that although most stars kept the same relative position night after night, as they appeared to move in circles around the earth, a few changed their position relative to the others. The latter are called planets (from a Greek word meaning wanderer).

**Aristotle** (4th century B.C.) and **Ptolemy** (2nd century A.D.) worked out complicated explanations for the motions of the planets on the assumption of a geocentric universe, i.e. on the assumption that the sun, moon, and stars move around the earth. They used intricate combinations of circles. **Aristarchus** (3rd century B.C.) suggested that the sun should be considered the center around which the planets move, but practically no attention was paid to this.

**Copernicus** (1473–1543) showed by calculations that the observed motions are simply and well described if one assumes that the earth and planets move in circles around the sun — a heliocentric universe. Most of his contemporaries would not think of accepting such a point of view. After all, it is not easy to stop thinking of the earth as something unique. It is difficult to adopt the point of view, that man's home is just another planet moving around the sun; it certainly is not obvious.

**Tycho Brahe** (1546–1601) was a superb astronomer and accumulated volumes of data on the motion of the planets. **Johannes Kepler** (1571–1630) became his assistant and eventually inherited the volumes. Kepler tried to fit the data into the Copernican system. He found it necessary to make an important modification: the path of a planet around the sun is an ellipse, not a circle.

Kepler, who was a gifted mathematician rather than an experimenter, was able to fit the tremendous amount of data into three laws. We have already mentioned the first one.

1. Planets move in elliptical paths. The sun is at one focus of the ellipse.

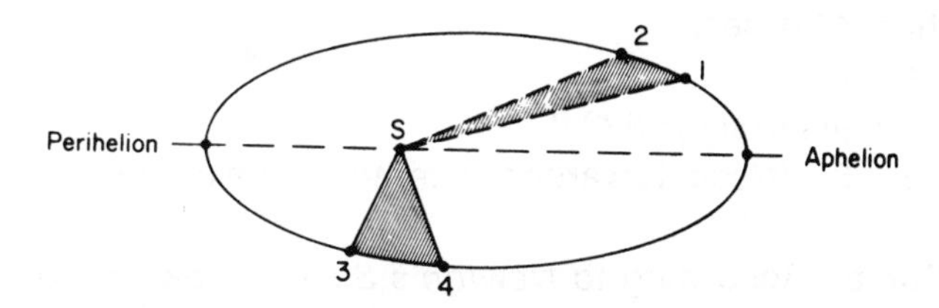


2. During equal time invervals, the line drawn from the sun to the planet sweeps out equal areas. In the diagram, it takes the planet the same time to go from position 1 to 2, as from position 3 to 4. Therefore the two shaded areas are equal. (Note that the planet moves faster when it gets closer to the sun.)
3. The periods or times ($T$) required for any two planets to complete their elliptical paths are related to their average distances from the sun ($R$) by the proportion

$$\frac{T_1^2}{T_2^2} = \frac{R_1^3}{R_2^3}$$

Notice that *Kepler's Laws* beautifully and rather simply summarize volumes of data. With their help it is possible to predict the position of planets. However, the Laws gave no reason for the motion. Why do planets move in elliptical paths?

This question was answered by Newton (1642–1727) when he formulated the more general *Law* (or *Theory*) *of Universal Gravitation*. Some more advanced books give the derivation. This Law was a powerful tool for explaining heavenly and terrestrial phenomena, but its validity was questioned in 1830. The planet Uranus had been discovered in 1781. On the basis of *Newton's Theory* and a few observations, the elliptical path of Uranus was predicted. But by 1830 it was obvious that Uranus was not following this predicted path.

Is there something wrong with the Theory? If so, it must be abandoned or modified. Two young astronomers, Adams and Leverrier, independently did some calculating and predicted that if a really good telescope were pointed in a certain direction, another new planet would be found. This was done, and Neptune was discovered. The Theory survived gloriously.

9 * **Centripetal Acceleration** If an object moves with constant speed around a circle, its velocity is not constant. Its velocity keeps changing because the direction of its

motion keeps changing. (If an automobile makes a turn at 20 miles per hour, it may have been going north before the turn and west after the turn. Although its speed remained at 20 mi/hr, its velocity changed because its direction changed.) Since its velocity changed, the object was accelerated. Acceleration required for motion around a circle is known as *centripetal acceleration*. As for all acceleration, it is a vector quantity; its direction is towards the center of the circular path. Its magnitude is proportional to the square of the speed of the object and inversely proportional to the radius of the path

$a = \frac{v^2}{r}$ where $a$ is the centripetal acceleration
$v$ is the speed around the circle
$r$ is the radius of the circle

**EXAMPLE:** An automobile moving at 30 miles per hour goes around a curve having a radius of 330 ft. What is its acceleration?

In order to have consistent units, we change the speed to feet per second.

30 mi/hr = 44 ft/sec.

$a = v^2/r$

$a = (44 \text{ ft/sec})^2/330 \text{ ft}$

$a = 5.9 \text{ ft/sec}^2$ towards the center of the curve.

**10** * **Centripetal Force** According to *Newton's Second Law of Motion*, a net force is needed to produce an acceleration; therefore, a force is needed to produce the centripetal acceleration. This is known as the centripetal force. The *centripetal force* is the force required to keep an object moving at constant speed in a circle. It is directed towards the center of the circle and is proportional to the mass of the object and its centripetal acceleration. It is therefore also proportional to the square of the object's speed and inversely proportional to the radius of the circle. If the speed of the object is tripled, the required centripetal force is multiplied by $3^2$, or 9.

$F = mv^2/r.$

The centripetal force on the moon and artificial earth satellites is supplied by the earth's gravitational pull. Strictly speaking, therefore, it is not correct to say that the satellite or the astronaut in space is weightless, as long as we define weight as gravitational pull. As was pointed out earlier, 4000 miles from the earth, gravity is one-fourth as great as on the surface of the earth. An astronaut orbiting 4000 miles above the earth weighs one-fourth as much as on earth and that is also the centripetal force on him.

$a = v^2/r$

$F = mv^2/r$

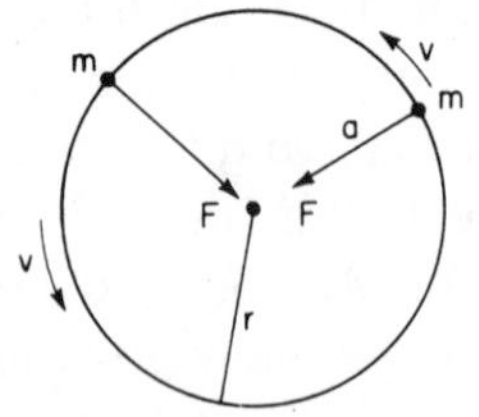

If we whirl a stone at the end of a string, the string pulls in on the stone; the string supplies the centripetal force. The stone reacts and pulls outward on the string. This equal and opposite force pulling on the string is sometimes called a centrifugal force. If the string breaks, both centripetal and centrifugal force disappear simultaneously, since one is a reaction to the other, and the stone tends to go off with constant velocity tangent to the circle in accordance with *Newton's First Law*. The centripetal force is the unbalanced force acting on the object moving in the circle.

**11** **Friction** is a force which always opposes motion or a tendency for motion. It acts between two surfaces in contact.

**EXAMPLE:** Imagine a block resting on a horizontal surface. If a one-pound force is applied toward the left and the block doesn't move, a one-pound force must be acting toward the right on the block. If the pull is increased to two pounds and the block doesn't move, the force to the right must have increased to two pounds;

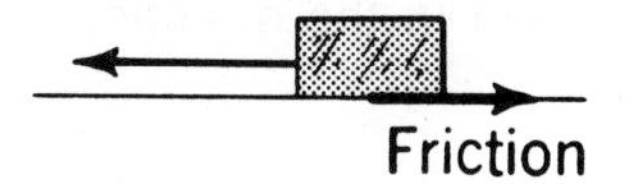


otherwise, there would be a net force and the block would accelerate. In each case the force to the right is friction, acting on the contact surface. If the force toward the left is made sufficiently large the block starts to slide. Once sliding has been produced the force of friction remains practically constant.

Sliding friction is practically independent of the speed with which the object moves and of the area of contact (unless it is point-like). Friction does depend on the nature of the surfaces in contact, on the material, and on the force pushing the surfaces together; this is the so-called *normal force*, or the force perpendicular to the surfaces in contact.

**EXAMPLE 1.** In the above figure, assume that the block has a mass of 10. *kg* and that it is being pulled by a force of 18 newtons. If the force of friction is 2.0 newtons, what is the acceleration of the object?

$$F = ma$$
$$(18 - 2.0)Nt = (10.\ kg)a$$
$$a = 1.6\ \text{m/sec}^2.$$

Note that in the *MKS* system, the force is in newtons and the mass in kilograms; the acceleration then is in meters/sec$^2$.

**EXAMPLE 2.** Assume that the block in Example 1 has been accelerated to a speed of 32 m/sec when the force of 18 newtons stops. What will be the motion of the object?

Since the block is moving, the force of friction still acts and is 2.0 newtons. It is now the net force and causes the block to decelerate.

$$F = ma$$
$$2.0\ Nt = (10\ kg)a$$
$$a = (2.0\ Nt)/(10\ kg)$$
$$a = 0.20\ \text{m/sec}^2$$

The deceleration of the block is 0.20 m/sec$^2$.

12 ** **Coefficient of Friction** The maximum force of friction between two materials is proportional to the normal force and depends on the materials used. This characteristic of the materials is described by a number called the coefficient of friction:

$$\text{coefficient of sliding friction} = \frac{\text{friction during motion}}{\text{normal force}}$$

EXAMPLE: A 30-pound object is dragged along a horizontal surface. The coefficient of friction is 0.20. How large is the force of friction?

The weight is the normal force.

$$0.20 = \text{friction}/30 \text{ lb};$$

$$\text{friction} = 0.20 \times 30 \text{ lb}$$

$$= 6.0 \text{ lb}$$

If the force dragging the object in the direction of motion is greater than 6.0 lb, accelerated motion results.

## Questions and Problems

1. The speed of an object keeps increasing. Explain why each of the following also keeps increasing. *a.* velocity *b.* momentum
2. A 10-pound object rests on a level table. What are the forces acting on the object? Why does the object stay at rest?
3. An artificial satellite moves around the earth at constant speed. *a.* What provides the centripetal force? *b.* Since the satellite is moving at constant speed, why is the acceleration not equal to zero?
4. An automobile weighs 4800 lb. Calculate the force needed to give it an acceleration of 10 ft/sec$^2$.
5. The force needed to give an automobile an acceleration of 10 m/sec$^2$ is *F*. What force is needed to give the automobile an acceleration *a.* twice as great? *b.* three times as great? *c.* one-half as great?
6. An object has a mass of 15 kilograms. What force is needed to give it an acceleration of *a.* 5 m/sec$^2$? *b.* 10 m/sec$^2$? *c.* 20 m/sec$^2$?
7. A force of 12 newtons gives an object an acceleration of 4.0 m/sec$^2$. Calculate *a.* the mass of the object *b.* the momentum of the object at the end of the third second.
8. A force of 30 newtons gives an object an acceleration of 8.0 m/sec$^2$. Calculate *a.* the mass of the object *b.* the momentum of the object at the end of the fourth second. *c.* the change in momentum during the fourth second.
9. Calculate the weight of a 15-kilogram object in *a.* newtons *b.* pounds.
10. What happens to the mass and weight of a 15-kilogram object when it is moved from Mt. Everest to the North Pole?
11. An object weighs 15 newtons. What force is needed to give it an acceleration of 98 m/sec$^2$?
12. A net force of 5.0 newtons acts on an object for 4.0 seconds. Calculate *a.* the impulse acting on the object *b.* the change in momentum of the object.
13. If the object in problem 12 has a mass of 100 grams, calculate *a.* its acceleration *b.* its speed at the end of 4.0 seconds.

14. A 40-gram bullet is fired from a 40-kilogram gun with a velocity of 300 meters/sec. Calculate *a.* the momentum of the bullet *b.* the speed of recoil of the gun.

15. Matilda weighs herself on a bathroom scale. As she steps off the scale she notices that the reading on the scale goes up before returning to zero. Why?

16. An elevator weighs 500. kilograms. Calculate *a.* its weight in newtons *b.* the force needed to move it up at constant speed, if friction is negligible. *c.* the additional force needed to accelerate it upward with an acceleration of 4.9 m/sec$^2$.

17. Calculate the force needed to give an electron an acceleration of 5.0 m/sec$^2$. (See the Physics Reference Tables in the Appendix for any information you may need.)

18. The speed of an object moving in a circular path of constant radius is changed from 4.0 feet per second to 8.0 feet per second. What happens to *a.* its centripetal acceleration? *b.* the centripetal force acting on it? *c.* its mass?

19. The gravitational force acting between two particles is *n* newtons. What is the value of this force, if the distance between the two particles is *a.* doubled? *b.* tripled? *c.* halved?

20. * A 2.0 kilogram object is moving in a circular path at a speed of 4.0 meters/sec. The radius of the path is 3.0 meters. Calculate *a.* the centripetal acceleration *b.* the centripetal force.

21. A net force of 60. newtons acts on a 40-kilogram object to change its linear velocity from 30. to 50. m/sec. Calculate *a.* the acceleration *b.* the initial momentum *c.* the change in momentum *d.* the required time.

22. A net force of 5.0 newtons acts on an object for 8.0 seconds. What is the object's change of momentum?

23. A 300-gram bomb rests on a table. It explodes into 2 pieces. One piece having a mass of 100. *gm* flies off with a speed of 10. m/sec. What is the speed and direction of the other piece?

24. A 500-gram block is projected along a horizontal table with an initial speed of 10 cm/sec. Friction brings the block to rest in 0.30 second. Calculate the force of friction.

25. A 15-kilogram mass is moving north with a constant velocity of 20 meters per second. A force of 20. newtons acting north is then applied to it for 2.0 seconds. Calculate: *a.* the original momentum of the object, *b.* its final momentum, *c.* its acceleration.

26. A 15-gram bullet is fired horizontally with a velocity of 820 meters per second. It strikes a wooden block which is suspended nearby and remains embedded in it. Calculate the velocity of the block immediately after it is struck by the bullet. The mass of the block is 1.5 *Kg*.

27. * Calculate the centripetal force on a 15-kilogram object moving in a circular path with a speed of 12 meters per second, if the radius of the path is 3.0 meters.

28. When the brakes are applied in a car moving with a speed of 30 Km/hour, it stops after moving an additional distance of 4 meters. How far will the car move if the same braking force is applied when the car is moving at 60 Km/hour?

## Test Questions

1. Which of the following is a unit of force? 1. meter 2. newton 3. slug 4. kilogram-second
2. As the mass of the contents of a box is increased, the force required to give all of it a constant acceleration 1. decreases 2. increases 3. remains the same.
3. If the distance between two objects of constant mass is doubled, the gravitational force between them becomes 1. one-fourth as great 2. one-half as great 3. twice as great 4. four times as great.
4. A force of 120 newtons acts on a mass of 60 *kg*. If no other force acts on the object, the acceleration, in meters/sec$^2$, is 1. 18 2. 1800 3. 0.50 4. 2.0.
5. A 4-kilogram mass moves with a velocity of 10 meters/sec for 5.0 seconds. Its momentum is ............
6. A 20-kilogram mass is raised 5 meters above the floor. Its weight, in newtons, is ............
7. An automobile weighs 3200 lb. The net force needed to give it an acceleration of 8.0 ft/sec$^2$ is ............
8. A man fires a rifle. The bullet from this gun has a momentum of 20 *kg*-m/sec. The magnitude of the momentum of the rifle under these conditions is, in *kg*-m/sec, 1. 20 2. less than 20 3. more than 20.
9. If an object with a mass of 6.0 *kg* experiences a force of 8.0 newtons for 5.0 seconds, its change in momentum, in kilogram-meters/sec, is 1. 30 2. 40 3. 48 4. 480.
10. * If the speed with which an object goes around a circle is doubled, its centripetal acceleration is 1. doubled 2. quadrupled 3. halved.
11. * If the speed with which an object goes around a circle of constant radius is doubled, the centripetal force acting on the object is multiplied by a factor of 1. 1 2. 2 3. 3 4. 4.
12. The product of a force and the time interval during which it acts is called ............
13. At a height of 16,000 miles above the earth's surface, a person who weighs 150 lb on the earth would weigh ............
14. The centripetal force acting on a stone moving with constant speed in a circular path varies inversely as ............
15. As a distant meteor approaches a planet with no atmosphere, its acceleration toward the planet ..........
16. As a constant unbalanced force acts on an object, its acceleration ............
17. When a kilogram of stainless steel is taken from sea level to the top of the Empire State Building, the weight of the kilogram ............
18. * The centripetal force which keeps a satellite in orbit about the earth is known as ............
19. As the force exerted by one object on another increases, the force exerted by the second object on the first ............
20. If the mass of one of two objects is doubled while the distance between them is also doubled, the gravitational force between them becomes ........... as great.

# Topic Questions

1.1 If an object is to accelerate, what must act on the object?

2.1 According to *Newton's First Law*, how large a force is acting on an object which moves with constant velocity?

2.2 What objects are in equilibrium other than those which remain at rest?

3.1 What are two terms which mean the same thing as *resultant force*?

3.2 To what is the acceleration of an object proportional?

3.3 How are mass and inertia related?

3.4 How are mass and weight related?

3.5 What is the weight of a one-kilogram box?

4.1 What is meant by momentum? Is it a vector or a scalar quantity?

4.2 What is meant by the impulse acting on an object?

5.1 What is the *Law of Conservation of Momentum*?

6.1 What is *Newton's Law of Gravitation*? To what part of the Universe does it seem to apply?

7.1 On what do the laws and theories of physics depend?

7.2 How does the gravitational pull depend on the distance between objects?

8.1* What is a heliocentric universe?

8.2* What is one important difference between the description of the universe given by Kepler and by Copernicus?

9.1* How does centripetal acceleration depend on speed?

10.1* In what direction does the centripetal force act?

11.1 In what direction can friction act on an object which is not moving?

11.2 On what does the friction between two solids depend?

# 6 Work, Energy, Power

## 1 Work

Most of us would readily agree that, in order to hold the earth, Atlas would have to do a great deal of work. However, the physicist says that Atlas would have to exert a great force but that he would do no work in holding the earth. It is a matter of definition.

*Work* is defined as the product of a force and the displacement in the direction of the force.

$$\text{work} = \text{force} \times \text{displacement}$$

**EXAMPLE:** If we pull a 50-pound object along the ground, we say that we are doing work on the object. How much we do depends on how far we move the object, and what force we exert. You might think that we have to use a 50-pound force to pull a 50-pound object. This is usually not so. If there were no friction we would have to exert no force to keep the object moving with constant velocity along the level ground. Suppose that a 20-pound force is used to drag the object 30 ft.

$$\text{work} = (20 \text{ lb})(30 \text{ ft}) = 600 \text{ foot-pounds}$$

In the British system the unit of work is the foot-pound (ft-lb). In the *MKS* system the unit is the *joule*. (In the *CGS*-system the unit is the erg.) If a force of 20 newtons is used to drag a 50-newton object along the level ground for a distance of 30 meters,

$$\text{work} = (20\ Nt)(30\ m) = 600 \text{ joules.}$$

In these examples it was implied that the force used to pull the object is in the same direction as the displacement. If it were not we would have to use its component in the direction of motion. For example, if a rope is used to pull a sled with a force

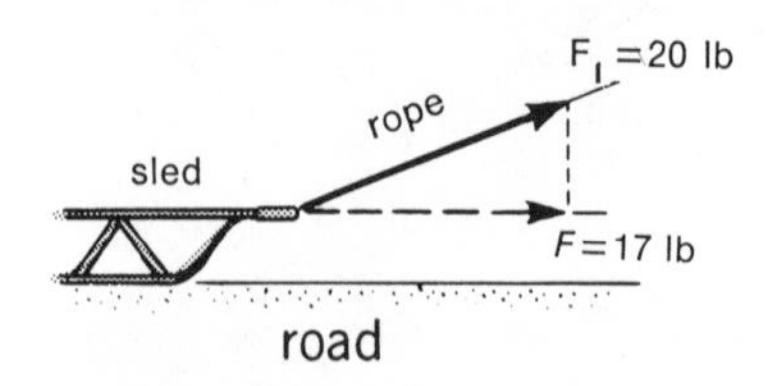


Work= 17 lb x 50 ft

($F_1$) of 20. pounds, as shown in the diagram, the sled moves along the level road, let us say, a distance of 50 ft. In order to calculate the work done on the sled we must multiply $F$, the component of the force in the direction of motion, by the distance of 50 ft. If the rope makes an angle of 30° with the road, the horizontal component is approx. 17. lb.

$$\text{Work} = 17 \text{ lb} \times 50 \text{ ft} = 850 \text{ ft-lb.}$$

** Work = force × displacement × cos $\theta$, where $\theta$ is the angle between the displacement and the direction of the force.

When we lift an object we do work on the object. If we lift the object with constant velocity, the force we have to exert is equal to the weight of the object. Therefore, when *lifting*,

$$\text{work} = \text{weight} \times \text{height}$$

**EXAMPLE:** If we lift a 100-pound box 3 feet, the work we do is:

$$\text{work} = (100 \text{ lb})(3 \text{ ft}) = 300 \text{ ft-lb.}$$

If we lift a box weighing 60 newtons to a height of 4.0 meters, the work we do is:

$$\text{work} = (60\ Nt)(4.0\ m) = 240 \text{ joules.}$$

Note that a *joule* is equivalent to a newton-meter. It is the work done when a force of one newton acts through a distance of one meter in the direction of the force.

Suppose you hold a 50-pound crate next to your chest and walk 100 feet; how much work do you do? You do no work on the crate if you walk along level ground! The force you exert on the crate is 50 pounds, but you exert it upwards (opposed to gravity); the displacement is at right angles to it. The horizontal component of the 50-pound force is zero. Zero times 100 feet is zero.

## 2 Power

*Power* is the time rate of doing work. It is the work divided by the time it takes to do the work.

$$\text{Power} = \text{work/time}$$

If the above work of 240 joules was done in 3.0 seconds, the power used is (240 joules)/(3.0 sec) = 80 joules/sec, or 80 watts.

$$1 \text{ watt} = 1 \text{ joule/sec}$$

$$1 \text{ kilowatt } (Kw) = 1000 \text{ watts.}$$

In the MKS system, the watt is a unit of power. In the British system we use the foot-pound per second or the horsepower (hp).

$$1 \text{ hp} = 550 \text{ ft-lb/sec}$$

$$1 \text{ hp} = 746 \text{ watts.}$$

Since work = force × displacement, the formula for power can also be written as:

$$\text{power} = \frac{\text{force} \times \text{displacement}}{\text{time}}$$

Since displacement over time is equal to velocity, another formula is:

$$\text{power} = \text{force} \times \text{velocity}$$

**EXAMPLE 1.** A motor is used to lift a 500-pound box to the roof. If the box is lifted 40 ft in 10 seconds, what are the work and power outputs of the motor?

$$\begin{aligned} \text{work} &= \text{weight} \times \text{height} \\ &= (500\ \text{lb})(40\ \text{ft}) \\ &= 20{,}000\ \text{ft-lb.} \\ \text{power} &= \text{work/time} \\ &= (20{,}000\ \text{ft-lb})/(10\ \text{sec}) \\ &= 2000\ \text{ft-lb/sec} \\ &= (2000/550)\ \text{hp} = 3.6\ \text{hp} \end{aligned}$$

**EXAMPLE 2.** A motor whose power output is 150 watts pulls a load to the roof with a force of 30 newtons. With what speed can the motor pull the load?

$$\begin{aligned} \text{power} &= \text{force} \times \text{velocity} \\ 150\ \text{watts} &= (30\ Nt)(\text{velocity}) \\ v &= (150\ \text{watts})/(30\ Nt) \\ &= 5.0\ \text{meters/sec} \end{aligned}$$

(Note that the units check. On the right side of the equation (line 2) we have: (newtons × meters/sec) which equals joules/sec or watts.)

# 3 Energy

In elementary physics *energy* is defined as the *ability to do work*. If we do work we have less energy left. One of the reasons we eat is to get energy to do work. In mechanics, work is done on an object, 1. to give it potential energy, 2. to give it kinetic energy, 3. to overcome friction, 4. to accomplish a combination of the above three. *Energy used to overcome friction is converted to heat.* Work, power, and energy are scalar quantities. *Units* of energy are the same as units of work: foot-pound, joule, erg, etc.

**Potential Energy** is the energy possessed by an object because of its position or distortion. Under ideal conditions it is equal to the work required to bring the object to this position or distortion. Instead of distortion we sometimes say con-

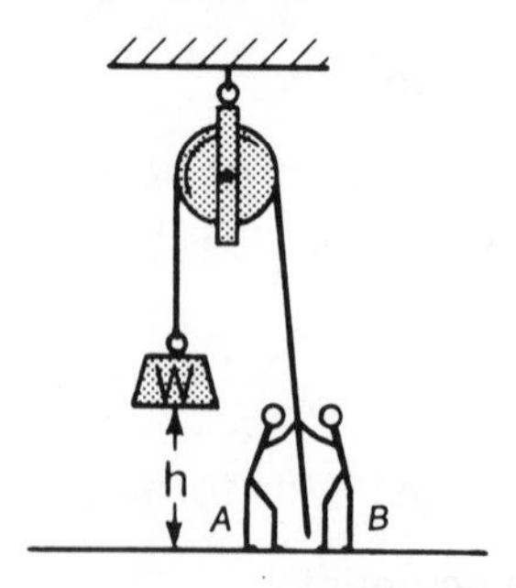


dition. For example, we have to do work to stretch a spring. As a result, potential energy is stored in the spring. (This is sometimes called elastic potential energy.) Now the spring can do work, such as lifting a weight or turning the gears in a watch.

If we lift an object we change its position with respect to the center of the earth; we do work against the force of gravity. As a result the lifted object has a greater ability to do work. This increased ability to do work is the object's potential energy with respect to its original position. (This is sometimes called gravitational potential energy.) This can be illustrated by referring to the diagram. Boys $A$ and $B$ have lifted the object of weight $W$ a vertical distance $h$. The object can now do work: if $A$ lets go, the object may now be able to lift $B$ thus doing work on $B$. If the weight is 80. newtons, and $h = 4.0$ meters, the object gained a potential energy of 320 joules.

$$\text{Potential energy} = \text{weight} \times \text{vertical height}$$

Note that this is a change in potential energy with respect to the given level, the original position. The boys may be lifting the object in the laboratory on the second floor. Then, with respect to street level, the object has an additional potential energy since someone had to do work to lift the object from street level to the second floor.

4 **Kinetic Energy** is the energy an object has because of its motion. We know that winds (moving air) do work which stationary air cannot do: turn windmills, raise roofs, etc.

The kinetic energy of an object is proportional to the square of its speed. If a car's speed is changed from 30 miles per hour to 60 miles per hour, its kinetic energy has become four times as great. Also, a truck moving at 30 miles per hour has a great kinetic energy when loaded than when empty; i.e. kinetic energy is also proportional to the mass of the object.

Kinetic energy $= \frac{1}{2}mv^2$ where $m$ is the mass of the object
and $v$ is its speed.

**EXAMPLE:** A 10-kilogram object falls from a bridge. What will be its kinetic energy at the end of 2.0 seconds?

$$\begin{aligned} v &= gt \\ &= (9.8\ \text{m/sec}^2)(2\ \text{sec}) \\ &= 19.6\ \text{m/sec} \end{aligned}$$

$$\begin{aligned} \text{kinetic energy} &= \tfrac{1}{2}mv^2 \\ &= \tfrac{1}{2}(10\ kg)(19.6\ \text{m/sec})^2 \\ &\doteq 2000\ \text{joules} \end{aligned}$$

(Note that a $kg\text{-m}^2/\text{sec}^2 = (kg\text{-m/sec}^2) \times \text{meter}$
$= \text{newton} \times \text{meter} = \text{joule}$)

5 **Work Done Against Friction** We know that friction is a force which opposes motion or a tendency for motion; the force acts on the surfaces in contact. When the two objects in contact are both solids, the friction is practically independent of the speed. When an object moves through a fluid (gas or liquid), friction on the object does depend on the speed with which it moves through the fluid. For example, as rain drops fall through the atmosphere, the frictional force acting on the rain drop increases as its speed increases, until friction equals the weight of the raindrop. Maximum velocity is then reached and is known as *terminal velocity*.

When an object moves against friction, work is done. The work done against friction does not increase the potential or kinetic energy of the object. The energy required to do the work against friction is converted to heat or internal energy. This will be discussed later.

work against friction = friction × distance object moves

**EXAMPLE:** In the above problem with the sled, p. 58, if the sled moves with constant velocity, the 17-pound force is just enough to balance the force of friction; the force of friction is 17 lb in the opposite direction. The work done against friction is 850 ft-lb. If a force $F$ greater than 17 lb were used ($F_1$ greater than 20 lb), the opposing force of friction would still be 17 lb. but now there would be a net force in the direction of motion, and the sled would accelerate.

**6 Principle of Conservation of Energy** Energy cannot be created or destroyed, but may be changed from one form into another. Kinetic energy may be changed to potential energy, and vice versa. Mass can be considered a form of energy as a consequence of Einstein's Theory of Relativity. When mass is converted to forms of energy, such as heat, the

$$\text{energy produced} = mc^2$$

where $m$ is the mass converted and $c$ is the speed of light. When $m$ is expressed in kilograms and $c$ in meters/sec ($c = 3 \times 10^8$ $m$/sec), the energy will be in joules. Note that this formula does not have a factor of one-half. Under ordinary conditions the energy-mass equivalence can be neglected.

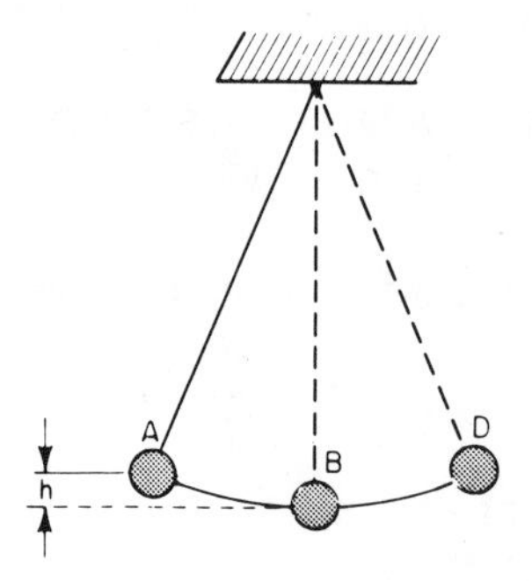


In a swinging pendulum kinetic energy is changed to potential energy, and potential energy back to kinetic energy. At the highest point reached by the pendulum bob, its speed is momentarily zero; its kinetic energy there is also zero. Let us assume this is point $A$ in the diagram. At the lowest point, $B$, the potential energy of the bob (weight $W$) is least. The loss in potential energy is $Wh$ (weight times vertical height), and all this has been converted to kinetic energy, if frictional losses are negligible. Because of inertia the bob keeps moving towards $D$; and as it rises it gains potential energy at the expense of kinetic energy: it slows down again. The highest point it can reach is again at the same height $h$, when all the kinetic energy will be changed back to potential energy. In practice the bob rises a smaller and smaller distance because of friction.

** It can be shown that the speed which the pendulum bob has at its lowest point, $B$, is given by the formula

$v = \sqrt{2gh}$ where $g$ is the acceleration due to gravity, and $h$ is the vertical height that it rises.

★★ The period ($T$) of a pendulum is the time it takes to swing from $A$ to $D$ and back to $A$. For a simple pendulum this time is given by:

$T = 2\pi\sqrt{L/g}$ where $L$ is the length of the string.

Suppose we pull an object up a flat, sloping surface (known as an *inclined plane*). When it reaches the top, the object has gained potential energy equal to the product of its weight and the vertical height.

potential energy $= Wh$.

If the object weighs 20. newtons and $h = 3.0$ meters, the gain in potential energy is 60. joules.

The work we do in pulling the object to the top of the plane is equal to the product of the force we use and the displacement in the direction of the force.

work done $= FL$, where $L$ is the length of the plane.

Ideally, if friction is negligible, that is, no work has to be done against friction, the energy we use to do the work is converted to potential energy gained by the object:

$FL = Wh.$

This can be written as a proportion:

$$\frac{F}{W} = \frac{h}{L}$$

(Check this by cross-multiplying.)

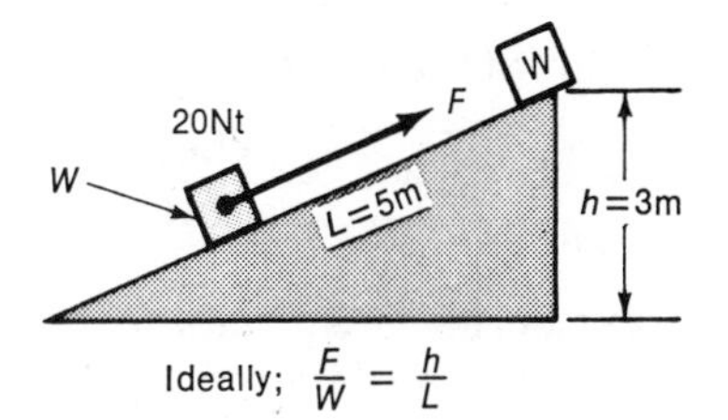


Therefore, the force we have to exert parallel to the plane to bring the object to the top is, in this problem, 20 Nt $\times$ 3/5, or 12 newtons.

If the object slides down the incline, and friction is negligible, what will be its kinetic energy when it gets to the bottom? According to the Principle of Energy Conservation, the kinetic energy at the bottom is equal to the potential energy at the top, or 60 joules.

## 7 Simple Machines

★★

Probably the most direct way of doing useful work on an object is to take hold of the object and lift it. This is often difficult or inconvenient, and we turn to machines to help us. A *machine* is a device which will transfer a force from one point of application to another for some practical advantage.

**FOR EXAMPLE:** Each morning a flag may have to be displayed from the top of a pole and taken down in the evening. It is inconvenient to climb to the top of the pole each day and pull the flag up. Instead, we keep a pulley at the top of the pole and with a rope apply a downward force at $A$ and produce an upward force on the flag at $B$. The pulley was used to change the direction of the force (from downward to upward) and the point of application of the force (from $A$ to $B$).

The pulley is an example of a *simple machine*; others are: the lever, wheel and axle, inclined plane, screw, and wedge. The force which we apply to the machine

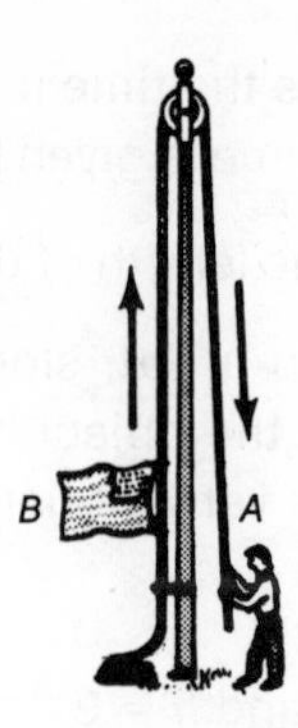


in order to do the work (such as at $A$ in the above case) is known as the *effort*, $E$. The force which we have to overcome (such as the weight at $B$) is known as the *resistance*, $R$.

8 **Mechanical Advantage** In the above case we used the pulley merely to change
** the direction of the force; the flag had to be pulled up, but we find it easier to pull down. Because of friction the actual effort we had to use may have been greater than the weight of the flag. Usually we use a machine because with it the effort we have to exert is less than the actual weight we are trying to lift, i.e. less than the resistance. The *actual mechanical advantage* (AMA) of a machine is defined as the ratio of the resistance to the effort.

$$AMA = \frac{\text{resistance}}{\text{effort}}; \quad AMA = R/E$$

**EXAMPLE:** If we want to lift a weight of 600 lb and, using a machine, need to exert an effort of only 100 lb, what is the actual mechanical advantage?

$$AMA = (R/E) = 600\text{ lb}/100\text{ lb}$$
$$AMA = 6.$$

Although a machine can make our work easier by changing the direction of the force or by reducing the required effort, *no machine allows us to do more work than the energy we supply.* This follows from the *Principle of Conservation of Energy*. How well the machine uses the energy we supply to it is a measure of its efficiency. We shall look at this and related ideas as we study some specific machines.

9 **The Pulley** Especially when heavy objects have to be lifted through a consider-
** able distance, the pulley is a convenient simple machine for the job. The *pulley* is a wheel so mounted in a frame that it may turn readily around the axis through the center of the wheel. If the frame and wheel move through space as the pulley is used, we have a *movable* pulley. If the axis of the wheel remains stationary as the wheel rotates, we call it a *fixed* pulley. In the diagram shown above for lifting the flag, we have a single fixed pulley. The wheel is usually grooved to guide the string or rope which is used with it. The term *block and tackle* is used for the commercial pulley arrangements.

In using a machine under ideal conditions we would have to waste no energy in overcoming friction or doing work in lifting parts of the machine. Although we cannot achieve such conditions, it is often desirable to know how close we are to the ideal conditions. If we are interested in lifting an object, the *useful work* done with the machine, or the *work output*, or work accomplished, is equal to the weight of the object multiplied by the distance the object is lifted. In general, work output is equal to the resistance times the distance the resistance is moved.

$$\text{Work output} = \text{resistance} \times \text{resistance distance}$$

In order to accomplish this work the effort must move a certain distance. The *work input* is the work done by the effort, and is the effort multiplied by the distance the effort moves.

$$\text{Work input} = \text{effort} \times \text{effort distance}$$

*Under ideal conditions* there is no useless work. Then,

$$\text{work output} = \text{work input;}$$

therefore, resistance × resistance distance = effort × effort distance. This can be written in a different form:

$$\frac{\text{resistance}}{\text{effort}} = \frac{\text{effort distance}}{\text{resistance distance}}.$$

The fraction on the right, the ratio of distance, is not affected by friction, but the one on the left is, since the effort has to increase if friction increases. The two fractions are equal only under ideal conditions. The one on the left we have already called the actual mechanical advantage. The one on the right is called the ideal mechanical advantage (*IMA*):

$$IMA = \frac{\text{effort distance}}{\text{resistance distance}}$$

$$IMA = \frac{\text{resistance}}{\text{ideal effort}}$$

**EXAMPLE:** A pulley system is used to lift a piano weighing 600 lb onto a platform 10 ft above floor level. The pulley system has an actual mechanical advantage of 4 and an ideal mechanical advantage of 5. *a.* What effort is required? *b.* What distance must the effort move?

*a.* $AMA = \text{resistance/effort}$

$4 = 600\ \text{lb/effort;}$

$\text{effort} = 600\ \text{lb}/4;$

$= 150\ \text{lb.}$

*b.* $IMA = \dfrac{\text{effort distance}}{\text{resistance distance}};$

$5 = \text{effort distance}/10\ \text{ft}$

$\text{effort distance} = 50\ \text{ft.}$

For many common pulley arrangements the ideal mechanical advantage can be determined visually by counting the number of rope segments supporting the movable pulley(s). In the diagrams below, of typical pulley arrangements, the rope segments are numbered to show how to count these rope segments. The weight to be lifted (60 lb) represents the resistance if we neglect the weight of the movable pulleys and friction. The effort shown is the value under these ideal conditions; it is the ideal effort.

Under actual conditions some work has to be done to overcome friction. This is

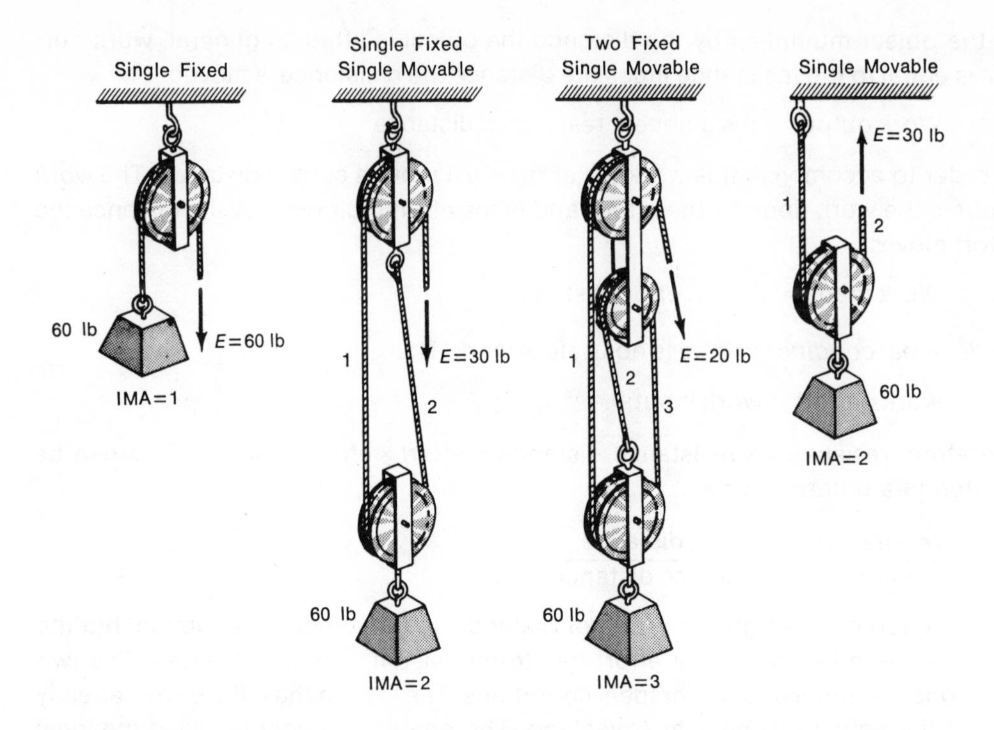


wasted work. The work done by the effort (input work) is then greater than the work done by the machine (output work).

$$\text{input work} = \text{output work} + \text{wasted work}$$

$$\text{efficiency} = \frac{\text{output work (or power)}}{\text{input work (or power)}}$$

It can be shown that this formula for the effiency of a machine is equivalent to the following:

$$\text{efficiency} = \frac{\text{ideal effort}}{\text{actual effort}}$$

It is customary to express the efficiency in percent. The fraction obtained with the above formulas is multiplied by 100. Under ideal conditions no work is wasted; the input work is equal to the output work; the efficiency is 100%.

**EXAMPLE:** A system of pulleys having an *IMA* of five is used to lift an object weighing 900 pounds. The effort required is 240 lb. What is the efficiency?

$$\frac{\text{resistance}}{\text{ideal effort}} = IMA;$$

$$\frac{900\text{ lb}}{\text{ideal effort}} = 5;$$

$$\text{ideal effort} = 180\text{ lb}$$

$$\begin{aligned}\text{efficiency} &= \text{(ideal effort)/actual effort}\\ &= 180\text{ lb}/240\text{ lb}\\ &= 75\%\end{aligned}$$

**10** ** **The Inclined Plane** When heavy objects have to be raised to a platform or put into a truck, it is often convenient to slide these objects up along a board. An *in-*

*clined plane* is a flat surface one end of which is kept at a higher level than the other. The effort to pull or push the object up is usually applied parallel to the plane and is represented in the diagram by $F'$. If $L$ is the length of the plane, the

input work = effort × effort distance;

$$\text{input work} = F'L$$
$$= (15\ Nt)(5\ m)$$
$$= 75 \text{ joules.}$$

$F'=15\text{Nt}$
$W=20\text{Nt}$
$L=5\text{m}$
$h=3\text{m}$

The output work is the useful work accomplished; that is the minimum work required to get the object to its new height; this is the same as the work required to lift the object vertically:

$$\text{useful work} = \text{weight} \times \text{vertical height}$$
$$= (20\ Nt)(3\ m)$$
$$= 60 \text{ joules.}$$

$$\text{Efficiency} = \frac{\text{output work}}{\text{input work}}$$
$$= \frac{60 \text{ joules}}{75 \text{ joules}}$$
$$= 4/5 \text{ or } 80\%.$$

As was shown earlier, under ideal conditions,

$$\frac{\text{ideal effort}}{\text{weight}} = \frac{h}{L};$$
$$\frac{\text{ideal effort}}{20 \text{ newtons}} = \frac{3\ m}{5\ m}$$
$$\text{ideal effort} = 12 \text{ newtons}$$

Note, as a check, that efficiency = (ideal effort/actual effort)

$$= (12\ Nt)/(15\ Nt)$$
$$= 4/5 \text{ or } 80\%$$

**11** ** **The Lever** In Chapter 3 we discussed some of the principles of the lever. Its practical value is apparent whenever we use crowbars, bottle openers, or oars. A *lever* is a rigid bar free to turn about a fixed point known as the *fulcrum* or pivot. As with other machines, its use makes it possible to overcome a large resistance by use of a small force. Of course, even under ideal conditions, we do not save any work by use of the lever; however, its use makes the work easier.

The principle of moments, discussed in chapter 3, shows why a small effort can overcome a large resistance. The diagram shows a horizontal lever 10 inches long, used like a crowbar to lift a 100-pound object. The resistance is then 100 lb. The effort, $E$, is applied at the other end of the lever, 8 inches from the fulcrum. The

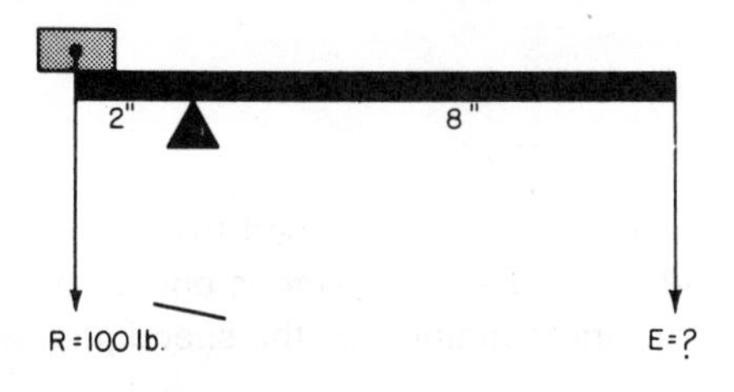

effort tends to produce a clockwise rotation, and has a moment arm (or lever arm) of 8 inches. The resistance tends to produce counterclockwise rotation and has a moment arm of 2 inches. For equilibrium,

$$\text{counterclockwise moment} = \text{clockwise moment}$$
$$\text{resistance} \times \text{resistance arm} = \text{effort} \times \text{effort arm}$$
$$100 \text{ lb} \times 2'' = E \times 8''$$
$$E = 25 \text{ lb.}$$

In other words, with the use of this lever, if we apply an effort of just a little more than 25 lb, we can lift a weight of 100 lb.

We can readily derive a formula for the mechanical advantage of the lever. Since, resistance × resistance arm = effort × effort arm,

$$\frac{\text{resistance}}{\text{effort}} = \frac{\text{effort arm}}{\text{resistance arm}}.$$

This shows that

$$\text{ideal mechanical advantage} = \frac{\text{effort arm}}{\text{resistance arm}}.$$

In the above problem, the ideal mechanical advantage = 8″/2″, or 4. Notice that as the effort moves down, the weight or resistance goes up; but the resistance goes up a much smaller distance than the effort does down (in the ratio of 2:8). Based on the principle of conservation of energy, effort × distance effort moves equals resistance × distance resistance moves.

In the above example the mechanical advantage is greater than one. In some applications of the lever we want a mechanical advantage of less than one. This, for example, is true in the use of the oar in rowing a boat. The oar can rotate around *P*. We pull on the oar handle at *A* and as a result a push is exerted on the water at *B* by the oar blade. The effort arm is smaller than the resistance arm. The effort is

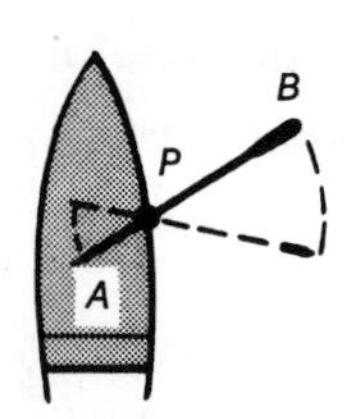


greater than the resistance of the water, but as a result the handle at *A* moves a shorter distance than the blade does during each stroke, as indicated by the dotted lines in the diagram. The mechanical advantage is less than one (effort arm divided by resistance arm), but we have a speed advantage since during each stroke the blade moves a greater distance than the effort does. (This description gives the point of view of the person in the boat. It is also possible to think of this with the fulcrum at B.)

## Questions and Problems

1. A mass of 0.20 kg is dropped from a building. Just before it hits the ground, the mass has a kinetic energy of 30. joules. Calculate *a.* the height of the building *b.* the speed just before hitting the ground.

2. Calculate the work done in pulling a wagon 100. meters along a level road with a force of 30. newtons applied to the handle which makes an angle of 60° with the ground.

3. A horizontal force of 100. lb is used to push a loaded wheelbarrow which weighs 200. lb a horizontal distance of 150. ft. Calculate the amount of work done.

4. Calculate the horsepower required to pull 200 tons of coal per hour from a mine which is 150. ft deep.

5. Calculate the kinetic energy of an object whose mass is 0.12 kg and moves with a speed of 15. m/sec.

6. A 2.0 *kg* pendulum bob attached to a string 0.50 m long is pulled aside until it has been raised a vertical distance of 0.15 m. Calculate *a.* the potential energy gained by the bob *b.* the work done on the bob.

7. A boy exerts a horizontal force of 20.0 newtons to push a 100. newton box along a horizontal surface at constant speed. It takes 10 sec to push the box a distance of 14. meters. Calculate *a.* the force of friction between the box and the surface *b.* the work done by the boy *c.* the power developed by the boy.

8. A machine can do 4000 joules of work in 20. seconds. What is its power, in watts?

9. ** An inclined plane is 10 ft long and 6.0 ft high. An object weighing 90. lb is to be pulled up the plane with a force parallel to the plane. If friction is negligible, calculate *a.* the minimum force needed *b.* the potential energy of the object at the top of the plane *c.* the work required.

10. ** A certain pulley system has an ideal mechanical advantage of 5.0. It is used to lift a 300-lb object a vertical distance of 20 ft. The effort which is required to do this is 120.0 lb. Calculate *a.* the ideal effort *b.* the actual mechanical advantage *c.* the distance moved by the effort *d.* the useful work *e.* the actual work *f.* the efficiency.

11. ** An inclined plane 12. ft long has one end on the ground and the other end on a platform 4.0 ft high. A man weighing 150. lb wishes to push a 90-lb object up this plane. The force of friction is 10. lb. Calculate *a.* the ideal effort *b.* the minimum force he must exert *c.* the potential energy gained by the object when it is at the top of the plane *d.* the minimum force required to keep the object from sliding down the plane.

12. A mass of 0.25 *kg* is dropped from a building. Just before it hits the ground, the object has a kinetic energy of 125 joules. Calculate *a.* the height of the building *b.* the object's speed just before hitting the ground. *c.* What is the potential energy of the object after it has fallen one-fifth of the height of the building?

13. A car having a mass of 2100 *kg* is moving with a speed of 14. m/sec when the brakes are applied. Assuming that the deceleration is constant, and that this brings the car to a stop in 2.0 seconds, calculate *a.* the deceleration *b.* the kinetic energy of the car before the brakes are applied *c.* the distance the car travels before it is brought to a stop.

14. How much work can be done in 1.0 minute by a motor whose power output is *a.* 5.0 watts *b.* 5.0 horsepower?

15. An object is allowed to fall freely. Compare its kinetic energy at the end of the third second with its kinetic energy at the end of the sixth second.

16. A mass of 2.0 *kg* is dropped from a height of 12 meters. After it has dropped 2.0 meters, what is its *a.* potential energy *b.* kinetic energy? *c.* What potential energy did it lose in dropping the 2 meters?
17. An electron and a proton move with a speed of $1.5 \times 10^4$ m/sec in opposite directions. *a.* Calculate the kinetic energy of the electron. *b.* How does the kinetic energy of the proton compare with that of the electron?
18. A 15-kilogram mass is moving north with a velocity of 10 meters per second. A force of 20 newtons acting north is then applied to it for two seconds. Calculate *a.* the original kinetic energy of the object *b.* the final kinetic energy of the object.
19. Jonathan raises a 50-kilogram object 2.0 meters in 0.50 second. What was the minimum power he required for this feat?

## Test Questions

1. The speed of an object is doubled. As a result its kinetic energy is 1. doubled 2. quadrupled 3. halved 4. quartered.
2. Jonathan weighs 100 lb. He exerts a horizontal force of 20 lb in pushing an object a horizontal distance of 30 ft at constant speed. The work done by Jonathan is, in foot-pounds, 1. 600 2. 2000 3. 3000 4. zero.
3. The force of friction in question 2 is, in pounds, 1. zero 2. less than 20 3. 20 4. more than 20.
4. A 40-newton object is released from a height of 10 meters. Just before it hits the ground its kinetic energy is, in joules, (air resistance is negligible) 1. 400 2. 3920 3. 12800 4. zero.
5. In question 4, the kinetic energy of the object at a height of 8 meters is, in joules, 1. 320 2. 80 3. 392 4. 790.
6. As an object falls toward the ground, its potential energy 1. increases 2. decreases 3. remains the same.
7. A box which weighs 10. newtons is dragged across a level floor by a horizontal force of 3.0 newtons for a distance of 5.0 meters. The work done, in joules, in 1. 15 2. 30 3. 50 4. 500.
8. As the time required to raise a given weight 6 ft decreases, the power needed 1. increases 2. decreases 3. remains the same.
9. If air resistance is negligible, as an object falls towards the ground, the sum of its potential and kinetic energies 1. increases 2. decreases 3. remains the same.
10. Paul carries his brother George on his shoulders along the level ground. If George weighs 60 lb and he is carried for a distance of 30 ft, the work done on him by Paul is, in foot-pounds, .............
11. A 5.0 kilogram mass is raised 8.0 meters above the ground. Its change in potential energy with respect to the ground is, in joules, 1. 40. 2. 80. 3. 200 4. 390
12. The speed of an object having a mass of 5.0 *kg* and a kinetic energy of 360 joules is, in meters/sec, 1. 12 2. 38 3. 72 4. 144.
13. A 30-newton object rests on a table 1.2 meters high. The work done by the table in supporting the object is, in joules, 1. zero 2. 30. 3. 36 4. 350.

14. A man exerts a force of 100 lb on a car that does not move. If the car weighs 3000 lb, the amount of work done by the man is ............

15. ** As the friction of an inclined plane increases, its ideal mechanical advantage ............

16. ** As the ratio of length to height of an inclined plane decreases, the ideal mechanical advantage of the plane ............

## Topic Questions

1.1 What is the physicist's definition of work?

1.2 What is the unit of work in the MKS system?

2.1 What is the definition of power? List three formulas that can be used to calculate power.

3.1 How can we calculate the potential energy which an object gains as a result of being lifted?

4.1 How do the mass and speed of an object affect its kinetic energy?

5.1 What happens to the energy which is used to move an object against friction?

6.1 How does a swinging pendulum illustrate the conservation of energy?

7.1 What is the purpose of machines?

8.1 What is the definition of mechanical advantage?

9.1 How can we calculate the useful work done by a machine?

9.2 How many pulleys are needed to get an ideal mechanical advantage of 2, if the effort is to be applied upward?

10.1 When using an inclined plane, in what direction is the effort usually applied?

11.1 What are some devices used around the house which operate on the principle of the lever?

11.2 What is the principle of the lever?

# 7 Temperature and Heat

## 1 Introduction

In a swinging pendulum, kinetic and potential energy are constantly being changed, one into the other. At the highest point of the swing, potential energy is maximum and kinetic energy is least, actually momentarily zero. At the lowest point, potential energy is least and kinetic energy is maximum. Potential energy was changed into kinetic energy. Ideally all of this kinetic energy is changed back to potential energy as the pendulum swings up on the other side. In practice the pendulum bob rises less and less, and finally stops. Why? What happens to the energy?

The moving pendulum bob encounters air resistance, a form of friction. Moving against friction requires work; the pendulum loses energy in doing this work against friction. Whenever work is done against friction, heat is produced. Energy is conserved, since heat is a form of energy.

What is heat? What is the difference between temperature and heat? How are they measured?

2 **Thermometers** When a person is feverish he feels hotter. We say his temperature has gone up. If we touch his forehead, our own hand gets hotter. At this stage we can define *temperature* as the degree of hotness or coldness; we can also say it is a measure of the relative hotness or coldness of an object. We also note that if a hot object is in contact with a cold object, the cold object gets hotter at the expense of the hot object. Some energy has gone from the hot object to the cold object. *Heat* is sometimes defined as the energy which goes from a hot object to a cold one because of their temperature difference.

All of us know that a thermometer is used to measure temperature. How does it work? When an object is heated, many of its characteristics may change. For example, when a solid gets hotter there may be a change in its color, dimensions, electrical resistance, etc. Any of these changes may be used to measure temperature. The more familiar thermometers depend on the fact that substances tend to expand when heated. Of these probably the most familiar is the mercury thermometer.

In manufacture a uniform, thick-walled capillary tube which has a glass bulb at one end is partly filled with mercury. The bulb is heated; the mercury expands and fills the tube, thus driving out the air. Then the top of the tube is sealed.

A temperature scale has to be put on the tube or stem of the thermometer. For both the Celsius and the Fahrenheit scales, two fixed points are located, the steam point and the ice point. If we insert the bulb of the thermometer into the water and heat steadily, the mercury level in the thermometer keeps rising until the water starts boiling. Thereafter, as long as the water keeps boiling, the level of the mercury does not change no matter how vigorously the water boils. This is, therefore, a readily obtained reference point, and is known as the *steam point* or boiling point of water. On the Celsius scale (also called centigrade scale) this is marked 100°. On the Fahrenheit scale it is marked 212°.

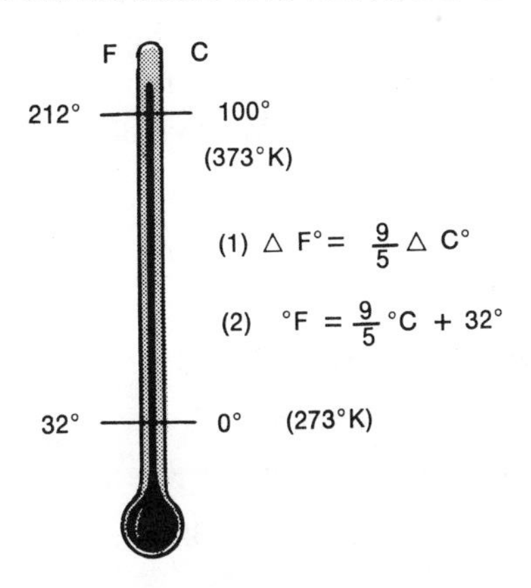


As the water is cooled, the level of the mercury drops and keeps dropping until ice starts forming. Further cooling results in the formation of more ice, but the level of the mercury remains the same as long as both ice and water are present and are mixed. This is the second fixed point and is known as the *ice point*—the temperature at which ice melts or water freezes. On the Celsius scale the ice point is marked 0°, on the Fahrenheit scale 32°. Both fixed points, especially the steam point, are affected by atmospheric pressure. The measurements should be made under conditions of standard pressure. This will be defined later.

Notice that between these two fixed points there are 100° on the Celsius scale, and 180° on the Fahrenheit scale. Both represent the same *temperature change*, the change from the temperature at which water freezes to the temperature at which it boils. If we want to compare other temperature changes on the two scales we can set up a proportion:

$$\frac{\text{change in Celsius degrees}}{100C^\circ} = \frac{\text{change in Fahrenheit degrees}}{180F^\circ}$$

$$\Delta F^\circ = \frac{9}{5}\Delta C^\circ$$

(In this book we distinguish between Celsius degress, C°, and degrees Celsius, °C. When we speak about actual temperatures, we say degrees Celsius, or degrees Fahrenheit. When we speak about temperature changes, we say Celsius degrees or Fahrenheit degrees.)

**EXAMPLE:** During the day the temperature changes from 56°F to 83°F. What change does this correspond to on the Celsius scale?

$$\Delta F^\circ = \frac{9}{5}\Delta C^\circ$$

$$(83^\circ - 56^\circ) = \frac{9}{5}\Delta C^\circ$$

$$\Delta C^\circ = \frac{5}{9} \times 27^\circ$$

$$\Delta C^\circ = 15.$$

The change on the Celsius scale is 15°, corresponding to a change of 27° on the Fahrenheit scale. We did not calculate the actual Celsius temperatures.

In order to calculate the actual Celsius temperature we can obtain another formula. We note that a Celsius temperature is a change from 0°. The corresponding change in Fahrenheit temperature is a change from 32°. To get the actual Fahrenheit temperature we must add 32° to this change. For example, as we saw above, a temperature change of 100C° is the same as a temperature change 180F°. A temperature of 100°C is the same as (180 + 32)°F. In general, for obtaining *actual temperatures*,

$$°F = \frac{9}{5}C + 32° \quad or \quad °C = (F - 32)5/9$$

**EXAMPLE:** Converting 83°*F*-

$$°C = (83 - 32)5/9$$
$$= (51)5/9$$
$$= 28°$$

Converting 56°*F*-

$$°C = (56 - 32)5/9$$
$$= (24)5/9$$
$$= 13°$$

This checks with our previous result: the change is 15°C.

**3 Expansion** Most solids expand when heated and contract when cooled. However, for the same temperature change, different solids of the same size expand different amounts. Most liquids, too, expand when heated and contract when cooled. This was made use of in the design of the mercury thermometer. Again, different liquids expand different amounts for the same temperature change. Water behaves peculiarly in this respect and is of special interest. It expands when heated from 4°C to 100°C, but it also expands when cooled from 4°C to 0°C. In other words, it is densest at 4°C.

If the pressure on a gas is kept constant, heating the gas will result in an increase in its volume. Strangely enough, all gases behave in nearly the same way, especially if the pressure is not too great. For example, if we start with a gas at 0°C, for each Celsius degree rise in temperature, the volume increases by 1/273 of whatever volume the gas occupied at *0°C*, provided the pressure is not allowed to change. If a gas occupying a volume of 273 cm³ at 0°C is heated to 10°C without letting its pressure change, the volume of the gas will become $273\ cm^3 + 273\left(\frac{10}{273}\right)$ $cm^3$ or 283 $cm^3$. This rule also operates when the gas is cooled, provided it is not cooled too much.

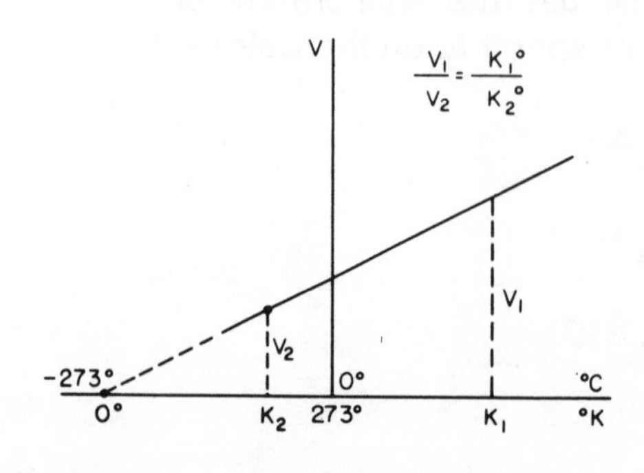

**FOR EXAMPLE:** If the above gas is cooled to −10°C, its volume will become 263 $cm^3$. *If* gases continued to obey this rule even at very low temperatures, the gas could disappear at −273°C.

This suggests using −273°C as the zero temperature on a gas scale of temperatures. No gas behaves this way. Fortunately, another theory developed by Lord Kelvin indicates that no temperature below approximately −273°C can exist. Therefore, this is taken as the zero temperature on the absolute scales. On the *Kelvin scale*, 0°K = −273°C; this is called absolute zero. It is not possible to reach absolute zero, but it has been possible to approach it within a small fraction of a degree. Also, each Kelvin degree is equal to one Celsius degree. Therefore we can change from one scale to the other by the use of the following relationship:

$$°K = °C + 273°.$$

**EXAMPLE:** What is the Celsius temperature when the absolute temperature is 20°K?

$$20°K = °C + 273°$$
$$°C = -253°.$$

## 4 Gas Laws
*

* **Charles' Law** As can be seen from the above diagram, the volume of a gas is proportional to its absolute temperature, if the pressure is kept constant. This is known as *Charles' Law*. It is true for ideal gases, and is practically true for many actual gases. The law can be expressed by the following formulas:

$$\frac{V_1}{V_2} = \frac{K_1^\circ}{K_2^\circ}; \quad \text{or,} \quad \frac{V_1}{V_2} = \frac{C_1^\circ + 273°}{C_2^\circ + 273°},$$

where $V_1$ is the volume of the gas at absolute temperature $K_1^\circ$ or temperature $C_1^\circ$ in Celsius degrees. The volume may be expressed in any volume units, such as cubic meters, cubic centimeters, liters, or cubic feet. Of course, both volumes must be expressed in the same units.

**EXAMPLE:** A sample of oxygen occupying a volume of 586 $cm^3$ at 20°C is heated to 40°C without letting its pressure change. What is the volume of the gas at 40°C?

$$\frac{586\ cm^3}{V_2} = \frac{20° + 273°}{40° + 273°};$$
$$\frac{586\ cm^3}{V_2} = \frac{293°}{313°}$$
$$V_2 = 586\ cm^3 \times 313/293;$$
$$V_2 = 626\ cm^3.$$

Notice that the volume is not proportional to the Celsius temperature: the volume of the gas is not doubled when the Celsius temperature is doubled. The volume is proportional to the absolute temperature: if the Kelvin temperature is doubled, the volume is doubled, if the pressure remains constant.

## 5 Boyle's Law
*

**Boyle's Law** In discussing the behavior of gases we have to keep talking about pressure. Let us review some of the basic facts about pressure. All of us are probably familiar with the fact that tires have to be inflated to certain pressures, and that our atmosphere exerts a pressure which can be measured with a barometer.

Pressure is defined as force per unit area.

$$p = F/A.$$

**FOR EXAMPLE:** The atmosphere exerts a pressure of about 15 lb per in$^2$. This means that the air pushes on every square inch of our body with a force of 15 lb. The air therefore exerts a force of about 100 lb on the palm of our hand. This pressure varies from day to day and also decreases as we go further away from the surface of the earth. It is measured by a barometer. The pressure in an automobile tire is about 25 lb/in$^2$ greater than atmospheric pressure.

In a mercury barometer the height of the mercury column varies with the atmospheric pressure. Therefore atmospheric pressure is often reported as the height of this mercury column. *Standard pressure* is taken as 76.00 cm at sea level and at 0°C. This pressure is about 14.7 lb/in$^2$ or $1.01 \times 10^5$ newtons/m$^2$. Sometimes this pressure is just referred to as "the pressure of one atmosphere"; a pressure twice as great would then be referred to as two atmospheres. This unit may be used to describe the pressure of any gas.

All of us know that if we squeeze a tire or balloon its volume becomes less. How does this change of volume depend on pressure? *Boyle's Law* tells us: If the temperature of a gas is kept constant, the volume of the gas varies inversely with the pressure. ($V = k/p$, where $k$ is a constant.)

**FOR EXAMPLE:** If the pressure on a gas is doubled without changing its temperature, the volume is reduced to half of its original volume.

It also follows that, for a given mass of gas, as long as the temperature remains constant, the product of its pressure and volume always remains the same:

$$p_1V_1 = p_2V_2$$

**EXAMPLE:** If 500 cm$^3$ of a gas, having a pressure of 76. cm of mercury, is compressed into a volume of 300 cm$^3$, the temperature remaining constant, what will be the new pressure?

$$p_1V_1 = p_2V_2$$

$$p_1(300 \text{ cm}^3) = 76. \text{ cm} \times 500 \text{ cm}^3$$

$$p_1 = \frac{76 \times 500}{300} \text{ cm} = 127 \text{ cm}.$$

**Note:** The pressure may be in any unit and the volume may be in any unit, but both pressures must be in the same units and both volumes must be in the same units.

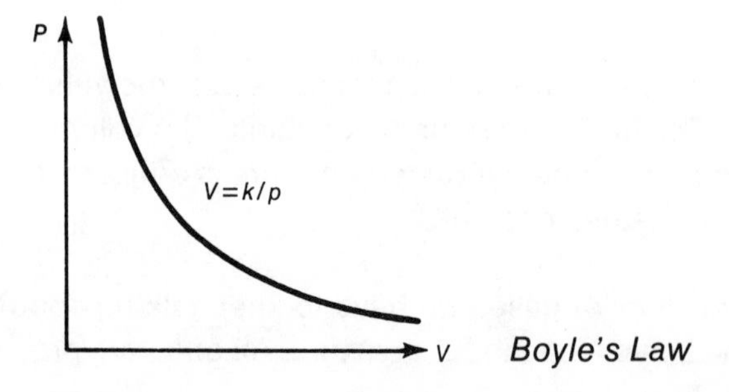


*Boyle's Law*

The graph for Boyle's Law, showing that the volume of a gas varies inversely with the pressure, is a hyperbola.

6 **Relationship Between Pressure of a Gas and Its Temperature** If we heat a gas
* and keep its volume constant, the pressure of the gas increases. In fact, the pressure may become so great that the container will explode. For a given mass of gas kept at constant volume, *the pressure is proportional to the absolute temperature.* If the absolute temperature is doubled, the pressure is doubled.

$$\frac{p_1}{p_2} = \frac{°K_1}{°K_2}$$

7 **General Gas Law** *Boyle's Law* and *Charles' Law* may be combined to give the
* *General Gas Law*:

$$\frac{p_1V_1}{K_1} = \frac{p_2V_2}{K_2}$$

## 8 Kinetic Theory of Gases

*Boyle's Law* and *Charles' Law* are empirical relationships, that is, the laws were not derived from any theory, but were obtained solely from experimental data. However, it is possible to derive these laws with the *Kinetic Theory*. This theory assumes that gases are composed of particles (molecules) in constant random motion. These molecules are relatively far apart and exert little force on each other except when they collide. The average distance between molecules is large compared with their diameters. In an ideal gas the total actual volume of the gas molecules is negligible, and so is the force between molecules. The pressure of a gas on its container is due to the collisions of the gas molecules with the walls of the container. (Note that when the molecule bounces back from the wall, there is a change in the molecule's momentum. Momentum is the product of mass and velocity; it is a vector quantity. Suppose the molecule has a momentum of 10 units when it hits the wall. If it bounces back with the same speed, the new momentum is −10 units. The momentum of the molecule changed 20 units. Force is rate of change of momentum. Pressure is force per unit area.)

The temperature of the gas is a measure of the average kinetic energy of its molecules; as the temperature of the gas goes up, its molecules move faster. They have greater momentum when they hit the walls of the container. This accounts for the increase in pressure of the gas when its temperature goes up, if the volume is kept constant. The sum total of the kinetic energy of all the molecules of an ideal gas is the internal energy of the gas. Sometimes this internal energy is referred to as the heat contents or thermal energy of the gas.

Many actual gases behave like ideal gases. This is especially true if the gas has a low density, is under moderate or low pressure, and is at a temperature considerably above that required to liquefy it at that pressure. In an actual substance, as we approach absolute zero the internal energy approaches a minimum but not necessarily zero. In an actual substance, potential energy is part of the internal energy.

9 *

**Specific Heat and Heat Measurement** Heat is a form of energy. We could use any units of energy to indicate the amount of heat, but for historical reasons special units are frequently used. The *calorie* (cal) is the amount of heat needed to raise the temperature of 1 gram of water 1 C°. The kilocalorie (kcal) is the amount of heat needed to raise the temperature of 1 kilogram of water 1 C°. Of course, we can convert from one unit to another:

1 kcal = 1000 cal. = 4185 joules

The British thermal unit (Btu) is the amount of heat needed to raise the temperature of 1 pound of water 1 F°.

**EXAMPLE:** Five kilograms of water are heated from 20°C to 60°C. How much heat is absorbed by the water?

The change in temperature is (60 − 20)°, or 40 C°. One kilocalorie is required to change the temperature of one kilogram of water one degree; therefore 40 kilocalories are required per kilogram for the 40° change. Five kilograms will require five times as much. Therefore the heat absorbed by the water is 5 × 40 kcal, or 200 kilocalories.

* **Specific Heat** From the definition of the calorie we know that if the temperature of 1 gram of water goes up from 15°C to 16°C, then one calorie was added to the water. However, we find that less heat is required to raise the temperature of 1 gram of most other substances 1 C°. We often define *specific heat* (sp. ht.) of a substance as the number of kilocalories required to raise the temperature of 1 kilogram of the substance 1 C°. From the above we can see that the specific heat of water is 1 cal/gm-C° or 1 kcal/kg-C°. (It also turns out to be 1 Btu/lb-F°.) Sometimes specific heat is defined in a slightly different way, so that it is not necessary to give any units.

**FOR EXAMPLE:** If we look at the *Reference Tables* in the back of the book, under *Heat Constants* we find that the specific heat of water is given as 1.00. Note that water in the solid state has a different specific heat; the table gives 0.50 for the specific heat of ice. The specific heat for iron is 0.11. This means that 0.11 calorie is needed to raise the temperature of 1 gram of iron 1 C°; or 0.11 kilocalorie is needed to raise the temperature of 1 kilogram of iron 1 C°.

If we want a larger temperature change, we must supply more heat. If we have a larger mass, we must supply more heat for the same temperature change. In general, if there is no change of phase,

heat required = mass × sp. ht. × temp. change

**EXAMPLE:** How much heat is required to change the temperature of 200 gm of iron from 30°C to 330°C?

METHOD 1: use the *MKS* system, mass in kilogram and heat in kilocal.
heat required = 0.200 × 0.11 × 300 = 6.6 kcal.

METHOD 2: use the *CGS* system, mass in grams and heat in calories.
heat required = 200 × 0.11 × 300 = 6600 cal.

The specific heat of a substance is often obtained in the laboratory by using the *method of mixtures*.

**FOR EXAMPLE:** A 100 gram-block of iron is moved from boiling water (100°C) to a beaker containing 110 gm of water at 18°C. Assume that this raises the temperature of the water in the beaker to 25°C and that the effect of the beaker is negligible. Heat passes from the iron to the water until both end up at the same temperature, in this case 25°C. On the basis of the Law of Conservation of Energy we expect:

heat lost by hot object = heat gained by cold object.

Heat lost by the iron = mass × sp. ht. × temp. change

= 100 × sp. ht. of iron × (100° − 25°).

Heat gained by water = mass × sp. ht. × temp. change

= 110 × 1.00 × (25° − 18°).

If we assume that no other substances are involved in the heat exchange,

100 × sp. ht. of iron × 75° = 100 × 1.00 × 7°;

sp. ht. of iron = 0.10. (The accepted value is 0.11)

10 * **Change of Phase** The three common states or phases of matter are solid, liquid, and gas. Sometimes we refer to plasma as a fourth phase. If we heat a solid, its temperature goes up until it starts melting. Further transfer of heat to the solid results in melting or *fusion*. There is no change in temperature during the melting if the solid and liquid are kept well mixed. Continued heating after all the solid has melted again leads to a temperature change; this time the liquid is the only phase being heated. The following graph illustrates this change for ice-water; assume that heat is supplied to this system at a constant rate, such as 20 cal/minute. Temperature (T) is plotted against Heat supplied ($\Delta Q$). Notice at the left of the graph: we have ice only. Assume it is at −20°C when the heating starts. Its temperature goes up steadily; the rise of temperature is in accordance with the relationship:

heat absorbed = mass of ice × sp. ht. of ice × temp. change.

Melting does not start until the ice reaches a temperature of 0°C. Further addition of heat does not result in a greater temperature of the ice. Instead the ice melts and the resulting water is at the same temperature as the ice, 0°C. When all the ice has melted, further addition of heat results in a temperature rise in the water. The curve again starts to rise, but notice that the slope is different from that for ice because the specific heats of ice and water are different. Notice that during the change of phase from ice to water the graph is horizontal, since the temperature remains constant during change of phase.

When a substance is changed from solid to liquid, heat must be supplied to pull the molecules away from each other against cohesive forces. This represents an

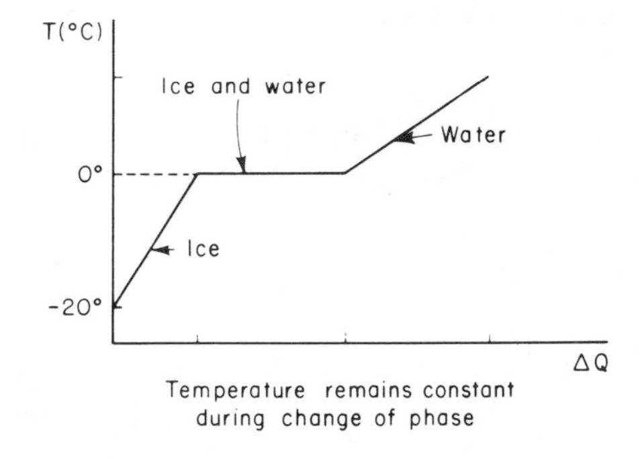

Temperature remains constant during change of phase

increase in potential rather than kinetic energy. Therefore temperature remains constant during change of phase. When the reverse change of phase takes place, when the liquid solidifies or freezes, this heat is released. For a given crystalline substance, such as ice, melting and freezing occur at the same temperature.

11 * **Heat of Fusion** Heat of fusion of a substance is the amount of heat required to melt a unit mass of the substance without change of temperature. For ice at 0°C this is 80 cal/gm or 80 kcal/kg. (Of course, on freezing the same amount of heat is liberated.) If we have twice as much mass, we need twice as much heat to produce melting. In general, for melting only,

$$\text{heat required for melting} = \text{mass} \times \text{heat of fusion.}$$

**EXAMPLE:** How much heat is required to melt 10 kg of iron at 1535°C? The Reference Tables in the back of the book gives the heat of fusion of iron at 1535°C as 7.9 kcal/kg.

$$\text{heat required} = 10 \text{ kg} \times 7.9 \text{ kcal/kg} = 79 \text{ kcal.}$$

If we want to melt a substance, frequently we must heat the solid first to get it up to the melting point. The required heat can then be calculated by combining the above two formulas;

$$\text{heat required} = \text{mass} \times \text{sp. ht.} \times \text{temp. change} + \text{mass} \times \text{heat of fusion.}$$

**EXAMPLE:** A block of iron is at a temperature of 35°C. If it weighs 10 kg, how much heat is required to melt it? Using the data from the previous example,

$$\begin{aligned}\text{heat required} &= 10 \times 0.11 \times (1535 - 35) + 10 \times 7.9\\ &= (1650 + 79) \text{ kcal} = 1730 \text{ kcal.}\end{aligned}$$

12 * **Vaporization and Heat of Vaporization** Vaporization is the process of changing a substance into a vapor. It includes evaporation, boiling, and sublimation. As explained above, the molecules of a substance are in constant motion. The speed of the different molecules is different. In a liquid, some of the fastest moving molecules at the surface can escape in spite of gravity and the attractive force between molecules. The molecules that escape are relatively far apart from each other, almost as in a gas, and form the vapor of the liquid. This escape of molecules from the surface is known as *evaporation*. Since the molecules with the greatest kinetic energy escaped, the average kinetic energy of the remaining molecules in the liquid decreased. That is, unless the liquid is heated, it becomes cooler when evaporation takes place. Therefore evaporation is a cooling process. This is the principle of alcohol rubs for feverish patients.

**Boiling** In evaporation the liquid is converted to vapor at the surface of the liquid. In *boiling* the liquid is converted to vapor within the body of the liquid. We observe the vapor in the form of bubbles. During boiling the temperature of the liquid doesn't change. The temperature at which this occurs is the boiling point of the liquid. Although the temperature of the liquid does not change during boiling, energy must be supplied for the vaporization (to do the work against the cohesive force between molecules). *Heat of vaporization* is the amount of heat needed to vaporize a unit mass of the liquid at its boiling point. For water at 100°C this is approximately 540 cal/gm or 540 kcal/kg. In general,

$$\text{heat required for vaporization} = \text{mass} \times \text{heat of vaporization}.$$

**EXAMPLE:** How much heat is required to change 10 gm of ice at 0°C to steam at 100°C?

To melt the ice: heat required = mass × heat of fusion

$$= 10\ \text{gm} \times 80\ \text{cal/gm} = 800\ \text{cal}.$$

To heat the resulting water from 0°C to 100°C:

$$\text{heat required} = \text{mass} \times \text{sp. ht.} \times \text{temp. change}$$
$$= 10 \times 1.00 \times (100° - 0°) = 1000\ \text{cal}.$$

To vaporize the water at 100°:

$$\text{heat required} = \text{mass} \times \text{heat of vaporization}$$
$$= 10\ \text{gm} \times 540\ \text{cal/gm}$$
$$= 5400\ \text{cal}.$$

$$\text{Total heat required} = (800 + 1000 + 5400)\ \text{cal}$$
$$= 7200\ \text{cal}.$$

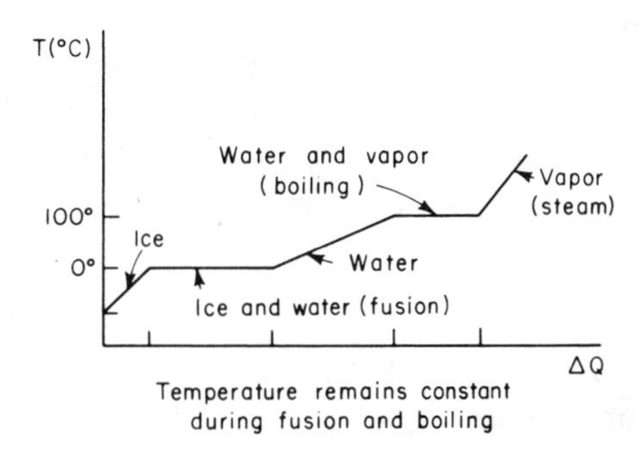

Temperature remains constant during fusion and boiling

This graph shows that there is no change in temperature while a change of phase is taking place. If we heat ice only, its temperature goes up until, at standard pressure, 0°C is reached. Further addition of heat results in the melting of ice. While we have ice and water in good contact, the temperature remains 0°C. When all the ice has melted, further addition of heat to the water results in its rise of temperature until 100°C is reached. Then, with further addition of heat, boiling takes place at 100°C. Rapid addition of heat results in more vigorous boiling, but not in a change of temperature. When all the water has been changed to steam, additional heating causes a rise in the steam's temperature.

13 ** **Effect of Pressure** When the pressure on a liquid is increased, its boiling point goes up; when the pressure on a liquid is decreased, its boiling point goes down. For example, at standard atmospheric pressure, (76 cm of mercury), the boiling point of water is 100°C. In mountainous areas the atmospheric pressure is usually considerably lower and water boils at temperatures significantly below 100°. Cooking at these lower temperatures proceeds rather slowly. To hasten the cooking, pressure cookers may be used; in these the pressure is allowed to build up by using special covers which do not allow the vapor to escape. The building up of the pressure raises the boiling point.

The melting point of ice also depends on pressure. An increase in pressure lowers the melting point of ice. However, the effect is small. If the pressure on the ice is twice the ordinary atmospheric pressure, the melting point is lowered to −0.0075°C. This is said to be important in ice-skating, where a person's weight is supported by a small area—the area of the bottom of the blades. This results in a high pressure (weight per unit area) on the ice—enough to melt it. Actually friction between the ice and blades is very important, too. When the skater passes, the water freezes again. *Regelation* is this melting under increased pressure and re-freezing when the pressure is reduced.

14 ** **Sublimation** Sublimation is the direct change from solid to vapor. For example, at ordinary room temperatures, solid carbon dioxide, *dry ice*, changes directly to carbon dioxide gas. *Condensation* is the changing of a vapor to a liquid. This, of course, is accompanied by the release of the same amount of heat that would be required to vaporize this quantity of heat at that temperature. The term condensation is also used occasionally to refer to the reverse of sublimation, that is, to the direct change of a vapor to a solid. This happens in the formation of frost on a cold night.

15 ** **Supercooling** At a given pressure, a liquid usually solidifies at the same temperature as the solid melts. For example, at standard pressure, ice melts at 0°C. Usually, when water is cooled it freezes at 0°C. However, under special conditions the liquid can be cooled below its ordinary freezing point without solidifying. This is apt to happen if the cooling is done slowly and the liquid is kept absolutely clean and undisturbed. This cooling of a liquid below its usual solidifying temperature is known as *supercooling*. Water can be supercooled readily to −20°C. If it is disturbed, some of the water quickly freezes and its temperature rises. The mixture of ice and water reaches a temperature of 0°C.

16 **Mechanical Equivalent of Heat** In paragraph 9 we stated that heat is a form of energy and that any unit of energy can be used to indicate the amount of heat. Much of the credit for establishing this goes to James P. Joule. The Physics Reference Tables gives the relationship between the kilocalorie, a unit devised to descripe amount of heat before its real nature was understood, and the joule, a unit of mechanical energy: $J = 4.19 \times 10^3$ joules/kilocalorie. In other words,

$$1 \text{ kilocalorie} = 4190 \text{ joules.}$$

**EXAMPLE:** Some engines burn fuel and use the resulting heat to drive machinery. Suppose such an engine converts 8 kilocalories of heat into work every second. Calculate the amount of power made available, in watts.

$$\begin{aligned} 8 \text{ kilocal/sec} &= 8 \times 4190 \text{ joules/sec} \\ &= 33{,}500 \text{ joules/sec.} \\ &= 33{,}500 \text{ watts} \end{aligned}$$

(Recall that 1 watt = 1 joule/sec.)

# Questions and Problems

1. What is standard temperature? Standard pressure?
2. Two hundred grams of water are heated from 30°C to 70°C. How much heat is required?

3. One hundred grams of water are cooled from 60°C to 40°C. How much heat is liberated by the water?
4. If 1400 calories are added to 150 grams of water at 30°C, what will be the final temperature of the water?
5. If 600 calories are added to 100 grams of copper, what will be the new temperature of the copper? Assume that the starting temperature of the copper is 20°C. (*See Reference Tables.*)
6. A boy has 200 grams of water at 40°C. How much ice at 0°C must he add to the water so that he will have only water at 0°C?
7. How much heat is required to change 20. grams of ice at 0°C to water at 50°C?
8. Thirty grams of steam at 100°C is changed to water at 40°C. How much heat is liberated?
9. A pot containing 6 pints of water is heated from 68°F to boiling. Assuming that "a pint is a pound the world around", how much heat is supplied to the water?
10. Convert the following Celsius temperatures to Fahrenheit temperatures: *a.* 30° *b.* −30° *c.* 220° *d.* −220°.
11. Convert the following Fahrenheit temperatures to Celsius temperatures: *a.* 80° *b.* 280° *c.* −80° *d.* −280°.
12. Convert the following Celsius temperatures to Kelvin temperatures: *a.* 30° *b.* 220° *c.* −30° *d.* −220°.
13. ★ A metal weighing 2200 grams requires 740 calories to raise its temperature from 20°C to 30°C. What is its specific heat?
14. ★ If 1.0 kg of water at 30°C is mixed with 2.0 kg of water at 60°, what is the final temperature of the mixture?
15. ★ A piece of aluminum having a mass of 0.4 kilogram and a temperature of 100°C is submerged in water having a mass of 2.0 kilograms and a temperature of 10°C. What is the final temperature of both?
16. ★ How much more energy is present in 100 grams of steam at 100°C than in 100 grams of water at 100°C?
17. ★ A piece of iron having a mass of 12. kilograms drops from a height of 10. meters into a puddle of water. Calculate *a.* the kinetic energy of the iron just before hitting the water, in joules and in kilocalories *b.* the rise in temperature of the water if there is 1.0 kilogram of water, and all the kinetic energy of the iron is used to heat the iron and water.
18. A certain heat engine converts 14 kilocalories of heat into work every second. Calculate the power output of the engine in watts.
19. ★ A gas is compressed into 0.2 of its original volume. If the temperature is kept constant, what happens to the pressure of the gas?
20. ★ Two hundred cubic centimeters of a gas are heated from 27°C and 700 mm of pressure to 227°C. Calculate the new volume if the pressure is kept constant.
21. ★ To what temperature must 120 $cm^3$ of a gas be heated, if its starting temperature is 40°C, and if *a.* its pressure only is to be doubled *b.* both its pressure and volume are to be doubled *c.* the pressure is to be doubled and the volume is to be tripled?
22. Newspapers reported that Lunar Orbiter 1 encountered temperatures of 275°F on the sunny side of the moon and minus 275° on the shady side. What is the temperature range on the Celsius scale?

## Test Questions

1. A change in temperature of 36 C° corresponds to a change in temperature of ........... F°.
2. A temperature of 290°C corresponds to a Fahrenheit temperature of ............
3. A temperature of 68°F corresponds to a temperature of ........... °C.
4. A temperature of −20°C is the same as a temperature of ........... °F.
5. A temperature of 280°C is the same as a temperature of ........... °K.
6. ★ To be at its maximum density, the temperature of water must be ............
7. A liter of water is heated from 30°C to 40°C. As a result, the volume of the water *a.* increases *b.* decreases *c.* remains the same.
8. In the water in question 7, the average kinetic energy of the molecules *a.* increases *b.* decreases *c.* remains the same.
9. The mass of the water in question 7 (liquid plus vapor) *a.* increases *b.* decreases *c.* remains the same.
10. The density of the water in question 7 *a.* steadily increases *b.* steadily decreases *c.* first increases and then decreases *d.* first decreases and then increases.
11. ★ A gas having a volume of 333 cubic centimeters at 70°C is heated at constant pressure to 140°C. The new volume is ........... cc.
12. If 150 calories are added to a liter of water at 100°C and standard pressure, the temperature of the water will *a.* increases *b.* decrease *c.* remain the same.
13. ★ Ice at 0°C is changed to water at 0°C. If the ice absorbs 400 calories, the number of grams of ice melted is ............
14. ★ When 5.0 grams of steam at 100°C are condensed to water at 100°C, the number of calories liberated is ............
15. ★ When a gas is compressed to one-half of its volume, and the temperature is kept constant, its pressure is *a.* constant *b.* doubled *c.* halved *d.* quadrupled.
16. ★ If the volume of a gas is to be doubled at constant pressure, we must double the ........... temperature.
17. ★ If the pressure on a gas is tripled, the temperature being kept constant, its volume *a.* triples *b.* becomes 9 times as great *c.* becomes one-third as great *d.* becomes one-ninth as great.
18. ★ If the temperature of a gas is raised from 200°K to 400°K at constant pressure, its volume *a.* doubles *b.* quadruples *c.* is halved.
19. ★ If the specific heat of lead is 0.03, the number of calories required to raise the temperature of 100 grams of lead from 20°C to 60°C is ............
20. When a closed tank of air is heated, its density (neglecting any expansion of the tank) *a.* increases *b.* decreases *c.* remains the same.

## Topic Questions

1.1 A swinging pendulum finally comes to rest. What happened to the kinetic energy it had?

2.1 What is one definition physicists use for heat?

2.2 What is meant by the steam point? by the ice point?

2.3 What formula can be used to convert Celsius temperature to Fahrenheit temperature?

3.1 What are two ways in which the expansion of gases is different from the expansion of solids?

4.1 * When is the volume of a gas proportional to the temperature?

5.1 * What does it mean when it is said that the atmospheric pressure is 76 cm?

5.2 * What must be kept constant, if the pressure of a gas is to vary inversely with its volume?

6.1 * If the pressure of a given mass of gas is proportional to its absolute temperature, what is kept constant in addition to the mass of the gas?

7.1 What is the *General Gas Law*?

8.1 What is meant by empirical relationship?

8.2 According to the *Kinetic Theory*, what causes the pressure on the walls of a cylinder which contains a gas?

8.3 How does the *Kinetic Theory* explain temperature and heat?

9.1 * What is meant by specific heat? How large is the specific heat of water?

10.1 * What temperature change takes place during a change of phase?

11.1 * What is meant by heat of fusion?

12.1 ** What is meant by heat of vaporization?

12.2 ** Evaporation produces cooling. How is this explained by the Kinetic Theory?

13.1 ** What is the effect of pressure on the boiling point? on the freezing point?

14.1 ** What is meant by sublimation?

15.1 ** Why is it unlikely that the water in a lake will supercool?

# 8 Wave Phenomena and Sound

## 1 Sound

Is there sound if a tree crashes in a deserted forest during a thunderstorm? Needless debates have developed over this question because of lack of initial agreement on the meaning of the word *sound*. In ordinary speech we usually mean by sound the sensation of hearing involving the use of auditory nerve and brain. In that sense there is no sound in the deserted forest because there is no organism to get the sensation. However, in physics sound is a certain kind of wave which can exist in the absence of a detecting organism.

**Origin and Speed of Sound** When we strike a tuning fork we hear a sound. We may not be able to see its prongs vibrating, but if we let the prongs touch the surface of water, we notice that the water is spattered. In general, the source of sound is a vibrating object.

We associate thunder with lightning. Thunder is produced by lightning, but frequently we have to wait a second or more after we see the lightning before we hear the thunder. It takes time for sound to travel. The *speed of sound* in air at 0°C is approximately 331 meters/sec or 1090 ft/sec. When the temperature of the air rises, the speed of sound in it increases. For each Celsius degree rise in temperature, the speed of sound in air increases approximately 0.6 meter/sec or 2 ft/sec.

FOR EXAMPLE: At 20°C, the speed of sound in air is $(331 + 20 \times 0.6)$ m/sec or 343 m/sec. In general, sound travels faster in liquids and solids than in air.

Sound does not travel through vacuum. This is usually demonstrated by suspending a vibrating bell in a bell jar. As the air is evacuated from the bell jar, the sound gets fainter.

## 2 Pulses and Periodic Waves

We say sound is a certain kind of wave. What does that mean? Wave motion in a medium is a method of transferring energy through a medium by means of a distortion of the medium; the distortion travels away from the place where it is produced while the medium itself moves only a little bit and then returns to its original position. (Later we shall speak about a wave in vacuum or space.) For example, a pebble dropped into quiet water disturbs it. The water near the pebble does not move far, but we can see a disturbance traveling through the water away from the spot where the pebble hit it. This disturbance is the wave. Some of the energy lost by the pebble is carried through the water by the wave, and a cork floating at some distance away can be lifted by the water; thus the cork gets some energy. The wave transferred energy through the water from where the pebble hit, to where the cork is.

A wave may be classified as a pulse or a periodic wave. The wave produced by the pebble in the previous paragraph is a *pulse*, a single vibratory disturbance which travels through the medium away from the source of the disturbance. If we push our finger down through the surface of the water, we produce a pulse. If we push our finger regularly and rapidly down and up through the surface of the water, we produce a periodic wave. A tuning fork produces a periodic wave in air. A *periodic wave* is a regularly repeated disturbance traveling away from the source; it is caused by regularly repeated vibratory disturbances of the medium. When the term *wave* is used without an adjective, usually a periodic wave is meant.

3 **Periodic Motion** This is motion which is repeated over and over again. The motion of a pendulum is periodic. As long as its arc of swing is small, the time for a back-and-forth swing is constant. This time is known as the period of the pendulum. (The formula for the period of the pendulum was given in an earlier chapter.)

The motion of the tuning fork prongs is periodic. As the prongs vibrate back and forth they push particles in the air, these push other particles, etc. When the prong moves to the right, particles are pushed to the right. When the prong moves to the left, the particles move back into the space left by the prong. As the prong moves back and forth, the particles of the medium move back and forth with the same frequency as the prong. The energy travels away from the vibrating source, but the particles of the medium vibrate back and forth past the same spot, their equilibrium position.

4 **Longitudinal and Transverse Waves** Two basic waves are the longitudinal wave and the transverse wave. The wave produced in air by a vibrating tuning fork illustrates the former. A *longitudinal wave* is a wave in which the particles of the medium vibrate back and forth along the direction in which the energy travels. We sometimes say that the disturbance of the particles is parallel to the direction of travel of the wave. A longitudinal wave is also referred to as a *compressional wave*.

A *transverse wave* in a medium is a wave in which the vibrations of the particles of the medium are at right angles to the direction in which the wave is traveling. A wave in a string or rope is transverse. A water wave is approximately transverse. Radio and light waves are transverse. (These are part of the electromagnetic spectrum which will be studied later.)

Simple longitudinal and simple transverse waves can be represented by a sine curve, a curve with troughs and crests (or peaks). Imagine a rope stretched horizontally with a tuning fork attached to the end at the left so as to produce a transverse wave in the rope. When the prong moves up, the rope moves up and the disturbance starts to travel to the right. As the prong moves down, the rope is pulled down, and this downward displacement moves along the rope, following the upward displacement. When the prong moves up again, it produces the second peak to follow peak 1. The periodic vibration of the prong results in a corresponding periodic wave along the rope. The whole pattern shown can be imagined moving to the right with a certain speed, the speed of the wave in the rope.

If we keep our eye on some particular spot on the rope, we see that spot going up and down. The frequency of vibration of each part of the rope is the same as the frequency of vibration of the source.

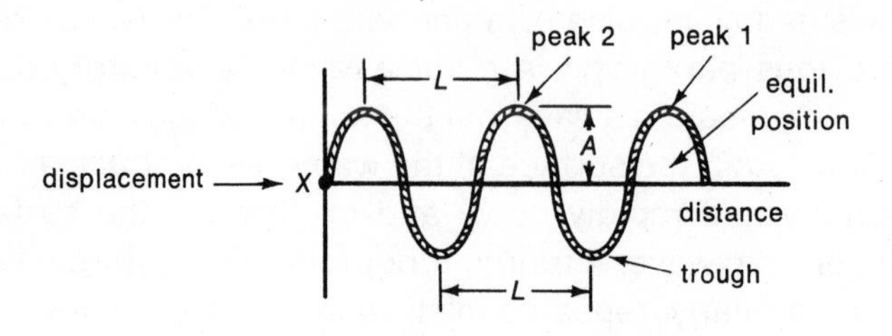


Amplitude *(A)*, Wavelength *(L)*

The *frequency of a wave* is the number of complete vibrations in unit time. Going upward produces the peak of the wave, going down the trough. Units for frequency are vibrations per second and cycles per second. The *hertz* (*Hz*) is also used: 1 hertz = 1 cycle per second.

The *amplitude of the wave* is the maximum displacement of a given part of the medium (rope) from its average or equilibrium position. This is represented by $A$ in the diagram.

The *wavelength* ($L$) is the distance in the medium between any two successive peaks, or between two successive troughs, or any two successive corresponding parts of the wave, such as $X$ and $Y$. The above graph shows $2\frac{1}{2}$ wavelengths. The Greek letter *lambda* ($\lambda$) is usually used instead of $L$ to represent the wavelength.

In the illustration, energy travels to the right along the rope, while the rope itself is displaced at right angles to this direction. As we said above, peak 1 and peak 2 are produced by two successive upward motions of the prong. The time between two such motions is the *period* ($T$) of the vibration: the time required for one complete vibration. During the next period, peak 2 will move to the position now shown for peak 1. In other words, the wave moves a distance equal to a wavelength $\lambda$ during the time equal to a period $T$. But for motion with constant speed, the speed $v$ is equal to the distance divided by the time required for the travel. Therefore, the speed of the wave $v = \lambda/T$.

Since the frequency ($f$) is the number of vibrations per second and the period ($T$) is the time for one of these vibrations,

$$T = 1/f$$

**FOR EXAMPLE:** If the frequency of the wave is 400 vibrations per second, the period is (1/400) second.

If we substitute $1/f$ for $T$ in the above equation ($v = \lambda/T$) we get the important equation for all waves:

speed of wave = frequency × wavelength.

$$v = f\lambda.$$

When something occurs over and over again, the term *cycle* is sometimes used to refer to the event which is repeated. For example, we can say a wave has 400 vibrations per second or 400 cycles per second.

**EXAMPLE:** A wave having a frequency of 1000 cycles/sec has a speed of 1200 meters/sec in a certain solid. Calculate the period and wavelength.

a. $T = 1/f = (1/1000)\ \text{sec} = 0.001\ \text{sec}.$

b. $\lambda = v/f = (1200\ \text{m/sec})/(1000\ \text{cy/sec})$
$= 1.2\ \text{meter}.$

Notice that in the calculation of wavelength, *sec* cancels. We could say that the wavelength is 1.2 meter per cycle, but the cycle is usually dropped.

The sine curve shown above can be used to represent a longitudinal wave as well as a sound wave. This will be described under the discussion of sound.

5 **Other Types of Waves** The two common classifications of waves are longitudinal and transverse. Other classifications include torsional and elliptical waves. A torsional wave is a twist wave; it may be set up in a thin metal rod by clamping one end and twisting the other end and then releasing the twisted end suddenly. Elliptical waves are found in ocean waves and are a combination of transverse and longitudinal waves.

6 **Sound Perception and Sound Waves** Earlier we described the formation of a sound wave in air by a vibrating tuning fork. The particles in the air (molecules of oxygen and nitrogen, etc.) are normally moving at random, but we can think of their average position. They are then equally spaced from each other. This is represented in the top diagram by a line of equally spaced circles to the right of a stationary tuning fork.

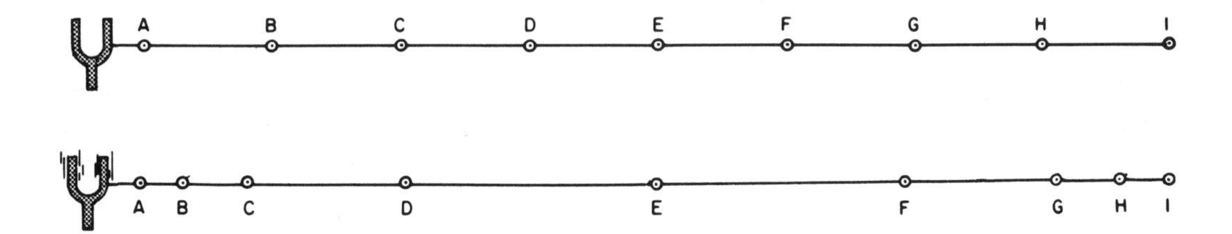


The second line shows the displacement of the particles at a specific instant after the tuning fork has been set into vibration. Some particles, such as *A* and *I*, are back in their average position. Others have been displaced to the left, such as *C*, and some others to the right, such as *F* and *G*. If we plot displacement (on the Y-axis) against original distance along the medium (on the X-axis), the graph is a sine curve just as for a transverse wave. The sound wave itself, however, is a longitudinal wave, since the particles have been displaced to the right and left while the wave itself travels to the right.

Note that near *A* and *I* the particles have been pushed closest together. In these regions, therefore, there is an increase in air pressure. Regions of higher pressure are known as *compressions* or *condensations*. Near *E* the particles are less crowded than average. The air pressure near *E* is therefore lower than normal. Regions of reduced pressure are known as *rarefactions*. If we plot this change in pressure against the original distance along the medium, we would also get a sine curve. Increase in pressure is positive, decrease in pressure is negative. (It is interesting to note that *A* is a region of zero displacement and a region of maximum pressure.)

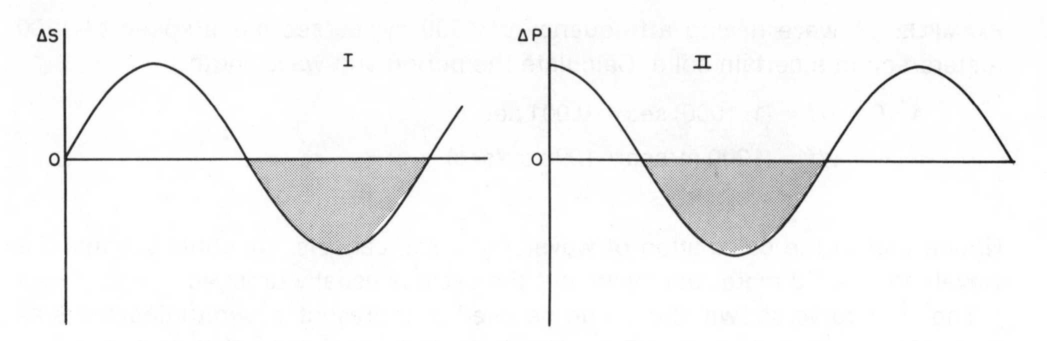


A longitudinal wave may be represented by a sine curve (or cosine)

I. by plotting displacement of particle (to left or right) against distance along the medium, or
II. plotting change in pressure at a point in the medium against distance along the medium.

7 **Musical Sounds** Sounds produced by regular vibration of the air are musical. Irregular vibrations of the air are classified as unpleasant sounds or noise. This is often demonstrated with a *siren* consisting of a wheel with concentric sets of holes. The wheel rotates at constant speed around an axis perpendicular to the wheel and passing through its center. The holes in the outermost circle only are irregularly spaced. When air is blown gently at the inner circles, a pleasant musical sound is heard. The wheel blocks the air at regular intervals, and the blasts of air going through produce compressions at these regular intervals. When these regularly recurring compressions (followed by rarefactions) reach the ear, pleasant sounds are heard.

The holes in the outermost circle are irregularly spaced. When air is blown at this circle, the air allowed through the holes produces successive compressions which are irregularly spaced. These compressions then reach the ear at irregular intervals and are perceived as noise.

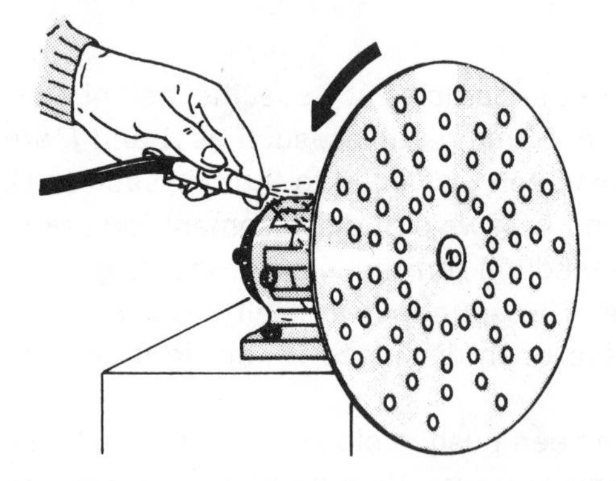

*Siren*

**Pitch** The *range of frequencies* of musical sounds is approximately 20–20,000 cycles per second. Some people can hear higher frequencies than others. When the frequency of the sound wave increases, we say that the *pitch* which we hear rises. The higher the frequency, the higher the pitch.

Longitudinal waves whose frequencies are higher than those within the human audible range are called *ultrasonic frequencies.* Dogs can hear frequencies which are out of the human range. Bats can produce and detect very high ultrasonic frequencies and use them for detecting obstacles to their flight in the dark. *Sonar* devices have been built for similar navigational purposes; they may use ultrasonic frequencies for such purposes as submarine detection and depth finding. Ultrasonic frequencies are also being tried for sterilizing food, since these frequencies kill some bacteria.

8 **Speed** Sound waves of all frequencies in the audible range travel with the same speed in the same medium. For example, in air at 0°C, they travel at approximately 331 meters per second. If the temperature of the air increases, the speed of sound increases. A change in pressure of the atmosphere does not affect the speed of sound unless the temperature also changes. The term *supersonic* refers to speeds greater than sound. An airplane traveling at supersonic speed is moving at a speed greater than the speed of sound in air at that temperature. *Mach 1* means a speed equal to that of sound. Mach 2 means a speed equal to twice that of sound, etc.

## 9 Characteristics of Musical Sounds

Musical sounds have three basic characteristics: pitch, loudness, and quality or timbre. As was indicated above, *pitch* is determined largely by the frequency of the wave reaching the ear. The higher the frequency, the higher the pitch.

**Loudness** Loudness depends on the amplitude of the wave reaching the ear. For a given frequency, the greater the amplitude of the wave, the louder the sound. In general, as the wave travels away from the source it spreads out, and the energy and amplitude of the wave in a given direction decreases. This agrees with the common experience that people who are far away from us have difficulty hearing us unless we shout. In this connection we also speak of intensity.

**Intensity** refers to the wave; it is the power transmitted by the wave through unit area of the medium. It is proportional to the square of the amplitude of the wave. Of course, the greater the intensity of the wave at the eardrum, the louder the sound.

**Quality** Sounds are produced by vibrating objects; if these objects are given a gentle push, they usually vibrate at one definite frequency producing a pure tone. This is the way a tuning fork is usually used. The graph of the wave, as it may appear on an oscilloscope, is a good sine curve. When objects vibrate freely after a force is momentarily applied, they are said to produce their *natural frequency.* If a pendulum or swing is pushed momentarily, it oscillates at a definite frequency. It doesn't make any difference who pushes it; the frequency is the same. It oscillates at its natural frequency. (Of course, we can hold on to the swing and force it to swing at other frequencies; but we have been talking about free oscillations.) Some objects, like strings and air columns, can vibrate naturally at more than one frequency at a time. The wave which is produced is somewhat complex, and its graph is not a pure sine curve. It might look something like curve *B*. The lowest frequency which an object can produce when vibrating freely is known as the object's *fundamental frequency.* Other frequencies that the object can produce are known as its *overtones.* The *quality* of a sound depends on the number and relative amplitude of the overtones present in the wave reaching the ear. In the above diagram, *A* and

*B* represent waves having the same fundamental frequency. Since *B* is a distorted sine wave, it represents a wave which has overtones present. Middle C on the piano and middle C on the violin have the same fundamental frequency (about 250 vibrations per second). They do not sound the same because they differ from each other in the number and relative amplitude of the overtones produced.

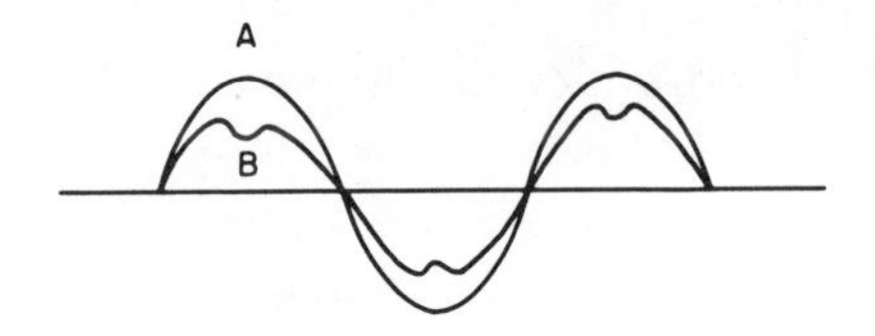


Tones with same fundamental frequency

## 10 Reflection of Waves; Echoes

**Reflection of Waves; Echoes** When a wave reaches another medium, part of the wave is usually reflected. When a sound wave is reflected, a distinct echo may be heard. Two sound waves reach the ear: one that goes directly from the source to the ear, and another wave that goes first to the reflecting surface and then to the ear. There is a distinct echo if the time between the two sounds is at least 1/10 of a second. Bats and porpoises use echoes to locate food and to maneuver around obstacles. The wave frequencies used by bats are about 100,000 cy/sec; they are ultrasonic.

Echoes may be used to determine the distance of reflecting objects, as in sonar. The same principle is used in radar, where electromagnetic waves are used. The total distance traveled by the sound wave is equal to the speed of the wave in the medium multiplied by the time between the start of the sound and the arrival of the echo:

$$\text{distance} = \text{speed} \times \text{time}.$$

If the reflected wave comes back along the same path as the incident wave, the distance to the reflecting wall is one-half of the total distance.

**EXAMPLE:** Six seconds after a boy yells in the country, he hears the echo. How far is the reflecting hill? (Assume an air temperature of 0°C.)

$$\begin{aligned} \text{distance} &= \text{speed} \times \text{time} \\ &= (331 \text{ m/sec}) \times 6 \text{ sec} \\ &= 1980 \text{ meters} \end{aligned}$$

This is the distance to the hill and back again. (Sound travels at the same speed in both directions.) Therefore, the distance to the hill is 1/2 of 1980 m, or approximately 990 meters.

## 11 Doppler Effect

**Doppler Effect** If there is relative motion between the source of the wave and the observer, the frequency of vibrations received by the observer, or the apparent frequency, will be different from the frequency produced by the source. If the source and observer are moving closer, the apparent frequency will be greater than the source frequency. If the source and observer are moving away from each other, the apparent frequency will be lower than the source frequency. Either one or both may be moving.

In the case of sound this will affect the pitch of the perceived sound, since pitch depends on the number of vibrations reaching the ear. When the source and ob-

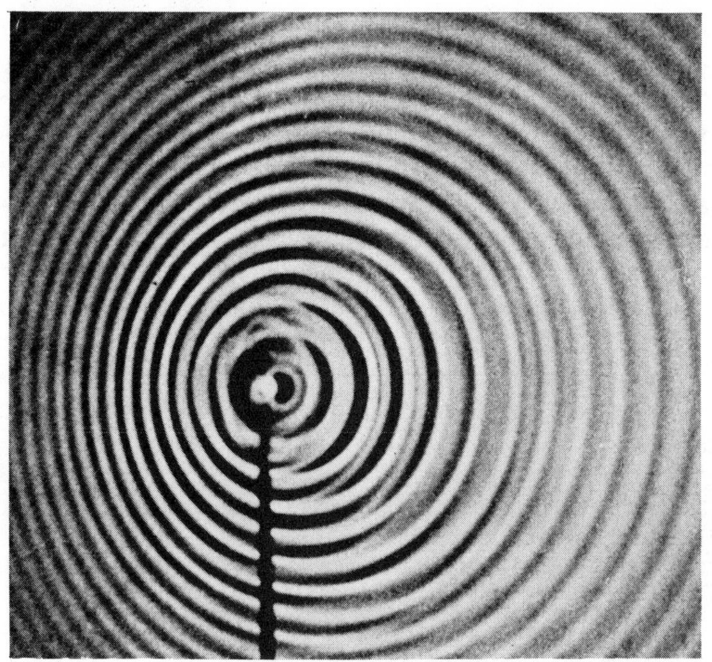

*Vibrating Source Moves to the Left*

*Doppler Effect*
Ealing Films

server approach each other, the pitch will be higher than if both were stationary. The pitch will be lower if source and observer are going further apart. (Note, however, that the pitch does not change *during* this motion, if the motion continues at constant velocity.)

(a) X S Y

(b) X′ 4′ 3′ 2′ 4 3 2 2′ 3′ 4′ Y′

The top diagram shows compressions going to the right and left away from a stationary source at S. The compressions are equally spaced, and the pitch will be the same for the stationary observers at *X* and *Y*. The bottom diagram shows the compressions going to the right and left as a result of motion of the source with constant velocity from left to right. Compression 4′ was produced when the source was at 4, compression 3′ when the source was at 3, etc. Notice that the compressions are more closely spaced in the direction towards which the source is moving; in this case towards the right. Therefore, the pitch for observer Y′ will be higher than for observer X′. The apparent frequency is higher than the source frequency when source and observer approach each other; it is lower when source and observer move away from each other. Notice that the wavelength changed as a result of the *source's* motion. Also note that during the motion itself, if it is with constant velocity, the apparent frequency remains constant.

12 **Natural and Forced Vibrations** As was stated above, objects tend to vibrate at their natural frequency. A simple pendulum has a natural frequency, and so does a swing. If given a gentle push the swing will vibrate at this frequency without any other push. Because of friction the amplitude of vibration decreases gradually. To keep the swing going, it is necessary to supply only a little push to the swing at its natural frequency (or at a sub-multiple of this frequency). The swing can be forced to vibrate or oscillate at many other frequencies. All one has to do is to hold on to the swing and push it at the desired frequency. Such vibration is known as *forced vibration*. For example, our eardrum can be forced to vibrate at any of a wide range of frequencies by the sound reaching the ear. We can think of a distant vibrating tuning fork forcing the eardrum to oscillate by means of a longitudinal wave traveling from the fork to the ear. Forced vibrations stop almost as soon as the push on the object stops.

*Strong winds set the bridge vibrating at its natural frequency in 1940 until it collapsed.*

*Tacoma Narrows*
Courtesy of the Tacoma Area Chamber of Commerce, Ken Whitmire Associates Photography

**Resonance** Resonance exists between two systems (or objects) if vibration of one system results in vibrations or oscillations of the second system whose natural frequency is the same as that of the first. In the case of sound, the two systems may be two tuning forks of the same frequency. We mount the two tuning forks on suitable boxes at opposite ends of a table. We strike one tuning fork, let it vibrate for awhile, and then put our hand on it to stop the vibration. We will then be able to hear the second fork. It resonated to the first one.

What caused the second fork to vibrate? Compressions starting from the first fork gave the second fork successive pushes at the right time to build up the amplitude of its oscillations. Thus energy was transferred from the first fork to the second fork by means of the sound wave. When resonance occurs with sound, the term *sympathetic vibration* is also used: The second fork is said to be in sympathetic vibration with the first one. Resonance is a very important phenomenon in radio and television.

13 ** **Vibrating Air Column** We can think of an air column in a tube as having a natural frequency. The natural frequency of an air column depends on its length. Tube *T*

can be raised or lowered to change the length $L$ of the air column above the water level inside the tube. If a vibrating tuning fork is held over the air column, the loudness of the observed sound will vary as we move the tube slowly up and down, keeping a constant distance between the tuning fork and the top of the tube. The sound will be loudest when the air column is in resonance with the tuning fork.

Since it is its natural frequency, the air column now readily absorbs energy from the vibrating tuning fork. It is then vibrating at the same frequency as the tuning fork and also emits energy. As a result the total energy is radiated by a larger body in a shorter time than by just the tuning fork alone. The sound is then louder.

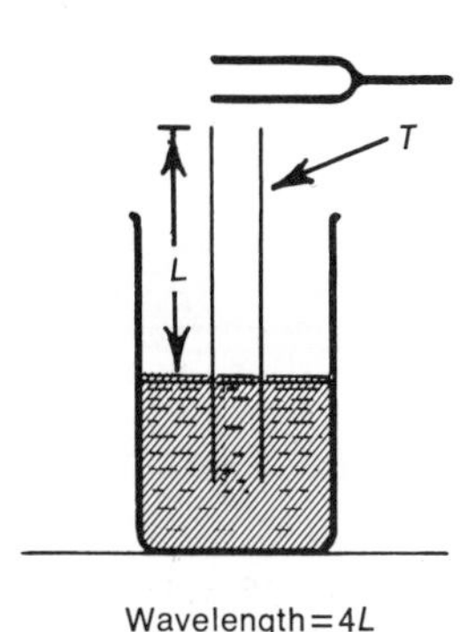


*Resonant Air Column*

At resonance, the length of the air column is approximately one-fourth of the wavelength produced by the tuning fork. In other words, an air column of length $L$ has the same natural frequency as the tuning fork used in the experiment. (Other lengths of air column will also resonate at this frequency, but this need not concern us here.) The principle of vibrating air columns is used in organs, clarinets, and similar musical instruments. If you blow gently into a bottle and get a musical sound, you have caused the air in the bottle to vibrate at its natural frequency. The longer the air column, the greater the wavelength; the greater the wavelength, the lower the frequency and the pitch.

** As we shall see, the principle of resonance is extremely important in many areas. One interesting application in the human body is hearing. The basilar membrane in the ear receives sound vibrations by way of the hammer, anvil, and stirrup. The end of the membrane nearest the stirrup resonates at the highest audible frequencies; the end furthest away is resonant to the lowest tones.

14 **Interference** When two waves go through the same portion of the medium at the same time, interference occurs. The two waves are superposed. Under the conditions that we consider, the resultant disturbance of the medium is the algebraic sum of the disturbances due to the individual waves. If, at a given place, the two waves tend to make the medium vibrate in the same direction, we have *reinforcement* or constructive interference.

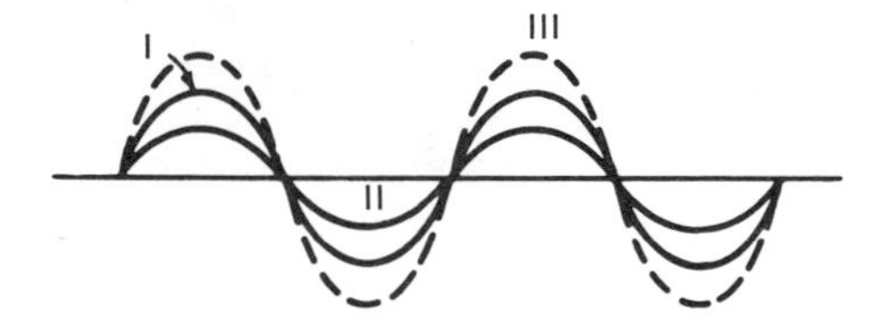


Waves I and II reinforce each other

**FOR EXAMPLE:** Waves I and II combine to give wave III. In this case, I and II were selected to be of the same wavelength or frequency, and to go through zero and maximum displacement at the same time. Such waves are said to be in step or *in phase*. They reinforce each other all along the wave. The amplitude of wave III is the sum of the amplitudes of I and II.

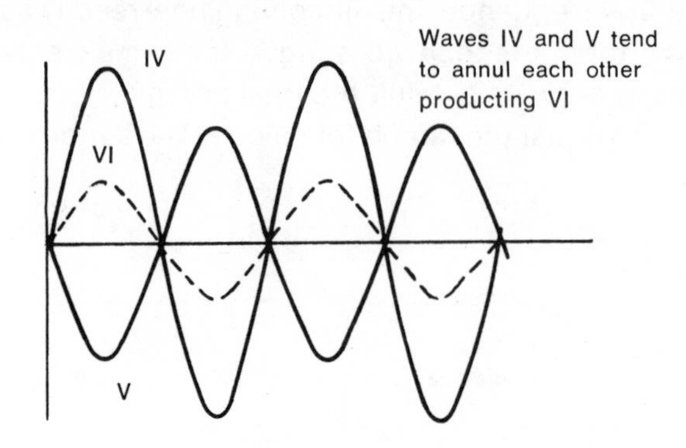


If two waves tend to make a medium vibrate in opposite directions, we have *destructive interference*; the two waves tend to annul or cancel each other. For example, waves IV and V produce wave VI. In this case, IV and V were selected to be of the same wavelength or frequency, and to go through zero displacement together. But when one tends to produce a peak, the other tends to produce a trough; one tends to produce a maximum displacement in one direction when the other tends to produce a maximum displacement in the opposite direction. Such waves are said to be of *opposite phase*; also one is one-half wavelength behind the other. Since a full cycle is considered to represent a variation of 360°, we can also say that two waves which are of opposite phase are 180° out of phase, or that their phase difference is 180°. Maximum destructive interference of two waves occurs when their phase difference is 180°.

Note, then, that *interference* is a phenomenon resulting from the superposition of waves and that the waves may reinforce or cancel each other. The resulting wave is the algebraic sum of the disturbances due to the individual waves.

When two pulses approach each other, interference results when they reach the same part of the medium. After they have passed each other, each pulse continues as though there had been no interference from the other.

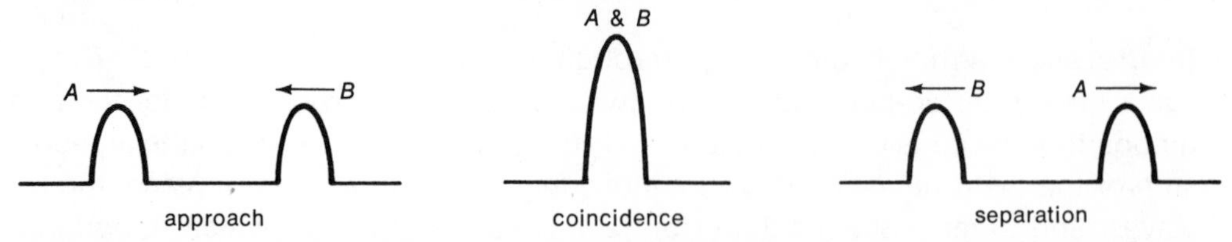


**Beats** If two sounds reach the ear at the same time, interference takes place. Beats are heard when two notes of slightly different frequencies and of approximately the same amplitude reach the ear at the same time. A *beat* is an outburst of sound followed by comparative silence. It turns out that the

number of beats = difference between two frequencies

**FOR EXAMPLE:** If two tuning forks are struck, and one produces 400 vibrations per second and the other 396, the number of beats heard is (400 − 396) or 4 per second.

15 **Standing Waves** Standing waves, also known as stationary waves, are produced when two waves of the same frequency and amplitude travel in opposite directions through the same medium. The two waves interfere with each other, and, under the right conditions, at some points of the medium the two waves always cancel each other. These points, where there is no vibration, are known as *nodes*. Midway between two adjacent nodes is the point where the two waves alternately reinforce and annul each other. As a result, at that point the displacement of the medium varies from zero to twice the amplitude of one wave. These points of maximum displacement in the standing wave are known as *antinodes*.

Standing waves may be produced in a stretched string. If one end of the horizontal string is attached to a wall while the other end is vibrated up and down regularly, a transverse wave will travel to the wall, and, on reaching the wall, be reflected and travel back. The reflected wave will be superposed on more waves traveling forward, and produce interference. The reflected wave has the same frequency as the original wave. If the string has the right length, a stationary wave is produced. *The distance between adjacent nodes is one-half of a wavelength.*

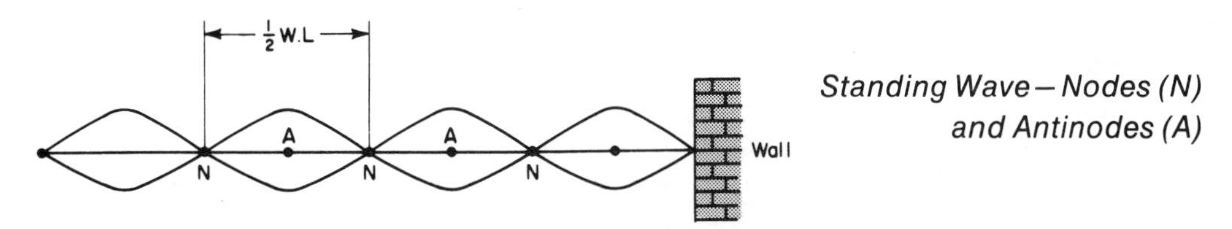


*Standing Wave—Nodes (N) and Antinodes (A)*

The resonant air column explained earlier on the basis of natural frequencies, can also be explained on the basis of standing waves. Water is practically incompressible. The sound wave traveling down the air column is reflected at the surface of the water. The reflected waves interfere with additional waves sent down by the vibrating tuning fork. If the air column has the right length, standing waves are set up in it. This standing wave is sketched in the usual transverse manner in the air column, although, of course, the sound wave is longitudinal. Since water is practically incompressible, the air near its surface cannot be pushed down. Therefore, at the surface of the water we get a displacement node, *N*. At the top of the column, the air can move freely. Near it an antinode (*A*) is set up. Since the distance between two adjacent nodes is one-half wavelength, and the antinode is midway between the nodes, the *distance from node to the adjacent antinode is one-quarter wavelength.* Therefore, the length of the resonant air column closed at one end is one-quarter wavelength.

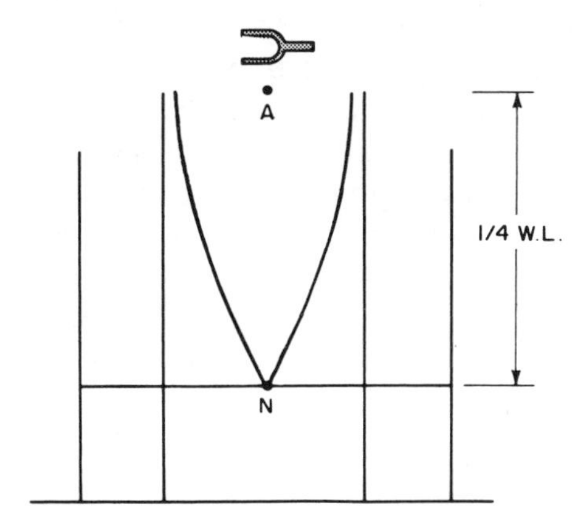


*Standing Wave in Air Column*

16 **Water Waves** Many characteristics of waves can be studied readily at home by producing water waves in a partly filled bathtub. If we put our finger into the water surface, we can see the pulse spreading out in a circle. We can see the pulse reflected by the sides of the tub. If we move our finger up and down regularly, we can see the reflected waves interfering with the direct waves.

Superficially water waves look like transverse waves. We have already pointed out that the actual water waves are more complex in structure, but for many laboratory purposes we may assume that the water wave is transverse.

** Another complication with water waves is their speed. We usually say that the speed of a wave remains constant as long as it remains in the same medium. This is true for sound in air, if the temperature of the air remains constant. It is true for water, except when the water is shallow. In shallow water, the speed of the water wave depends on the depth. The shallower the water, the lower the speed. As a wave moves into shallower water, its frequency does not change. Since its speed decreases, its wavelength must decrease also; speed equals the product of wavelength and frequency.

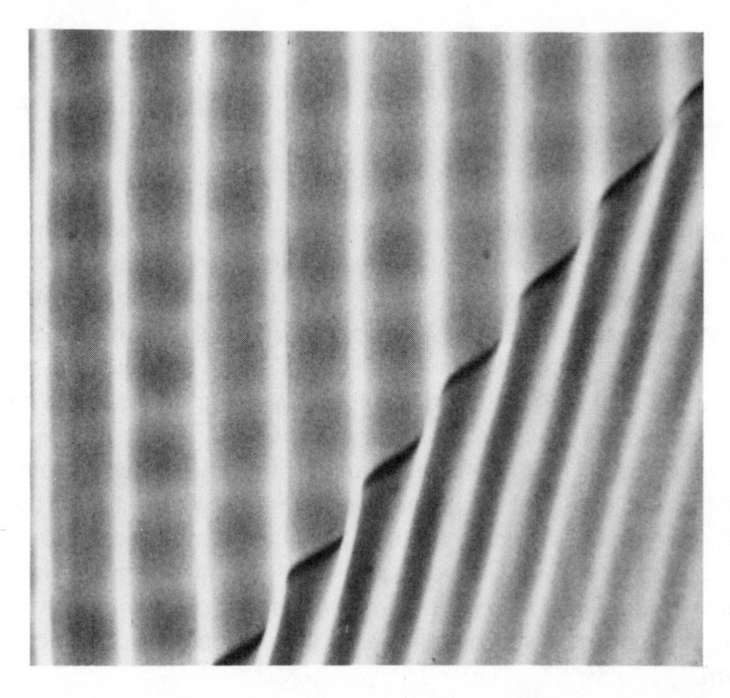

*Refraction of a Water Wave*
From PSSC *Physics*, D. C. Heath and Company, Lexington, Mass. 1965

17 ** **The Ripple Tank** At home some information can be gained from studying water waves in an aquarium or bathtub. In school, observations are more readily and precisely made with a ripple tank. This is a shallow container with a transparent bottom. Water is poured in to a convenient depth, and waves of various kinds can be produced by dipping suitably shaped objects into the surface of the water. Periodic waves are usually produced by attaching the objects to a motor-driven arm. The arm of such a wave generator moves up and down and, as a result, the object dips into the water surface with the same frequency as the vibration frequency of the arm. This frequency can usually be varied over a wide range by adjusting the speed of the motor.

When the object dipping into the water is a small bead, circular waves are produced. When the object is a straight bar, straight waves are produced. The direction of propagation of the straight waves is perpendicular to the straight lines.

A light can be placed so that the image of the water surface appears on a convenient "screen" such as the floor, ceiling, or wall. On this screen the periodic wave will then appear as a series of moving bright and dark lines. The bright lines are produced by the crests of the waves, the dark lines by the troughs.

18 ** **The Stroboscope** The study of moving objects is made easier by the use of a stroboscope. In industrial use the *stroboscope* is usually an intense light which can be flashed on periodically at a variable, known rate. The object which is to be studied can be seen only because it reflects this light. If the frequency of the stroboscope light is the same as the frequency of rotation of a fan which it illuminates, the fan will appear to be stationary. The stroboscope is said "to stop the motion". This makes it possible to find some strains and defects which develop in moving objects.

*Stroboscope*
Sargent-Welch Scientific Company

In school, the stroboscope for student use is often a disk with one or more slits in it. The student looks at a moving object through a slit as he rotates the disk close to his eye. He can see the object only when the slit is in front of his eye. Suppose the object is a rotating wheel, and the disk has only one slit. The first time around he may see spoke *A* at the top. If he sees spoke *A* again at the top the second time around, he is rotating the stroboscope disk at the same frequency as the wheel, or one-half the frequency, etc. It will seem as though the wheel had not turned. However, if he were rotating the disk at twice the wheel frequency, spoke *A* would be at the bottom when he can look through the slit. The view through the slit shows a moving wheel. The stroboscope "stops motion" at certain special fre-

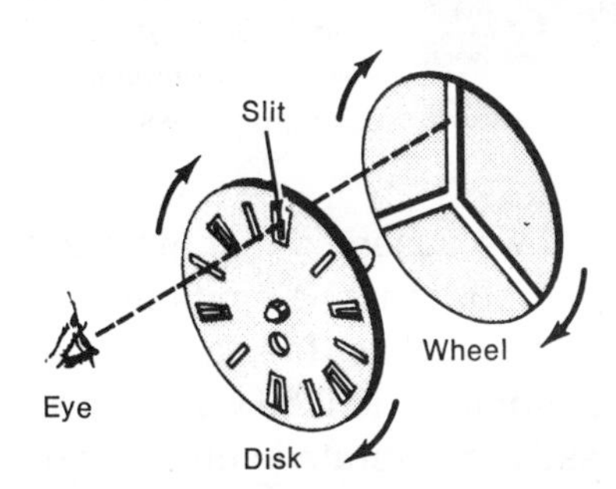


*Stroboscopic Disc*

quencies only; for example, when the frequency of rotation of the disk is the same as the frequency of the wheel.

If the disk has two slits, 180° apart, then we get a view of the wheel twice as often. If we rotate the disk at the same frequency as the wheel, we see the spokes every 180° of their motion. If we rotate the disk at one-half the frequency, the wheel appears to stand still.

In general, when the frequency with which slits pass the eye is equal to the frequency of rotation of the observed object, the object appears to stand still. If the disk is used to observe a water wave in a ripple tank, the wave motion can be "stopped" in a similar way. Again, as a special case, when the frequency with which the slits pass the eye is equal to the frequency of the wave, the wave appears to be stationary. This makes it easy to measure the wavelength, the distance between two adjacent bright lines, or two adjacent crests. The wave's speed can then be calculated:

$$\text{speed of wave} = \text{frequency} \times \text{wavelength}.$$

19 **Diffraction** It is common experience that we can hear a telephone ringing even if the instrument is blocked from our view. It seems that somehow the sound wave can bend around an obstacle while light does not, or at least not as much. The bending of a wave around an obstacle is known as *diffraction*. It can be observed in a ripple tank. Assume that a straight wave is produced and travels towards a small opening produced by the two barriers. The dotted lines enclose the region in which we might have expected a wave if the wave-energy traveled in straight lines only. In fact, we observe that the water wave spreads around the sides, and that the wave becomes nearly circular. The exact effect depends on the ratio between the wavelength and the width of the opening. This will be discussed more under Diffraction of Light.

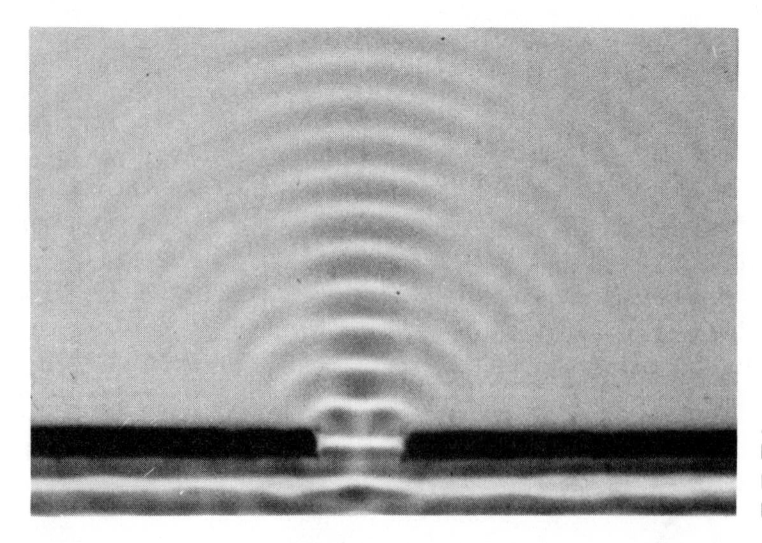

*Diffraction of Waves*
From PSSC Physics, D. C. Heath and Company, Lexington, Mass. 1965

20 ** **Interference of Water Waves** In the study of sound we observed that two waves interfere with each other when they reach the same portion of a medium at the same time. The interference may be constructive or destructive. The same thing is true in water waves. Suppose we have a wave generator in which the vibrating arm has two beads attached, a distance *d* apart. The beads will vibrate in phase and at

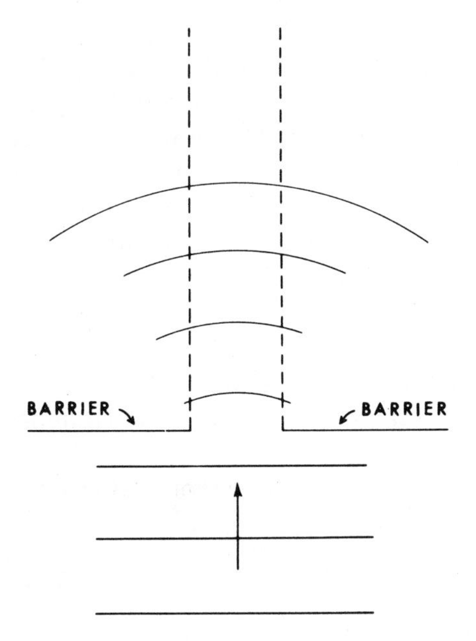


*Diffraction*

the same frequency. As they dip into the water, they act as two point sources, $S_1$ and $S_2$, and produce circular waves of the same frequency. The waves are in phase. The diagram shows these two waves.

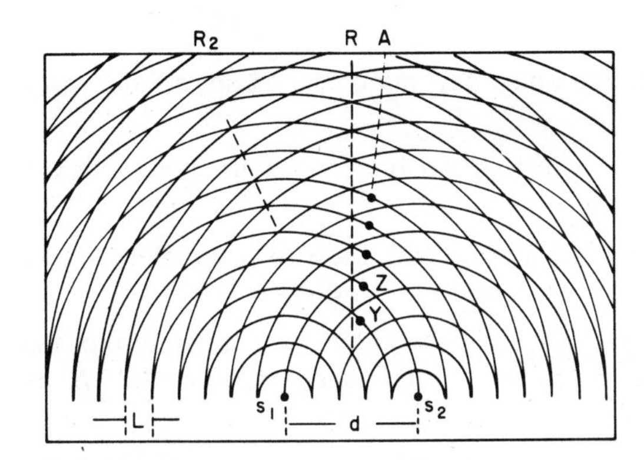


*Waves from Two Point Sources in Phase*

The circles represent the crests of the waves. They are a wavelength *L* apart. Where the crests from the two waves coincide, we get maximum reinforcement. The dotted line *R* down the center of the diagram shows a series of such points of maximum reinforcement. Other such lines of reinforcement can be drawn on either side of the center line.

In between the crests are the troughs. Consider point *Y*. It is on the crest of a wave produced by $S_1$ and on a trough produced by $S_2$. At point *Y* the waves annul each other. As successive waves reach point *Y* from the two sources, the waves will always be out of phase and will annul each other. Such a point of no disturbance in the medium we have called a *node*.

One period later, the crest and trough have moved one wavelength, to point *Z*. During this motion, the trough and crest keep cancelling each other, producing a *nodal line*: a line of zero disturbance of the medium. The dotted line marked *A* shows the more complete nodal line. There are many such nodal lines on both sides of the dotted center line *R*. The first one on either side of the center line is called the first nodal line.

Let us look at point $Y$ again. The wave from $S_1$ traveled 4 wavelengths to $Y$. The length of path $S_1Y$ is 4 wavelengths. Similarly, for the wave starting from $S_2$, the length of path $S_2Y$ is 3 1/2 wavelengths. The difference in paths for the two waves to $Y$ is 1/2 wavelength. This is true for every point on the first nodal line: the difference in path to it is 1/2 wavelength. If we study the second nodal line, we find that the difference in path from the two sources to any point on the line is 3/2 wavelengths. In general, for the $n$th nodal line,

difference in path = $(n - 1/2)$ wavelengths.

For example, for the 5th nodal line, the difference in path = (5 − 1/2) w.l., or 9/2 wavelengths. For destructive interference, the difference in path is an odd number of half wavelengths.

Interference produced by two point sources will be studied in more detail when we take up the subject of light.

## Questions and Problems

*(Unless the problem states otherwise, assume that the air temperature is 0°C)*

1. If you see lightning but don't hear any thunder, were any sound waves produced?
2. Why does a vibrating tuning fork sound louder when its stem is pushed against a desk top?
3. A pulse is observed to travel the length of a wire (2.0 meters) in 0.1 second. What is the speed of the pulse in the wire?
4. A tuning fork vibrates 440 times per second. If the temperature of the surrounding air is 20°C, calculate *a.* the speed of the sound *b.* its wavelength.
5. What is the frequency of a sound in air at 0°C if its wavelength is 1.8 meters?
6. In a certain medium, sound travels 4 times as fast as in air. What is the wavelength of a sound wave in such a medium if its frequency is 440 vibrations per sec?
7. An echo is heard on ship 5.0 seconds after its fog horn blows. If the temperature is 10°C, how far is the reflecting surface?
8. When two tuning forks are sounded together, 4 beats per second are heard. If the frequency of one tuning fork is 256 vibrations per sec, what is the frequency of the second tuning fork? What experiment could you perform to reduce your number of choices?
9. As a motorist starts his engine, the automobile horn gets stuck. What change in pitch, if any, do you observe as he drives away from you to get the trouble fixed? How does this compare with what the mechanic hears as the motorist drives up and stops?
10. A bullet traveling at 700 meters per second hits a wall 1400 meters away. How soon after it hits the wall, does the sound arrive which it made as it left the muzzle?

11. A tuning fork has a frequency of 440 vibrations per sec. What is the length of a closed column resonant to it, if the air temperature is 0°C?
12. A ship's sonar sends a signal of 5000 vibrations per second down to the bottom of the water. An echo is heard 3.0 seconds later. The speed of the sound wave is 1500 meters per sec. Calculate *a.* the depth of the water *b.* the wavelength of the sound wave in water.
13. A pebble is dropped into a dry well. The well is 100. meters deep. Assuming that the speed of sound is 350. meters per sec, how long does it take before he hears the pebble strike bottom?
14. In a certain standing wave pattern in air, the distance between successive nodes was observed to be 3.0 cm. Calculate *a.* the wavelength of the sound *b.* the frequency of the wave.
15. Giant ocean waves called tsunamis accompany earthquakes. They may travel at 500 miles per hour and cause tremendous damage on shores thousands of miles away. *a.* How does this speed compare with the speed of sound? *b.* How long would it take such a wave to reach a shore 3000 miles away?

## Test Questions

1. In wave motion, a single disturbance which is not repeated is known as a 1. peak 2. pulse 3. transverse wave 4. shoot.
2. A wave in which the particles of the medium vibrate at right angles to the direction in which the energy travels is known as 1. a pulse 2. electromagnetic 3. transverse 4. compressional.
3. A sound wave in air is 1. longitudinal 2. transverse 3. torsional 4. electric.
4. Paul holds on to one end of a stretched spiral spring, Jonathan to the other. George pinches several coils close together. This distortion is known as a 1. trough 2. crest 3. rarefaction 4. compression.
5. A tuning fork vibrates in air. The number of peaks (crests) going past a given point per second is the 1. wavelength 2. period 3. frequency 4. amplitude.
6. Steve produces a certain note when he strikes a tuning fork gently. If he strikes the fork harder, the most probable change will be that the sound wave will 1. travel faster 2. have a higher frequency 3. have a greater amplitude 4. have a longer period.
7. If a standing wave pattern is set up in a spring, certain parts of the spring never move from their equilibrium position. These parts are known as 1. nodes 2. antinodes 3. loops 4. crests.
8. The production of echoes depends on the wave phenomenon known as 1. reflection 2. refraction 3. interference 4. annulment.
9. At a given temperature, all sound waves in air have the same 1. frequency 2. speed 3. period 4. intensity.
10. If the frequency of vibration of an object in air is increased, the characteristic of the resulting wave which must change is the 1. amplitude 2. speed 3. wavelength 4. intensity.
11. Two waves having the same amplitude and frequency travel in phase through the same medium and are superposed. The resulting wave will have an increased 1. speed 2. frequency 3. amplitude 4. period.

12. The speed of a certain water wave is 5.0 meters/sec. If its wavelength is 2.0 meters, its frequency is ............
13. The lowest audible sound frequency for most people is, in cycles per sec, about 1. 6 2. 16 3. 60 4. 256.
14. A sound has a certain frequency (1000 cy/sec) and amplitude. If the amplitude is doubled, the 1. loudness increases 2. the loudness is quadrupled 3. the pitch goes up 4. the distortion increases.
15. In order for two sound waves to produce beats, it is most important that the two waves have 1. the same amplitude 2. the same frequency 3. slightly different amplitudes 4. slightly different frequencies.
16. If a fire engine moves away from us, while sounding its siren, the pitch we would hear, as compared with that heard by the fireman, is 1. the same 2. lower 3. higher.
17. When two sound waves of the same frequency are superposed on one another, the resulting phenomenon is known as 1. interference 2. resonance 3. beats 4. an echo.
18. A sound wave in air has a frequency of 500 cycles per sec. Its period is ............
19. If the speed of a sound wave in air is 350 m/sec, and its frequency is 1000 cy/sec, the wavelength is ............,
20. and the temperature of the air is about ............
21. In question 19, if an echo is heard in 3.0 sec, the distance of the reflecting wall is ............

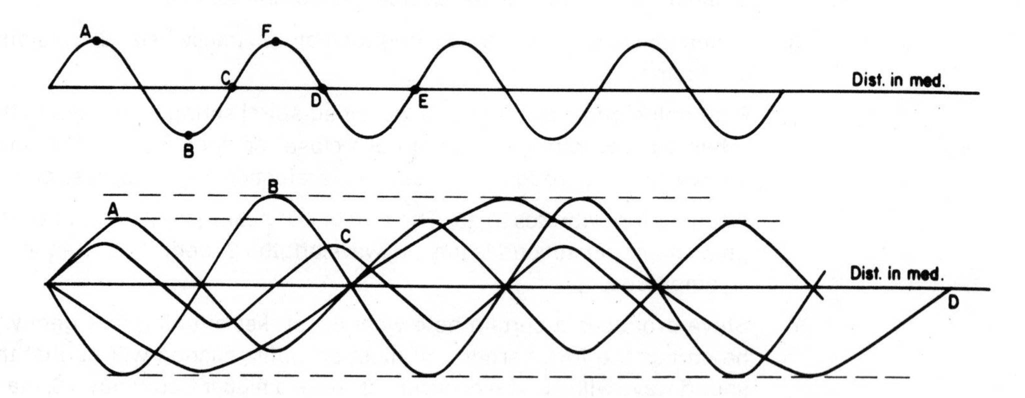


22. One wavelength is the distance between point A and point ............
23. The wave with the greatest amplitude is ............
24. The wave with the longest period is ............
25. Two pulses are shown moving towards each other. When the two pulses coincide, the phenomenon which takes place is known as 1. reflection 2. dispersion 3. diffraction 4. interference.

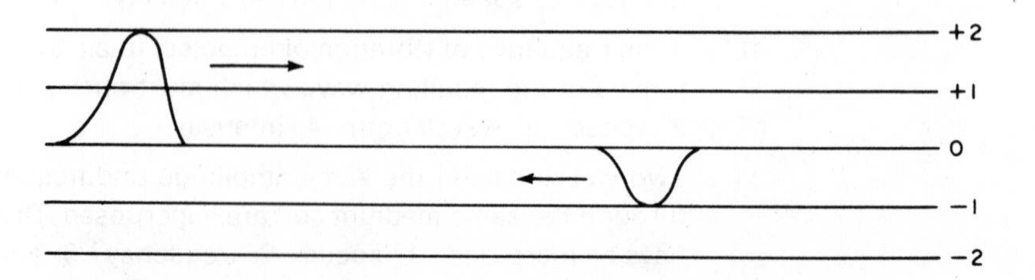

26. When the above two pulses coincide, the amplitude of the resulting pulse will be 1. +1 2. +2 3. +3 4. −1.

## Topic Questions

1.1 What is the source of sound?

2.1 What evidence can you think of to show that a wave transmits energy?

2.2 What is one difference between a pulse and a periodic wave?

3.1 A tuning fork vibrating in air sets the molecules in the air in motion. How does the frequency of vibration of these molecules compare with the frequency of the tuning fork?

4.1 What is the chief difference between a longitudinal wave and a transverse wave?

4.2 What are the definitions of *a.* frequency *b.* amplitude *c.* wavelength *d.* period?

5.1 What are the two common classifications of waves?

6.1 What kind of a wave is sound? What variables can be used to represent a sound wave as a sine curve?

7.1 What is the difference between musical sounds and noise?

7.2 How is pitch related to sound frequency?

8.1 What is meant by *a.* Mach 1 *b.* supersonic?

9.1 How are the following related to characteristics of the sound wave: *a.* pitch *b.* loudness *c.* quality or timbre?

9.2 Middle C can be produced by a violin and a piano. How do the resulting waves differ from each other?

10.1 What is the common principle used by sonar and radar? How do they differ?

11.1 What is the relation between the emitted frequency and the observed frequency, if the source approaches the observer?

12.1 If two tuning forks are in resonance, what is true of their natural frequencies?

13.1 In a pipe organ, how does the natural frequency of a long pipe compare with that of a short pipe?

14.1 Under what conditions do two waves interfere with each other constructively?

14.2 What are two other ways of saying that two waves are of opposite phase?

15.1 What are nodes and antinodes?

19.1 What is meant by diffraction of a wave?

20.1 When will two waves from two point sources annul each other?

# 9 Light—Reflection and Refraction

## 1 Nature of Light

Light is radiation that can produce the sensation of sight. Sometimes the term is used to include ultraviolet and infrared radiation which do not directly produce this sensation.

Light is a form of energy. We can demonstrate this in different ways. When light is absorbed by a black oil, the temperature of the oil rises. Light has been converted to heat, another form of energy. Also, if light shines on the vanes of a Crookes' radiometer, the vanes of the radiometer start turning. The required work was somehow done by the light, indirectly. Energy is the ability to do work, and the light had this ability.

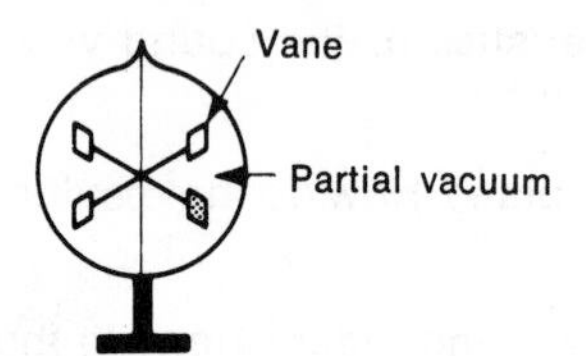


*Crookes Radiometer*

Sir Isaac Newton thought that light is a stream of very small particles. He was able to explain reflection and refraction of light with his theory, and predicted that the speed of light in water is greater than in air. Huygens, who lived at the same time, thought that light is a wave, and was also able to explain reflection and refraction. He predicted that light travels more slowly in water than in air. At that time the measurement could not be made. Most people favored *Newton's Theory.*

In 1801, Young performed an experiment that showed clearly that two light beams can interfere with each other. The explanation required that light be considered a wave. In 1864 Maxwell developed a theory which indicated that light is an electromagnetic wave; what this means will be explained later. Measurement of the speed of light in water turned out to be in agreement with Huygens' prediction; it does travel more slowly in water than in air. The wave theory of light seemed well established.

About 1900 the photoelectric effect was observed and studied: when light shines on certain metals, electrons are emitted by the metal. The wave theory of light could not explain the observations. In 1905 Einstein explained the photoelectric effect by use of ideas of the quantum theory proposed by Planck in 1900. Einstein said that we have to think of light as traveling in little bundles of energy. These bundles are known as *photons*. The amount of energy in each bundle is known as a *quantum* (plural: quanta).

*Thomas Young*
Niels Bohr Library, A.I.P.

We are faced with the *duality* or dual nature of light. In some respects it behaves like a wave and in some respect it behaves like little particles, the photons.

We can make use of interference effects to measure the wavelength of light. Its wavelength is about $5 \times 10^{-7}$ meter. Its exact length depends on its color. X-rays are also electromagnetic waves; their wavelength is about $10^{-10}$ meter. When dealing with such small wavelengths it becomes convenient to introduce another unit:

$$1 \text{ Angstrom (Å)} = 10^{-10} \text{ meter.}$$

2 **Speed of Light** The speed of light and other electromagnetic waves is fantastically high. In vacuum it is about 186,000 miles per second or about $3 \times 10^8$ meters per second. In air the speed of light is only slightly less. In transparent liquids and solids, the speed of light is considerably less than in air.

For light, as for any wave,

$$\text{speed} = \text{frequency} \times \text{wavelength}$$
$$\text{distance} = \text{speed} \times \text{time}$$

**EXAMPLE:** *a.* The sun is about 93,000,000 miles away. How long does it take for sunlight to reach us? *b.* What is the frequency of light whose wavelength is $5.0 \times 10^{-7}$ meter?

*a.* distance = speed × time
93,000,000 miles = (186,000 mi/sec) × time
time = 500 sec. (approx. 8 minutes)

*b.* speed = frequency × wavelength
$3 \times 10^8$ m/sec = frequency × ($5 \times 10^{-7}$ m)
frequency = $0.6 \times 10^{15}$ cycles per sec
= $6 \times 10^{14}$ cycles per sec.

(Note that we used the speed in meters per second because the wavelength was given in meters.)

3 **Standard of Length** The wavelength of light can be measured with great precision. The meter is now defined in terms of the wavelength of a special isotope known as Krypton-86.

1 meter = 1,650,763.73 wavelengths

The *Standard of Time* now recommended is the duration of a certain number of periods of the radiation from another isotope, Cesium-133.

1 second = 9,192,631,770 periods.

4 **Light Travels in Straight Lines** Many phenomena of light can be explained without considering whether light is a wave or a stream of photons. We merely make use of the common observation that light seems to travel in straight lines as long as it remains in the same medium. For example, this explains why we have sharp shadows of objects in sunlight, including eclipses. The direction in which the light energy travels is known as a *ray*. Sometimes the term ray is also used for the energy which travels along this line. A *beam* is then a narrow bundle of such rays. We shall make use of the idea that light travels in straight lines to explain the pinhole camera, reflection, and refraction.

5 **Pinhole Camera** A pinhole camera is a closed box with a small hole near the center of one side and film or a screen on the other side. These two sides are shown below. Let us assume that the right side is wax paper, acting as the screen, so that we can look at it from the outside during the experiment. Light radiates in all directions from every point of the candle flame. Consider point *A* of the candle

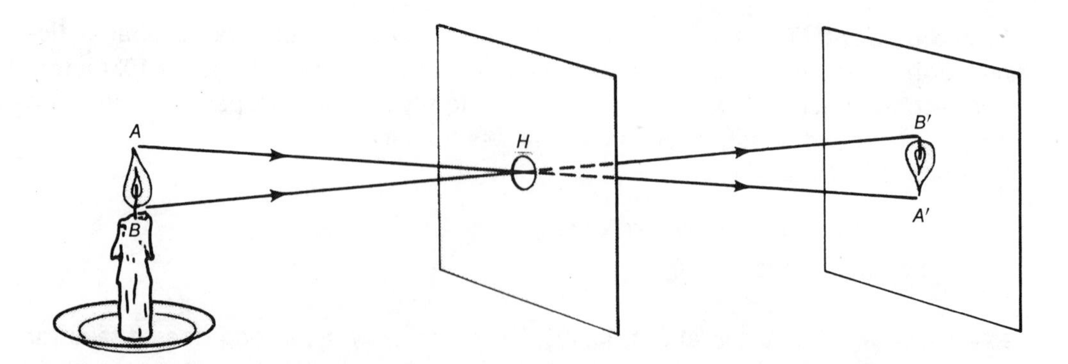


*Real, Inverted Image Formed on Screen*

flame. The only light which can get from it to the screen passes through the pinhole at $H$, the rest is blocked. The light arrives on the screen at $A'$. Similarly, light from $B$ arrives at $B'$. Every point of the flame is thus reproduced as a spot of light on the screen. We thus get an image or likeness of the flame, but notice that this image is upside down, or inverted. The image formed by the pinhole camera is a *real image*: it is actually formed on a screen by light reaching it from the object. If we make the hole smaller, the image gets fainter but sharper. (Under the right conditions, if the hole is small enough, we observe diffraction. This will be discussed later.)

# 6 Reflection

**Law of Reflection** When light hits a surface, usually some of the light is reflected. The operation of mirrors, of course, depends on this. In order to understand the characteristics of our image in a mirror, it is necessary to introduce some terms.

The light that goes towards the surface is known as the *incident* light. It is represented by the incident ray. The *normal*, $N$, for this light is the line drawn perpendicular to the surface at the point where the incident ray touches the surface. The *angle of incidence*, $i$, is the angle between the incident ray and its normal. The reflected light is represented by the reflected ray. The *angle of reflection*, $r$, is the angle between the reflected ray and its normal. Every incident ray has its own normal.

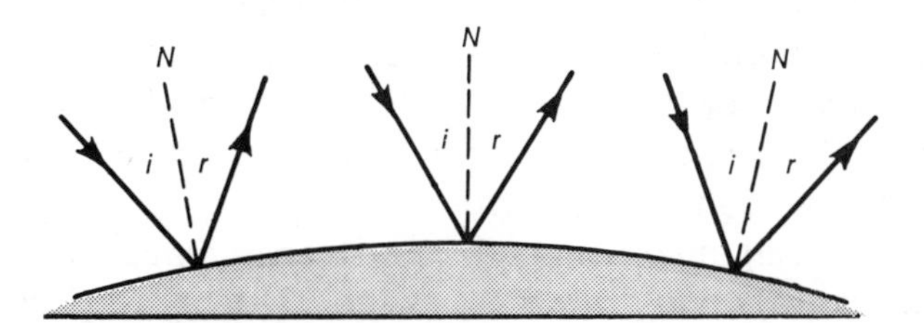


The *Law of Reflection* states that when light is reflected, the angle of incidence equals the angle of reflection, and the incident ray, the normal, and the reflected ray lie in one plane. Theory and experiment show that this is true for all waves. It is true for rough as well as for smooth surfaces.

Reflection obtained from mirror-like surfaces, such as highly polished metals, is known as *specular reflection*. In such reflection the reflected beam is sharply defined. If a surface is flat and highly polished, parallel incident rays are reflected as parallel rays. Most surfaces are rough to a degree. This is true even of this paper. From it we get *diffuse reflection*: reflected light goes off in all directions even when the incident light rays are parallel.

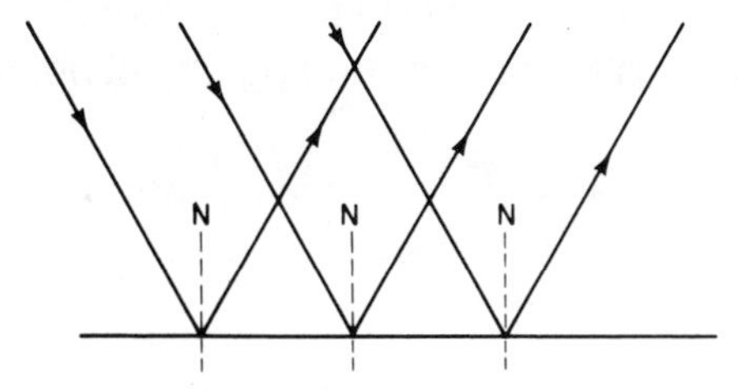


*Specular Reflection*

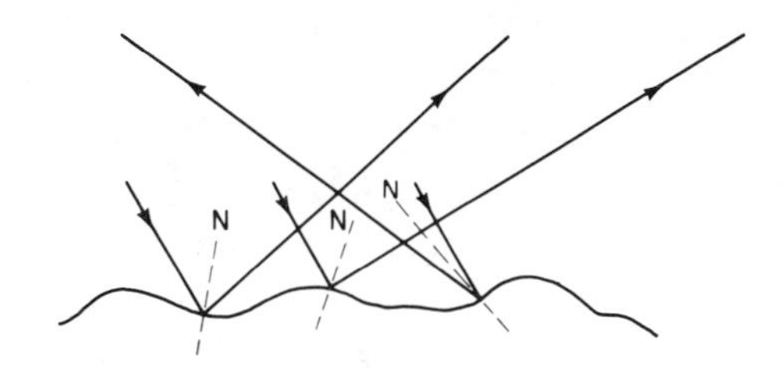


*Diffuse Reflection*

*Luminous* objects are objects which emit light. We see them because some of the light which they emit enters our eyes. Non-luminous objects are visible because light is diffusely reflected from their surfaces and enters our eyes. A perfectly smooth reflecting surface cannot be seen because it gives specular reflection. That is why people sometimes walk into plate glass doors.

**7 Image in Plane Mirrors** A plane mirror is a perfectly flat mirror. The characteristics of an image produced by such a mirror can be determined by means of a *ray diagram*. We represent the object in some suitable way; often an arrow is used. We take two rays from *A*, the top of the object, and draw them as the incident rays, *1* and *2*. We draw their normals, and then the corresponding reflected rays, *1′* and *2′*. Of course, for each ray, the angle of incidence is equal to the angle of reflection. (When you do it, use a protractor.)

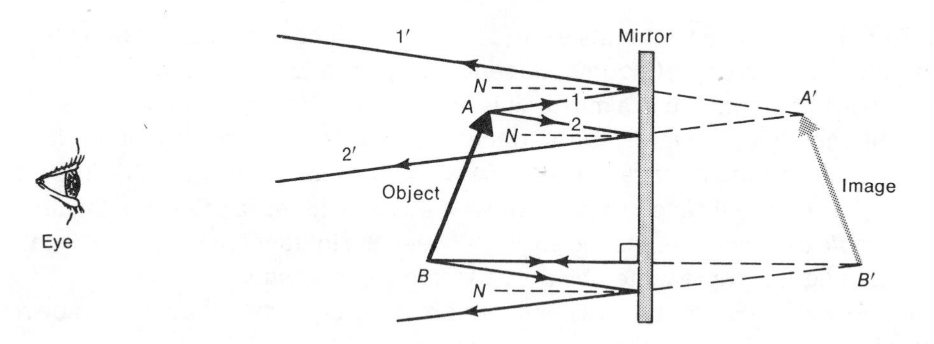


The observer into whose eyes these reflected rays go, imagines them to come from some point behind the mirror. We get this point by extending the reflected rays backwards until they meet, point *A′*. This is the image point corresponding to the top point of the object. The extensions are shown dotted because the light does not actually get behind the mirror.

We do the same with the other end of the object, point *B*, to get its image point. Again we take two rays and draw them as the incident rays. This time we took an incident ray which is perpendicular to the mirror; the reflected light goes right back along the same path. We extend the two reflected rays backwards behind the mirror. Where they meet we get image point *B′*.

We could do this for every point on the object. This is not necessary. We merely connect *A′* and *B′* by a straight dotted line. *A′B′* is the image of *AB*. We draw it dotted because the light we use to see it does not actually come from there.

From the diagram we can determine the *Characteristics of an Image in a Plane Mirror:*

1. The image is the same size as the object.
2. The image is erect (if the object is erect).
3. The image is as far behind the mirror as the object is in front. (Every point of the image is as far behind the mirror as the corresponding object point is in front.)
4. The image is *virtual*. This means that the image is formed by rays which do not actually pass through it. If a screen were placed at the position of the virtual image, no image would be formed on the screen.
5. The image is laterally reversed. The left part of the object becomes the right part of the image. This can be thought of in terms of two people

shaking hands; their arms extend diagonally between them. When a person holds his hand out to the mirror, the arm of his image comes out straight at him.

## 8 Images in Curved Mirrors

** **Images in Curved Mirrors** Spherical surfaces are commonly used as the reflecting surfaces of curved mirrors. A *convex* mirror is one whose reflecting surface is the outside of a spherical shell. A *concave* mirror is one whose reflecting surface is the inside of a spherical shell. The *center of curvature*, *C*, is the center of the spherical shell. The *principal axis* is the line through the center of curvature and the midpoint of the mirror.

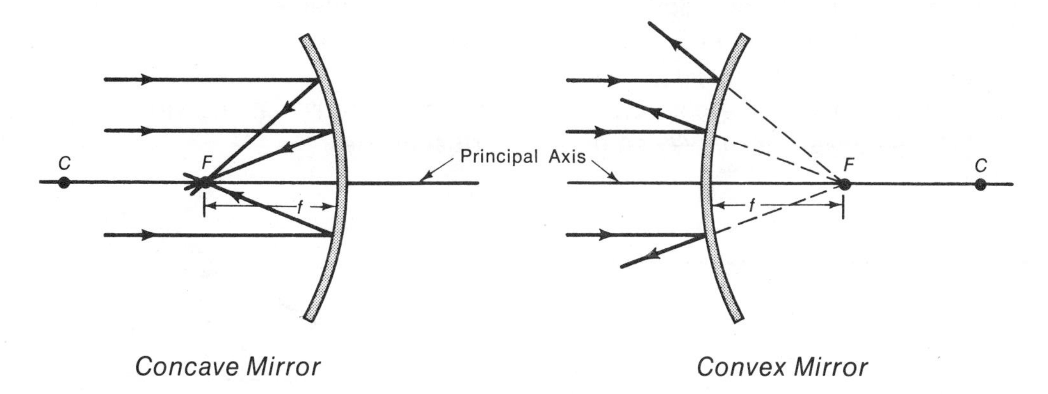


*Concave Mirror* *Convex Mirror*

Incident rays parallel to the principal axis of a concave mirror, after reflection pass through a common point on the principal axis. This point is the *principal focus*, *F*, of the mirror.

A convex mirror also has a principal focus. Incident rays parallel to the principal axis are reflected by a convex mirror so that the reflected rays seem to come from a common point on the principal axis. In the above diagram the reflected rays are extended backwards by means of dotted lines until they intersect at *F*. This principal focus is a virtual focus because the rays of light don't really meet there.

For both mirrors, the *focal length*, *f*, is the distance from the principal focus to the mirror. For a spherical mirror the focal length is equal to one-half of the radius of the spherical shell.

$$f = R/2$$

**The Normal of Spherical Mirrors** We need one more bit of information to draw ray diagrams of spherical mirrors: a radius is perpendicular to its circle, and is the direction of the normal. Therefore any ray directed along the radius of the spherical shell is perpendicular to the spherical mirror. The angle of incidence is then equal to zero, and so is the angle of reflection. In other words, a ray going through the center of curvature of the mirror is reflected right back upon itself.

**Ray Diagrams, Concave Mirror** Some of the basic ideas used for drawing the ray diagram of a plane mirror also apply to the spherical mirror. We try to find the points of the image that correspond to selected points of the object. We often use an arrow as the object. If we place the object so that its tail touches the principal axis, we need to use rays only from the head of the arrow. The tail of the image will also touch the principal axis. We select two special rays from the head of the arrow; we already know their path after reflection. Ray 2, parallel to the principal axis, after reflection passes through the principal focus. Ray 3, whose direction passes through the center of curvature, is normal to the mirror, and is reflected right back upon itself. The reflected rays, 2′ and 3′, intersect in the left diagram. The point where they intersect is the image point, the head of the arrow in the image. We can draw in the rest of the image, remembering that the tail of the arrow is on the principal axis.

In the diagram on the right the reflected rays don't meet. Therefore no real image is produced. However, a person into whose eye the reflected light goes will see an image behind the mirror, at the place where the extended reflected rays meet. This is a virtual image, as in the plane mirror. Rays extended behind the mirror are always drawn dotted. The virtual image is often shown dotted, too. Any other ray starting from the head of the arrow, after reflection, will seem to come from the head of the image.

Notice that with the concave mirror we can get both real and virtual images. With the convex mirror we can get only virtual images.

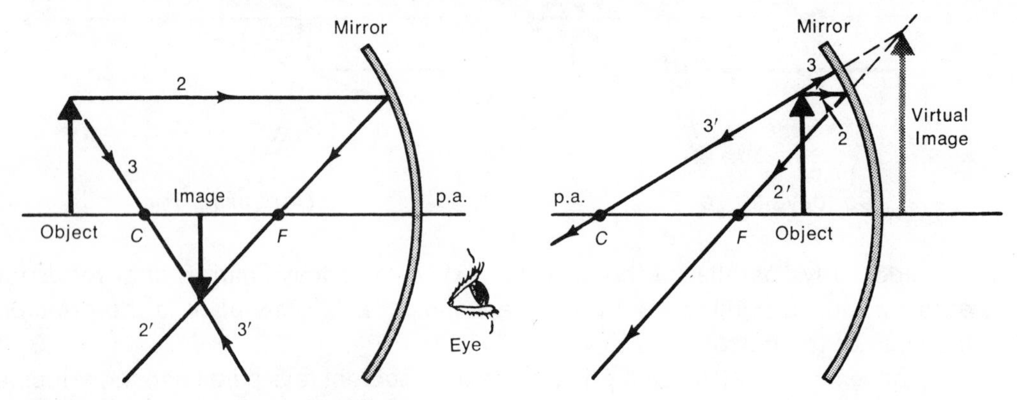


*Ray Diagrams — Concave Mirror*

## 9 Ray Diagram, Convex Mirror

**Ray Diagram, Convex Mirror** The procedure for drawing the ray diagram is similar to the above. Notice that the center of curvature and the principal focus are on the inside of the spherical shell. The parallel ray is reflected so that it will seem to come from the principal focus. The ray directed towards the center of curvature is reflected right back upon itself. The rays originally starting from the same point on the object don't meet after reflection and therefore cannot produce a real image.

The observer sees a virtual image behind the mirror. The virtual image formed with a convex mirror is smaller than the object; the virtual image formed with a concave mirror is larger than the object.

*Numerical calculations* will be discussed under the heading of lenses.

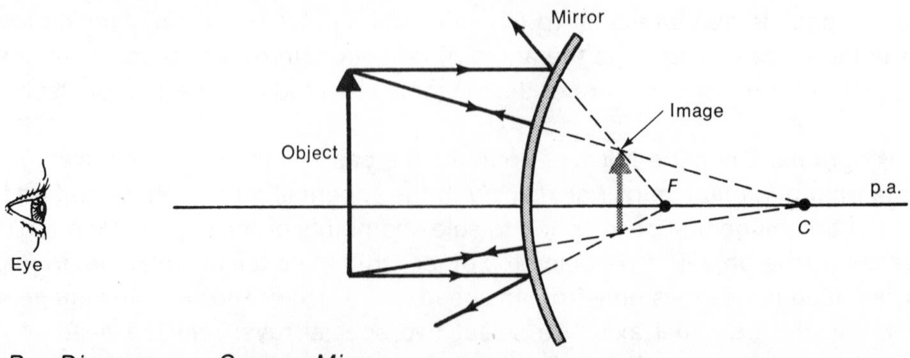


*Ray Diagrams — Convex Mirror*

# 10 Refraction

**Law of Refraction** A spoon in a cup of tea seems bent. A lens can make things look larger or smaller. These and other interesting phenomena depend on the refraction of light. *Refraction* is the bending of light on going into a second medium. Again, as in reflection, we can make some observations and generalizations without having to decide whether light is a wave or a stream of particles.

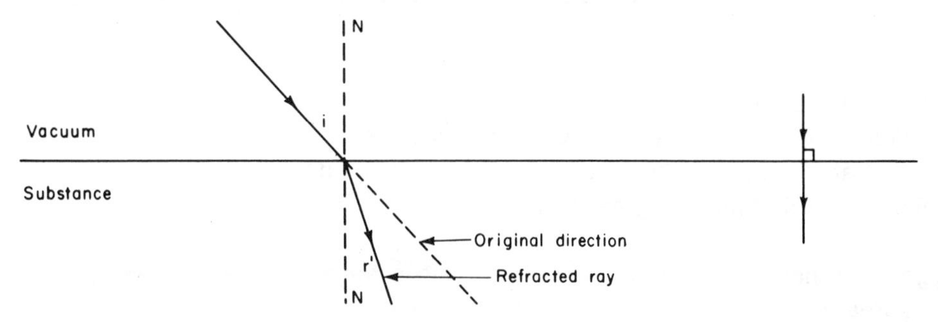


*Oblique ray bent towards normal*

When a ray of light passes obliquely from vacuum into a transparent substance, the light is bent towards the normal. The ray in the substance is known as the *refracted ray*. The angle which the refracted ray makes with the normal is known as the *angle of refraction*, $r'$. Snell discovered in the 1600's that no matter what the angle of incidence, the ratio of the sines of the angle of incidence and the angle of refraction remains constant. This is known as *Snell's Law*. The constant ratio is known as the index of *refraction*, $n$, of the substance. It is also known as the absolute index of refraction. It is different for different substances. For most purposes we may use air instead of vacuum. If, as shown on the right side of the diagram, the incident ray is perpendicular to the substance, the ray enters the second medium without being bent in either direction. (The angles of incidence and refraction are both zero.)

$$\text{index of refraction} = \frac{\text{sine of angle of incidence}}{\text{sine of angle of refraction}},$$

$$n = \sin i / \sin r'.$$

If a ray of light goes obliquely from a substance into vacuum, the refracted ray is bent away from the normal. If we reverse the arrows in the above diagram, we would get the diagram for this situation. It is, therefore, sometimes more convenient to think of the absolute index of refraction in a more general way as:

$$\text{index of refraction} = \frac{\text{sine of angle in vacuum (or air)}}{\text{sine of angle in substance}}.$$

**11 Explanation of Refraction** As mentioned in the introduction to this chapter, both Newton and Huygens were able to explain refraction. Newton used his theory that light is a stream of particles or corpuscles. This led him to the prediction that in water light travels faster than in vacuum. Huygens used his wave theory. He predicted that in water light travels more slowly than in vacuum. When it became possible to measure the speed of light in water and glass, Huygens' prediction turned out to be correct. The speed of light in water is less than in vacuum, and it is even less in glass. When a wave obliquely enters a medium in which its speed is decreased, the water is refracted towards the normal. The less the speed in the medium, the greater the amount of refraction. When a wave enters a new medium and there is an increase in speed, the wave bends away from the normal. The speed of the light wave in a medium like water and glass is the same in all directions.

$$\text{index of refraction} = \frac{\text{speed of light in vacuum } (c)}{\text{speed of light in substance } (v)};$$

$$n = c/v.$$

**EXAMPLE:** If we look at the *Reference Tables* in the back of the book, we see that the index of refraction of air is 1.00. This means that the speed of light in air is close to that in vacuum. The index of refraction of lucite is 1.50. The speed of light in vacuum is 1.50 times as great as in lucite.

12 **Huygens' Principle** If we dip our fingers into a water surface, we see a circular wave spreading out. At every instant we get a new circle. Every point on this circle gets the energy at the same time, setting these points of the medium into vibration in phase. A *wave front* is the locus of adjacent points of the wave which are in phase. The circular waves we see spreading out in the water are wave fronts.

If we dip a ruler into a water surface instead of our finger, we get straight wave fronts.

For us, the important part of *Huygens' Principle* is: Every point on a wave front may be considered a source of wavelets with the same speed. Let us see what this means by applying it to the refraction of a wave. Assume that we have a straight wave front, *AB*, traveling from air into water. The speed of the wave in air is *c* and in water it is *v*. The direction of the wave is *RA*, perpendicular to the wave front.

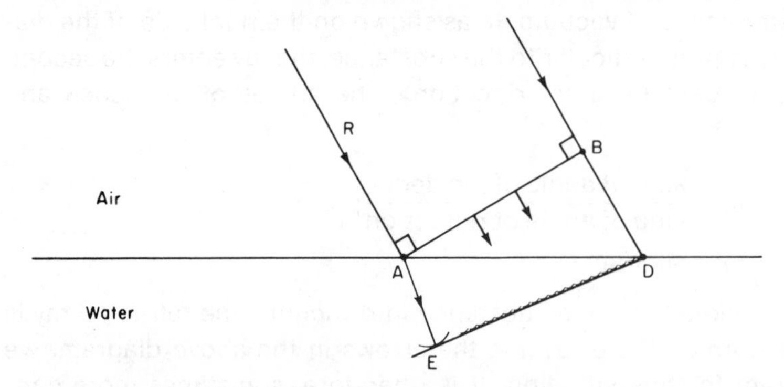


*AD* is on the surface of the water. Assume that it takes time *t* for the wave to travel from *B* to *D*. (The distance *BD* is equal to *ct*, since distance equals speed times time.) During the same time, the disturbance set up by point *A* on the wave front has traveled out as a circular wavelet with radius *AE*. (The length of this radius is *vt*.) As the other parts of the wave front *AB* reach the water surface, they set up little disturbances which travel out as spherical wavelets; they are represented in the diagram by little circular arcs. The new wave front in the water is obtained by drawing the line tangent to these circular arcs. This is *DE*. With a little geometry it can be shown that

$$\frac{c}{v} = \frac{\text{sine of angle of incidence}}{\text{sine of angle of refraction}}.$$

We have already mentioned that either ratio is the index of refraction.

**13 Another Form of Snell's Law** Suppose light goes obliquely from water into glass; which index of refraction do we use? We may use both, applying a more general form of Snell's Law:

$$\frac{\sin \theta_1}{\sin \theta_2} = \frac{n_2}{n_1}; \text{ or: } n_1 \sin \theta_1 = n_2 \sin \theta_2,$$

where $\theta_1$ = angle in medium 1; $\theta$ is a Greek letter, pronounced *theta*
$\theta_2$ = angle in medium 2
$n_1$ = index of refraction of medium 1 with respect to vacuum
$n_2$ = index of refraction of medium 2 with respect to vacuum

(The ratio, $n_2/n_1$, is called the relative index of refraction of medium 2 with respect to medium 1.)

Notice that if medium 1 is vacuum or air, its index of refraction $n_1$ is 1.00, and we get back to the more familiar form of *Snell's Law*:

$$\frac{\sin \theta_1}{\sin \theta_2} = n_2$$

N
Water
(Medium 1)
$\theta_1$
Glass
(Medium 2)
$\theta_2$
N
$n_1 \text{Sin } \theta_1 = n_2 \text{Sin } \theta_2$

Notice that light travels more slowly in glass ($n = 1.5$) than in water ($n = 1.3$). Therefore light going obliquely from water into glass is refracted towards the normal. Light going obliquely from glass into water is bent away from the normal.

Sometimes we use the term *optically dense* to refer to the speed of light in the substance. The medium in which the light travels more slowly is known as the optically denser medium. Glass is optically denser than water. Water is optically rarer than glass. Using these terms we can say that light entering an optically denser medium obliquely, is bent towards the normal.

**14 Refraction by Rectangular Glass Plate** If light goes through parallel surfaces of a transparent substance, such as a flat glass plate, the emerging ray is parallel to the entering ray, provided the two rays are in the same medium. In the diagram we assume that we have a glass plate in air. (A similar diagram would be drawn for a fish tank with a transparent bottom.) The light is *monochromatic*: light of a single

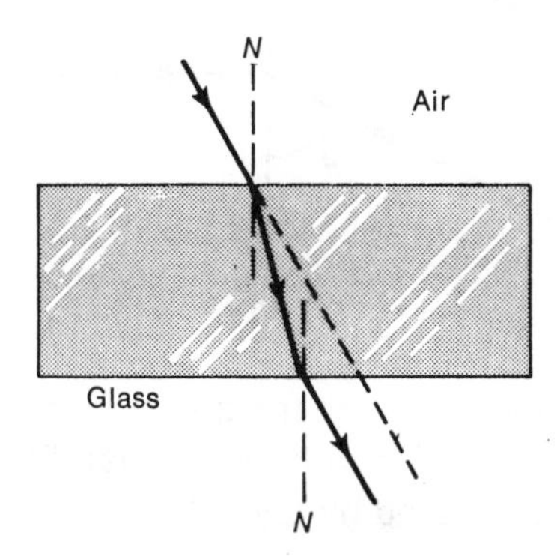

wavelength. The effect of ordinary light will be considered later. Notice that the light entering the glass is bent towards the normal; light entering the air is bent away from the normal. The entering ray is parallel to the emerging ray.

## 15 Refraction by Triangular Glass Plate

**Refraction by Triangular Glass Plate** Again we assume that the surrounding medium is air, and that we are using monochromatic light. Notice that in the triangular plate, the two surfaces at which refraction occurs are not parallel. At the first surface, light enters glass obliquely; because it is slowed up, it is bent towards its normal. At the second surface, the light enters the air obliquely; since it travels faster in air than in glass, the ray bends away from its normal. The overall effect is that the light has been bent towards the base.

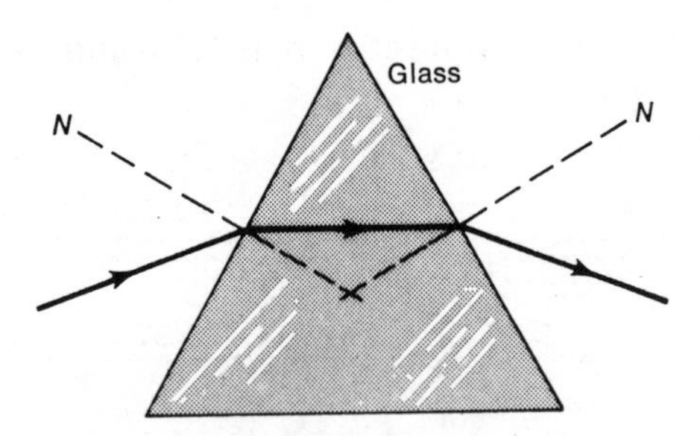


## 16 Critical Angle and Total Reflection

**Critical Angle and Total Reflection** Whenever light enters a second medium whose index of refraction is different from the first medium's, some of the light is reflected. The greater the angle of incidence, the greater the amount of reflection. This is called partial reflection and was deliberately omitted in the above diagrams on refraction. In some cases all of the light is reflected.

Consider this diagram in which rays are shown making different angles of incidence; notice that they are all directed from a medium with a greater index of refraction (1.3) towards one with a smaller index of refraction (1.0). The ray in air is bent further away from the normal. As we go from ray 2 to ray 4, the angle of incidence in water keeps increasing; therefore the angle of refraction in air keeps increasing. For each ray, the angle in air is greater than the corresponding angle in water. As the angles of incidence increase, the refracted rays 2′, 3′, etc. lie closer and closer to the surface.

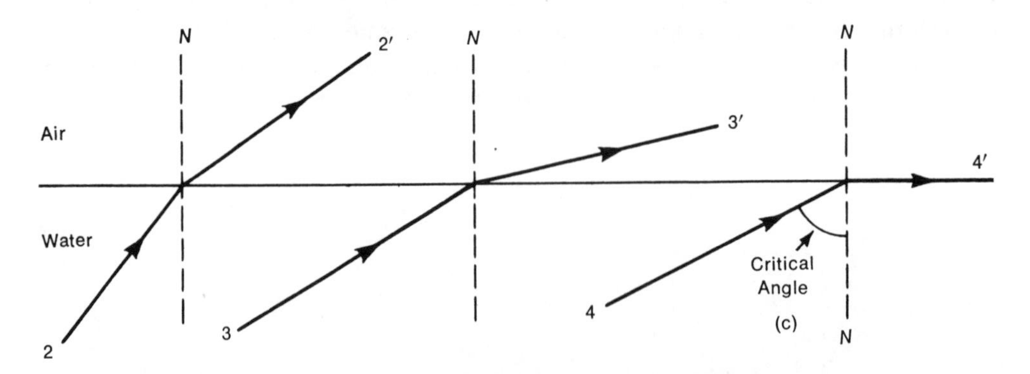


The *critical angle* is that angle of incidence in the medium with the greater index of refraction for which the refracted ray makes an angle of 90° with the normal. The refracted light then goes along the surface. This is illustrated by ray 4-4′.

Let us apply *Snell's Law* to this special case:

$$\text{index of refraction} = \frac{\text{sine of angle in air}}{\text{sine of angle in substance}};$$

$$n = \frac{\sin 90°}{\sin c};$$

$$n = 1/\sin c.$$

For water, $n = 1.33$. Therefore, $1.33 = 1/\sin c$. Sin c = 0.75. The critical angle for water is approx. 49°.

If we increase the angle of incidence beyond the critical angle, the light reaching the surface has to return to the water; in other words, it is reflected. *Total reflection* occurs when the angle of incidence in the medium with the greater index of refraction is greater than the critical angle.

Total reflection is applied in prism binoculars and in curved transparent fibers and in lucite rods. The light stays inside the curved rod because of successive internal reflections. For all reflections the angle of incidence equals the angle of reflection. Let us see how this works out in the case of a triangular isosceles glass prism, such as is used in prism binoculars. One angle is a right angle. The two acute angles are 45°.

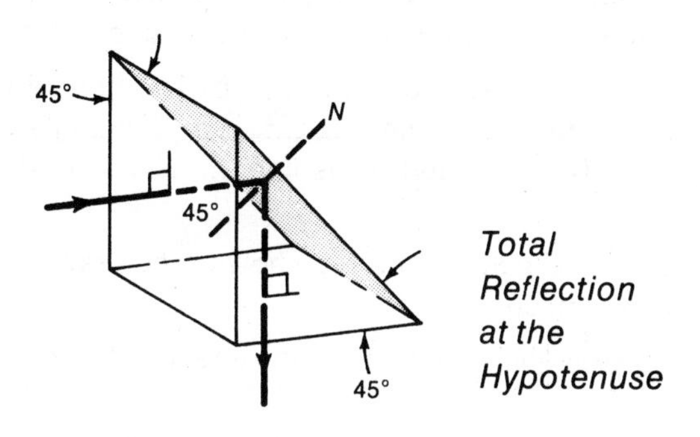


*Total Reflection at the Hypotenuse*

The ray entering the prism is shown perpendicular to the first surface. Therefore it is not refracted and goes on to the hypotenuse where it makes an angle of incidence of 45°. Assume that the critical angle for this glass is 42°. Therefore the angle of incidence is greater than the critical angle and total reflection takes place at the hypotenuse. You should be able to show that the reflected ray is perpendicular to the second surface; it then emerges without being refracted. Of course, in the air the light travels faster than in the glass.

# Questions and Problems

(Assume that the critical angle for glass is 42°.)

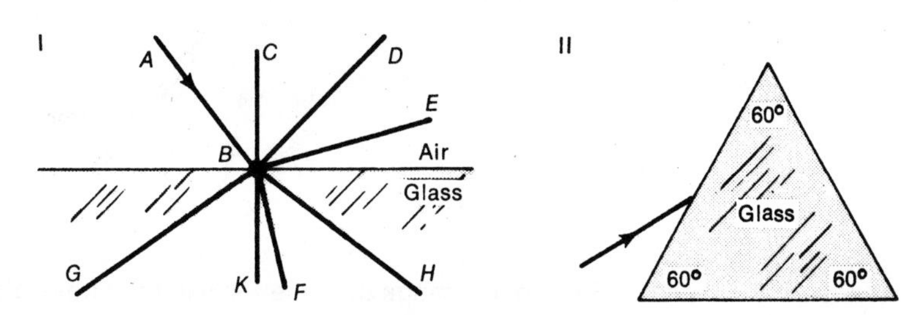

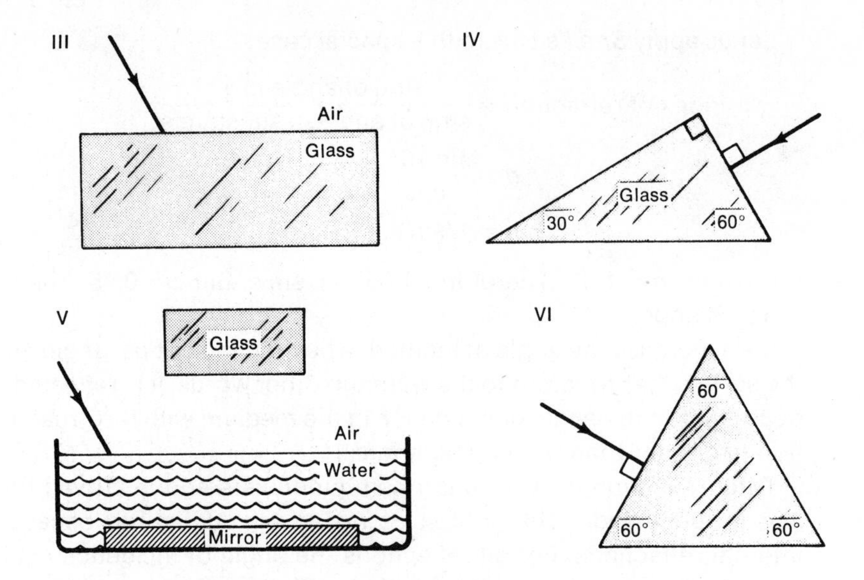


1. *a.* Complete the diagram to show how the *Law of Reflection* and two rays from point A can be used to locate the image of point A in the mirror. *b.* Compare image distance with object distance on your diagram. *c.* How does this compare with theoretical prediction?

2. In diagram I, *a.* the incident ray is line ............ *b.* The normal is line ............ *c.* The reflected ray is line ............ *d.* The angle of incidence is ............ *e.* The refracted ray is line ............ *f.* The angle of refraction is ............

3. *a.* In diagram II, complete the path of the ray shown. *b.* Draw and label the normals at each surface.

4. In diagram V, continue the ray shown to indicate the effect of the water, mirror, air, and glass.

5. *a.* In diagram III, complete the path of the ray shown. *b.* Draw and label the normals at each surface.

6. *a.* In diagram IV, complete the path of the ray shown. *b.* Draw and label the normals at each surface. *c.* Calculate all angles.

7. Calculate the index of refraction of a substance in which the speed of light is $2.5 \times 10^8$ m/sec.

8. When light goes from air into water, what happens to *a.* its frequency *b.* its speed *c.* its wavelength?

9. *a.* In diagram VI, complete the path of the ray shown. *b.* Draw and label the normals at each surface. *c.* Calculate all angles with the normals produced by the rays.

10. By means of a ray diagram, show how the image of the arrow is formed by the plane mirror.

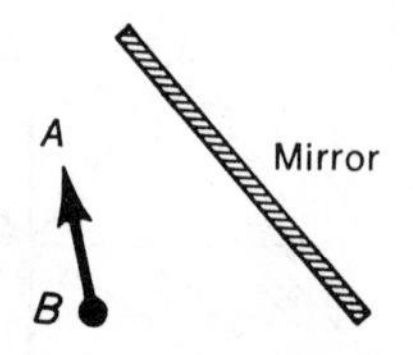


11. Sketch the image of the letter *B* in the above diagram.

12. Flint glass and carbon disulfide have almost the same index of refraction. How does this explain the fact that a flint glass rod immersed in carbon disulfide is nearly invisible?
13. When a straight stick is placed in water at an oblique angle to its surface, the part under the surface appears bent upwards. Explain this effect with the aid of a diagram.
14. A small source of light is at the bottom of a tank filled with water. Use a ray diagram to show the conditions for *a.* partial reflection of the light *b.* total reflection *c.* refraction at 90°.
15. Show by means of a diagram how atmospheric refraction affects the apparent position of the sun at sunset.
16. Calculate the angle of refraction for light entering water from air with an angle of incidence of *a.* 20° *b.* 30° *c.* 45° *d.* 60°.
17. Using the values in the *Reference Tables*, calculate the speed of light in *a.* benzene *b.* diamond *c.* quartz *d.* water.
18. For light leaving each of the following substances and entering air, calculate the critical angle. *a.* benzene *b.* diamond *c.* quartz *d.* water.
19. Calculate the relative index of refraction for *a.* water with respect to crown glass *b.* carbon tetrachloride with respect to flint glass *c.* crown glass with respect to water.
20. Make a ray diagram for the formation of an image by a concave mirror whose focal length is 6 cm, if the object distance from the mirror is *a.* 12 cm *b.* 18 cm *c.* 10 cm *d.* 3 cm.
21. *a.* A light year is the distance that light travels in one year. Express the distance in miles using standard notation and two significant figures. *b.* The planet Neptune is about $2.8 \times 10^9$ miles from the earth. Light from the nearest star other than the sun takes about 4 years to get to the earth. About how many times as far as Neptune is this star from the earth?

## Test Questions

**Given:** 1 Å = $10^{-10}$ m; critical angle for glass = 42°.

1. When parallel rays are reflected from a smooth, plane surface, the reflected rays will be 1. parallel to the incident rays 2. parallel to each other 3. non-parallel 4. part of a spherical surface.
2. When light is diffusely reflected, the angle of reflection is 1. greater than 2. less than 3. equal to the angle of incidence.
3. A person approaches a mirror with a speed of 7 ft/sec. His image approaches him with a speed, in feet per second, of 1. 0 2. 7 3. 14 4. 21.
4. As a person approaches a plane mirror, the size of his image 1. increases 2. decreases 3. remains the same.
5. A length of 0.047 cm is equal to ........... Angstrom units.
6. Twenty Angstrom units is equal to ........... centimeters.
7. If the index of refraction of a transparent substance is 1.7, the speed of light in it is ........... m/sec.

8. The bending of light when it passes from one medium to another is known as 1. reflection 2. refraction 3. diffraction 4. dispersion.
9. If a ray of light goes from one medium to another with a different index of refraction without being bent, the angle of incidence must be ........... degrees.
10. The index of refraction of alcohol is 1.36. The speed of light in it is ........... m/sec.
11. If a ray of light falls on a rough surface, the angle of reflection of this ray 1. is less than the angle of incidence 2. is greater than the angle of incidence 3. is equal to the angle of incidence 4. may be any of the previous three.
12. The glass of some electric light bulbs is frosted so that the glass will be 1. opaque 2. transparent 3. translucent 4. non-refracting.
13. The shadow of an eclipse containing no light is called ............
14. This paper does not act as a mirror because the light falling on it is 1. transmitted 2. absorbed 3. diffusely reflected 4. regularly reflected.
15. In New York and California the sun is never where it appears to be because of the earth's revolution around the sun and which one of the following: 1. smog 2. the atmosphere absorbs light 3. atmospheric refraction 4. formation of ozone.
16. Of the materials listed in the Reference Table, Indices of Refraction, the speed of light is lowest in ............
17. If the index of refraction for glass is 1.61, the sine of the critical angle for light going from this glass to air is ............
18. Light goes from medium 1 to medium 2. The angle of incidence in medium 1 is 45°, the angle of refraction in medium 2 is 35°. The speed of light is 1. the same in both 2. greater in medium 1 3. greater in medium 2.
19. ** An object is 30 cm from a concave spherical mirror whose focal length is 20 cm. The image is 1. virtual and enlarged 2. virtual and reduced in size 3. real and reduced in size 4. real and enlarged.
20. ** The radius of curvature of the mirror in question 19 is ........... cm.

## Topic Questions

1.1 What measurement supported the wave theory of light?

1.2 What is an Angstrom unit?

2.1 What is the speed of light in vacuum?

3.1 The light emitted by an element is now used to provide a universal standard for length. What is the name of this element?

4.1 What simple explanation can we give for the fact that objects in sunlight cast a sharp shadow?

5.1 What are the characteristics of the image formed by a pinhole camera?

6.1 What is the law of reflection?

6.2 What is meant by *a.* diffuse reflection *b.* luminous?

7.1 What are the characteristics of the image formed by a plane mirror?

8.1 What kind of images are formed by concave mirrors?

9.1 What are the characteristics of the image formed by convex mirrors?

10.1 *a.* What is *Snell's Law*? *b.* What is meant by the index of refraction?

11.1 How does the speed of a wave in a medium affect the amount of refraction of a wave?

12.1 What is Huygens' Principle?

13.1 What does it mean to say that one substance is optically denser than another?

14.1 Light enters a parallel glass plate and leaves it obliquely. What is the direction of the emerging ray compared to that of the entering ray?

15.1 What is the path taken by light which enters a triangular glass plate?

16.1 What is meant by the critical angle?

16.2 Under what conditions do we get total reflection?

# 10 Light—Lenses and Color

## 1 Introduction

Many of us use eyeglasses to improve our vision; telescopes give us a closer view of distant objects; microscopes permit a detailed examination of nearby objects; cameras provide a pictorial record for pleasure, business, and science. All these, and other devices, depend on lenses. How does a lens work?

**2 Types of Lenses** A *lens* is a transparent device shaped to converge or diverge a beam of light transmitted through it. Lenses described in this book are thin spherical lenses; one or both of the surfaces are spherical. If there is a non-spherical surface, it is plane. In addition, unless we say otherwise, it will be assumed that the lens is made of glass and is surrounded by air. Lenses are usually divided into two types, converging and diverging.

**Converging Lenses** A converging lens is one that converges parallel rays of light. It is thicker in the middle than at the edges. It is also called a convex lens, and

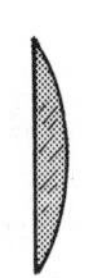

*Shapes of Converging Lenses*

comes in three shapes. In the middle lens, the left surface is plane; all other surfaces are spherical. In our diagrams we use the shape shown on the left as typical of all three.

The *principal axis* (p.a.) of a lens is the line connecting the centers of its spherical surfaces. A narrow bundle of rays parallel to the principal axis of a converging

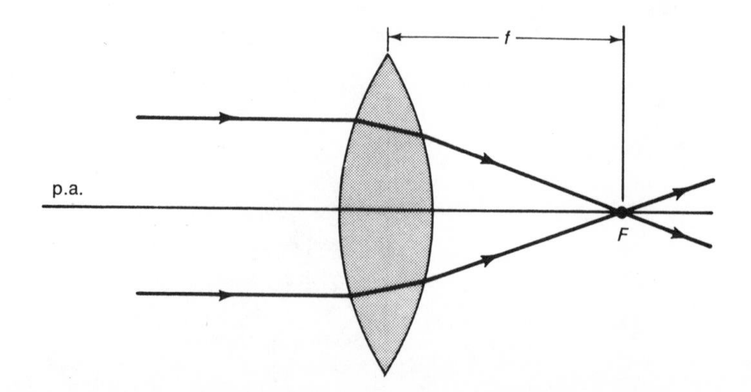

lens is refracted by the lens so that these rays, after refraction, pass through a common point on the principal axis known as the *principal focus* (F). The distance from the lens to the principal focus is known as the *focal length* (f) of the lens. The lens is drawn quite thick, but remember that we are discussing thin lenses only and therefore we may show the focal length as a distance to the center line of the lens.

The focal length of a lens depends on its shape and on the index of refraction of the glass. The greater the index of refraction the smaller the focal length. You probably expected this, since the rays will be bent more if the index of refraction is greater.

## 3 Images Formed by Converging Lens

A converging lens can be used to form both real and virtual images. The formation of these images, and their characteristics, can be explained well by means of a *ray diagram.* In the diagram the formation of a real image is shown. This happens with the converging lens whenever the object is more than one focal length away from the lens.

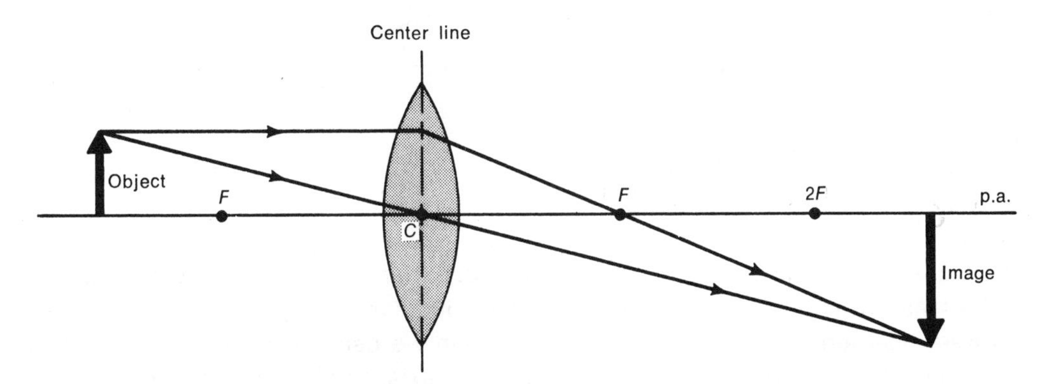


*Real Image Formed*
*Object Distance Greater than* **f**

The procedure for making the ray diagram is as follows. Draw the lens and its principal axis. Then draw the center line of the lens and note the center of the lens (C) where the principal axis intersects the center line. Mark off a convenient length from the lens along the principal axis to represent the focal length; this gives the location of the principal focus (F) on both sides of the lens. A point twice as far is represented by 2F. We usually measure these distances from the center line; since the lens is thin we can think of the center line as representing the lens with the arcs merely drawn in to remind us of the type of lens with which we are working.

The object is usually represented by an arrow. Place the object at some appropriate distance from the lens, depending on the situation. In this case, the object was placed between one and two focal lengths away from the lens; in short, the object distance is between one and 2f. We draw two special rays from the head of the object. One ray, parallel to the principal axis, is drawn to the center line and from there is drawn straight through the principal focus, *F*. The second ray is drawn as a straight line through the center of the lens. Where the two rays meet, the head of the image is formed: on a screen placed there, the top point of the object is sharply reproduced. We draw in the rest of the image.

Note the characteristics of the image for the special case shown in the diagram. When the object is between one and two focal lengths from the lens, the image is real, inverted, enlarged, and more than two focal lengths away from the lens on the opposite side from the object.

The above procedure for making a ray diagram does not give a straight line for the ray inside the lens. Of course, all rays inside the lens are straight. Although it is permissible to leave the diagram as drawn above, a slight change is sometimes made in the drawing of the first ray. This is exaggerated in the next diagram, where we use a rather thick lens. First we proceed as above, drawing the ray parallel to the principal axis to the center line and from there through the principal focus; now, however, the portion inside the lens is drawn as broken lines. The two points of intersection with the lens surfaces, *A* and *B*, are then connected by a solid line to represent the actual path of the light inside the lens.

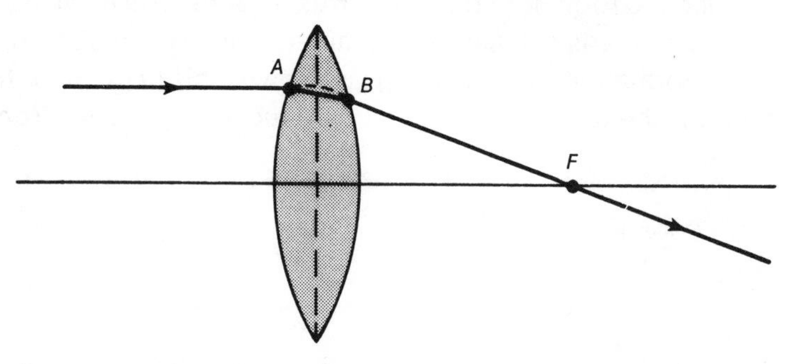


*Connect points A and B on lens surfaces by a straight line.*

Consider the case when the object is more than two focal lengths away from the lens. Draw a ray diagram as in the previous case, but place the object about three focal lengths away from the lens. Again we use two special rays from the head of the object, one parallel to the principal axis, and the other straight through

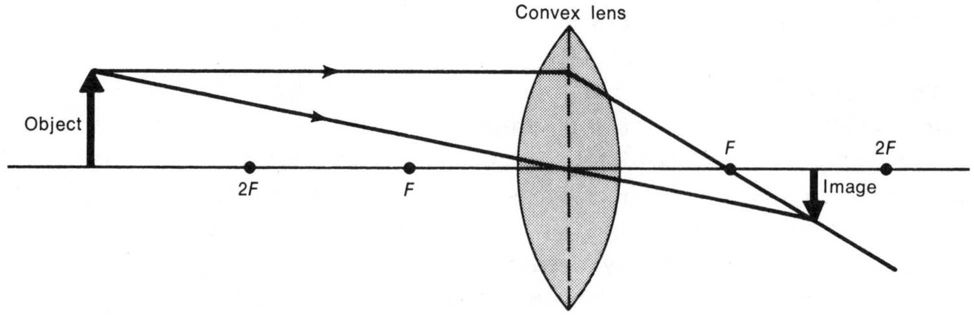


*Real Image Formed*
*Object Distance Greater than* **2f**

the center of the lens. Where the two rays meet the head of the object is sharply reproduced on a screen placed there. A real image is formed. We draw in the rest of the image. Again the real image is inverted and on the other side of the lens, but this time it is smaller than the object, and between one and two focal lengths away from the lens.

When the object is two focal lengths away from the lens, a real image is formed which is of the same size as the object and two focal lengths away from the lens and on the opposite side from the object.

4 **Object Distance Less Than One Focal Length** When the object is less than one focal length away from a converging lens, a virtual image is formed. This is shown in the next ray diagram. We start as in the previous cases after drawing the object about one-half a focal length away from the lens. We draw the two special rays

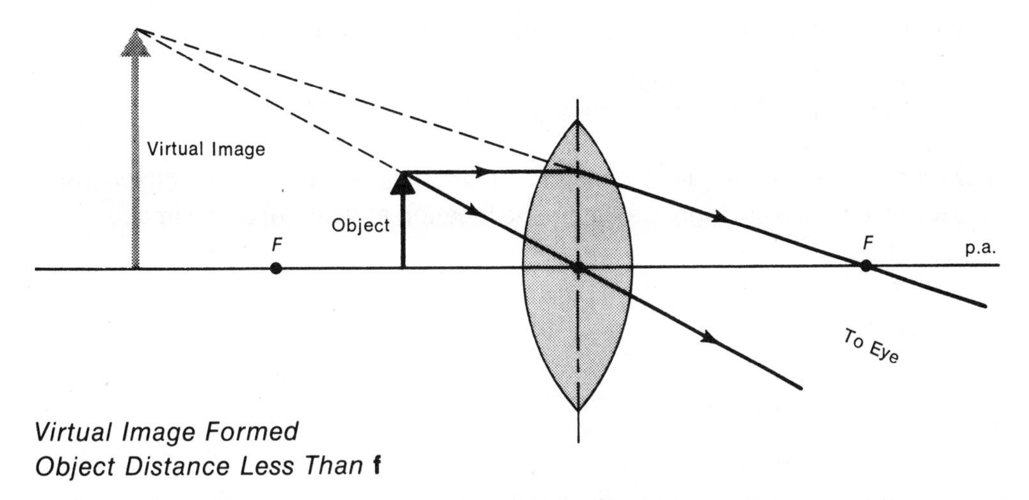


*Virtual Image Formed*
*Object Distance Less Than* **f**

from the head of the object. The ray parallel to the principal axis is refracted by the lens to pass through the principal focus. When we draw the second ray through the center of the lens, we notice that it does not meet the first ray. The two rays diverge. No image can be formed on a screen in this case. However, a person looking through the lens towards the object will see a virtual image. If we extend the diverging rays beackwards, as shown by the dotted lines, we get the location and characteristics of this virtual image. It is erect, larger than the object, and further away from the lens than the object, and on the same side of the lens as the object. Notice, as with the virtual image in the case of the plane mirror, that a virtual image is not formed on a screen; no rays come to a focus where the virtual image is. A virtual image is formed by rays which do not actually pass through it.

Notice that real images are formed on a screen. Every object point is reproduced as a sharp image point. In our diagrams we have drawn only two rays from the head of the object to obtain the corresponding image point. Actually, all rays starting from the head of the object will pass through this image point after refraction by the ideal lens. Note that virtual images are not formed on a screen. They are the result of refraction in the observer's eye. The diverging rays are focused on the retina, and the observer interprets the rays as coming from the point where the rays, extended backwards, meet.

5* **Numerical Calculations** The location of the image can be calculated with the lens equation:

$$\frac{1}{\text{object distance}} + \frac{1}{\text{image distance}} = \frac{1}{\text{focal length}};$$

$$\frac{1}{d_o} + \frac{1}{d_i} = \frac{1}{f}$$

Observe the following rules when using the lens equation.

1. Distances are measured from the lens; object distance is distance of object from lens, etc.
2. The object distance ($d_o$) is always positive.
3. The image distance ($d_i$) is positive for real images, negative for virtual images.
4. The focal length is positive for converging lenses, negative for diverging lenses.

* The height of the image can be calculated from this equation:

$$\frac{\text{height of image}}{\text{height of object}} = \frac{\text{image distance}}{\text{object distance}} = \text{magnification of lens.}$$

**EXAMPLE:** A converging lens whose focal length is 20 cm is placed 30 cm from an object which is 4.0 cm high. Calculate the location and height of the image.

$$\frac{1}{d_o} + \frac{1}{d_i} = \frac{1}{f};$$

$$\frac{1}{30} + \frac{1}{d_i} = \frac{1}{20};$$

$$d_i = 60 \text{ cm.}$$

$$\frac{\text{height of image}}{\text{height of object}} = \frac{\text{image distance}}{\text{object distance}};$$

$$\frac{\text{height of image}}{4.0 \text{ cm}} = \frac{60 \text{ cm}}{30 \text{ cm}};$$

$$\text{height of image} = 8.0 \text{ cm.}$$

The image is 60 cm from the lens and is enlarged to a height of 8.0 cm.

** Students of courses such as PSSC Physics should be careful to avoid confusion in applying the above formulas. In these, object and image distances are measured from the lens. In PSSC Physics distances are calculated by a different formula:

$$S_i S_o = f^2,$$

where $f$ is the focal length of the lens
$S_i$ is the distance of the image from the principal focus
$S_o$ is the distance of the object from the corresponding focus on the left side of the lens.

Notice that the distances are not measured from the lens.

Also, $H_i/H_o = f/S_o$, where $H_i$ is the height of the image
$H_o$ is the height of the object

**6** ** **Spherical Mirrors** The same equations apply to spherical mirrors, but what has been said in the rules about convex lenses applies to concave mirrors; and what has been said about concave lenses applies to convex mirrors. Also, since a mirror depends on reflection, not refraction, the real images formed by a concave mirror are formed on the same side of the mirror as the object.

# 7 Diverging Lenses

A diverging lens always produces a virtual image which is smaller than the object and erect (if the object is erect). This will be illustrated with a ray diagram, but first we must explain what is meant by the principal focus and focal length of a concave or diverging lens. As for the converging lens, the principal axis is the straight line connecting the centers of curvature. However, a narrow bundle of rays parallel

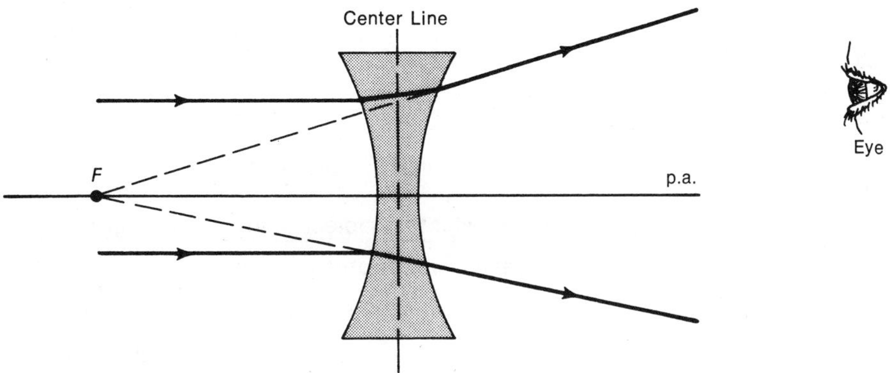


*Focal Length of a Concave Lens*

to the principal axis is refracted by the diverging lens so that after refraction the rays diverge away from the principal axis. If these diverging rays enter the observer's eye, they will seem to come from a point on the principal axis on the other side of the lens. This point is located in the above diagram by extending the diverging rays backwards by dotted lines. It is called the principal focus of the concave or diverging lens, F. It is a virtual focus because the actual rays do not meet there. The focal length ($f$) is the distance from the principal focus to the lens.

Now we can draw the ray diagram for a diverging lens. Again we draw two special rays from the head of the object after selecting a suitable point F as the principal focus. The ray parallel to the principal axis is diverged on the right side of the lens so as to seem to come from F. It is extended by a dotted line backwards to F. The second ray from the head of the object is drawn straight through the center of the lens. The image point of the head of the object is located at the intersection of this ray and the dotted extension of the first ray. The rest of the image (I) is drawn in. Since the rays do not actually meet there, no image will be formed on a

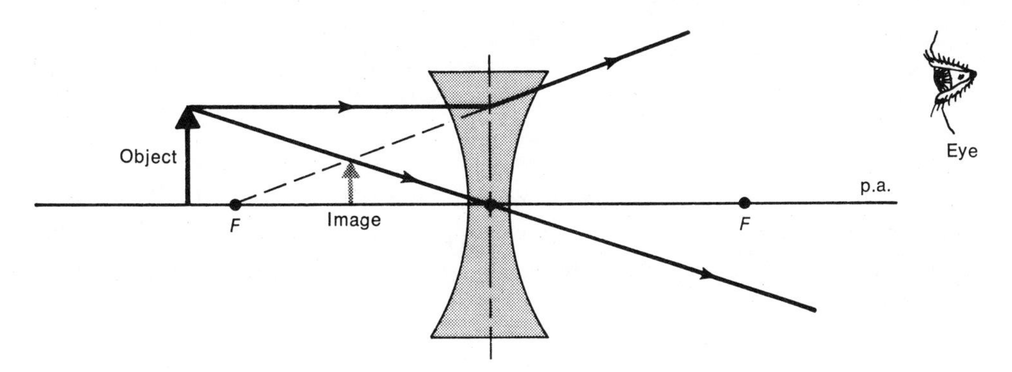


*Virtual Image Formed by Concave Lens*

screen placed in this location. It is a virtual image; on a diagram it is usually represented by a dashed line.

Notice from the diagram that the image produced by a diverging lens is always virtual, erect, smaller than the object, and closer to the lens than the object.

The characteristics of the image produced by converging and diverging lenses under different conditions are summarized below:

## 8 Characteristics of Image Produced by a Lens

| | Object Distance | Image Characteristics |
|---|---|---|
| Converging Lens | greater than $2f$ | real, smaller than object, inverted, between $f$ and $2f$ |
| | $2f$ | real, same size as object, inverted, at $2f$ |
| | between $f$ and $2f$ | real, larger than object, inverted, greater than $2f$ |
| | less than $f$ | virtual, larger than object, erect, $d_i$ greater than $d_o$ |
| Diverging Lens | any distance | virtual, smaller than object, erect, $d_i$ less than $d_o$ |

## ** Optical Instruments

**9 Eye and Camera** The camera and eye have many points of similarity. The camera has a shutter to admit the light from the object at the right time. In the eye this corresponds to the eyelid. In the camera the light has to pass through an opening, known as an *aperture*, in a diaphragm. The diameter of the aperture is usually

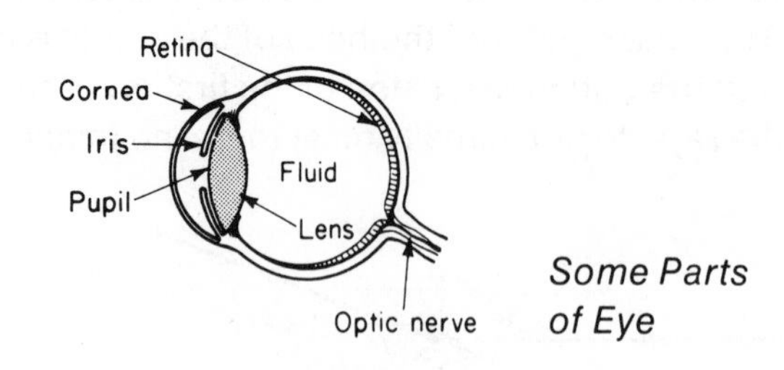


*Some Parts of Eye*

adjustable. In the eye, the corresponding opening is the *pupil*, a hole in the iris. The iris is the colored portion of the eye. The pupil appears black but is actually a hole. In the camera the light passes through a glass lens (or combination of lenses) and is brought to a focus on the photographic film. In the eye the light goes through a complicated lens system consisting of the *lens* and some liquids with curved surfaces. The light then falls on the retina where the image is formed. From the retina appropriate electrical impulses are transmitted along the optic nerve to the brain.

In both the camera and the eye, the image is real, inverted, and smaller than the object. In most cameras it is possible to adjust the focusing by changing the position of the lens with respect to the film. In the eye, the equivalent is accomplished automatically by changing the focal length of the lens; muscles attached to the lens change its thickness.

**10 F-number** A lens is sometimes described by an *f*-number, such as *f*/8. This means that the diameter of the lens, measured along the center line, is 1/8 of the focal length. In a camera this *f* number means that the aperture has been adjusted to 1/8 of the focal length. This is related to the *speed of the lens*. If the *f*-number is changed from *f*/8 to *f*/4, the diameter of the aperture is doubled. Therefore the area of the opening is quadrupled and, if the illumination of the scene is constant, four times as much light goes through the lens and only 1/4 of the exposure time is required.

**11 Some Defects of the Eye** A *nearsighted* person can see nearby objects well. However, distant objects seem blurred because his eyes focus the light in front of the retina. When the light reaches the retina, it has diverged again and results in a blurred image. In order to prevent this focusing in front of the retina, a nearsighted person uses an appropriate diverging lens.

Similarly, a *farsighted* person's eye does not focus the light soon enough when the object is close. The light would be focused if the retina were a little further away. Farsightedness is corrected by a suitable converging lens.

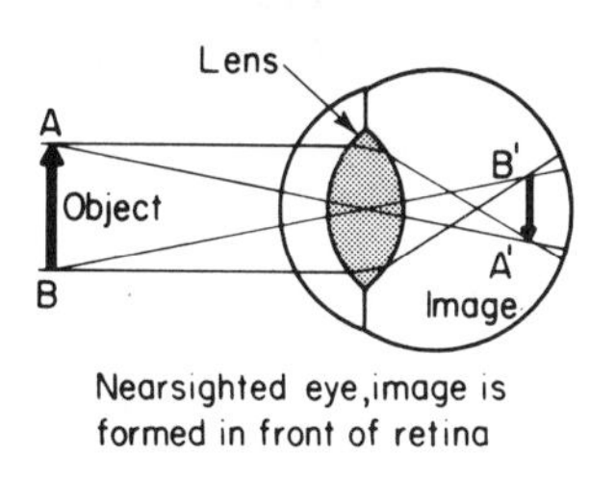


Nearsighted eye, image is formed in front of retina

**A Uncorrected**

Lens
A
Object
B
Image
B'
A'

Concave lens focuses light on retina

**B Corrected**

*Nearsightedness corrected by Diverging Lens*

**12 Magnifying Glass** We can use a converging lens to make this print seem larger. When we do so, we hold the print ("the object") a little less than a focal length away from the lens through which we look at the print. The result is a virtual, erect, enlarged image. We can change the distance for different magnification. The maximum magnification for the magnifying glass is approximately

$$\frac{25}{f}+1$$

where $f$ is the focal length; it must be expressed in centimeters. The magnifying glass is sometimes called a simple microscope.

**13 Microscope** To get greater magnification than can be obtained with a single lens, we use a compound microscope. For short it is usually referred to as a microscope. The basic parts are two converging lenses. The lens closer to the object is called the *objective*, the one closer to the eye is called the *eyepiece*. (In practice each lens is usually a combination of lenses.) The object is a little more than one focal length away from the objective, which therefore gives an enlarged real image, the "first image" in the diagram. The eyepiece is used as a magnifying glass to

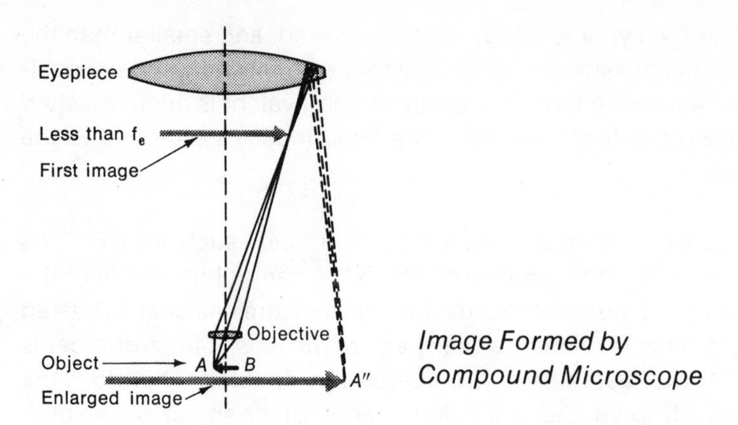


*Image Formed by Compound Microscope*

enlarge this real image. It is, therefore, placed a little less than one focal length away from where the first image is focused by the objective.

14 **Refracting Telescope** There are many types of telescopes. We shall describe briefly a simple type, the refracting telescope. Like the compound microscope, it has two lenses; the one closer to the object is known as the *objective*, the one closer to the eye is the *eyepiece*. In the telescope, the object is far away; usually much more than two focal lengths. The image formed by the objective is real, inverted, smaller than the object, and only a little more than one focal length away from the objective. As in the microscope, the eyepiece is used as a magnifying glass to enlarge this real image. The magnification of such a telescope is approximately the focal length of the objective divided by the focal length of the eyepiece.

$$m = f_o/f_e.$$

## 15 Color and Light

**Dispersion** In describing refraction by means of a triangular glass prism, we specified the use of *monochromatic light*, which literally means light of a single color, but often is used to mean light of a single wavelength. Newton discovered that if sunlight is used instead, beautiful colors are obtained. In the laboratory, instead of sunlight we often use ordinary incandescent tungsten filament bulbs, whose light is described as *white light*. When we allow this light to be refracted by a triangular glass prism, we get a beautiful display of colors. By means of another prism held upside down, these colors can be recombined to give back white light. *Dispersion* is the breaking up of light into its component colors.

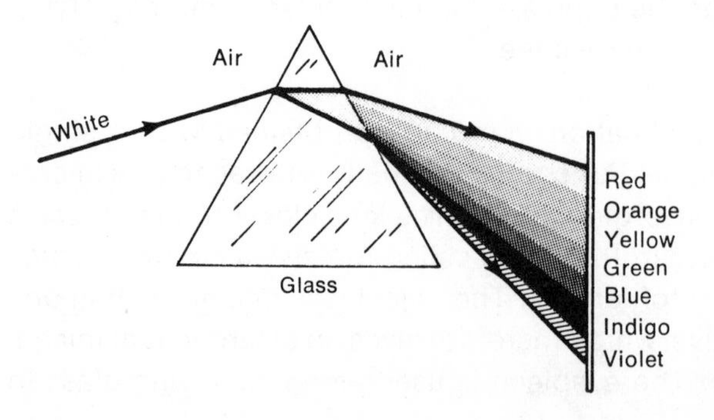


*Visible Spectrum of White Light*

Seven distinct colors may be observed if white light is used. If a glass prism is used, the order of the colors is, from the one bent the least to the one bent the most: red, orange, yellow, green, blue, indigo, violet. Some students like to remember this sequence by using the first letters to spell a name. Light which is a mixture of different colors is known as *polychromatic light.*

The term spectrum is used in many different ways. Here we shall define the *spectrum* of visible light as the array of colors produced by the dispersion of white light, or the range of wavelengths capable of stimulating the sense of sight. This range of wavelengths is approximately 4000–8000 Angstroms, or $4 \times 10^{-7}$–$8 \times 10^{-7}$ meters. In this spectrum, the shortest wavelength is violet, the longest is red.

We have already said that light is an electromagnetic wave. Energy radiated by the sun includes electromagnetic waves other than light. *Infrared* radiation has a greater wavelength than red light. *Ultraviolet* radiation has a shorter wavelength than violet light. Both are invisible to the human eye.

Dispersion is produced because the different wavelengths of light travel with different speeds in a material medium. In vacuum all electromagnetic waves travel with the same speed. In water and glass, violet light is slowed up more than red light, and therefore is refracted more. Therefore the index of refraction of glass is greater for violet than it is for red light. It is intermediate for the other colors. The index of refraction for yellow light is usually taken as the typical value. The term *dispersive medium* is sometimes used to refer to a medium in which different frequencies travel with different speeds.

The frequency of light does not change when it goes from one medium to another or vacuum. Its speed does depend on the medium. Since, for any wave,

$$\text{speed} = \text{frequency} \times \text{wavelength},$$

when an electromagnetic wave goes from vacuum into glass, its wavelength decreases, because its speed decreases. Also note that the different colors of light are produced by different frequencies. In the case of sound, different frequencies produce a different pitch.

16 **Color of Objects** The color of an opaque object depends on the color of light it reflects. A red object reflects mostly red light; it absorbs the rest. If an object reflects no light, it is said to be black. If a red object is exposed only to blue light, which it absorbs, the red object would appear black.

If an object reflects all the light, it is said to be white. A white object exposed to sunlight or ordinary artificial light appears white. Such an object exposed to red light only will appear red, because that is the only light reflected to our eyes.

Similarly, the color of a transparent object is determined by the light it transmits. If we hold a piece of red glass in front of an incandescent bulb, the glass appears red because it transmits only red light.

17 ** **Primary and Complementary Colors** If the above seven colors of light are recombined, white light is produced. However, it is possible to produce white light with fewer colors. With the proper choice of three colors, known as the *primary colors*, it is possible to produce not only white light, but nearly all colors of light. The three primary colors of light are red, green, and blue-violet. This principle is applied in color television. For example, if we mix red and green light, we get yellow light. Don't confuse this with the effect of mixing paints.

If we mix all three primary colors of light, in proper intensity, we get white light. Now, the red and green alone give us yellow. Therefore you may not be too surprised to hear that yellow mixed with blue-violet, the third primary color, gives us white light.

*Complementary colors* are two colors of light whose combined effect on the eye is that of white light. One pair, as shown in the above paragraph, is yellow and blue-violet. Yellow is the complement of the blue-violet. We get the complement of each primary by using the color resulting from the mixture of the other two primaries. The complement of red is blue-green; the complement of green is purple (also called magenta).

Paints, like other opaque objects, are seen by the color of light which they reflect. If we mix two paints, each paint subtracts the color of light which it absorbs. The color of the mixture is the color of light which neither absorbs. Color obtained by the mixing of *lights* is known as an additive process.

18 ** **Aberrations** Even a perfectly-made thin, spherical lens doesn't function exactly as described above if white light is used. The operation of a lens depends on refraction. We saw that refraction by the triangular prism results in dispersion.

In a lens, the different colors are brought to a focus at different points. The focal length of a lens depends on the color of light used, just as the index of refraction of a substance depends on the color of light used. The result is that images produced by lenses have a slight color fringe, and photographs with such lenses may be slightly fuzzy. This is known as *chromatic aberration*. It can be minimized ("corrected") by using a suitable combination of lenses known as an *achromat*. A simple lens also suffers from other aberrations. These will be ignored here.

# 19 Types of Spectra

**Spectroscope** Newton was able to disperse sunlight by letting a small beam of light pass through a glass prism. For a more careful examination of the colors produced by a light source we use a special instrument, the spectroscope. The *prism spectroscope* is illustrated here. The light to be examined goes through a narrow

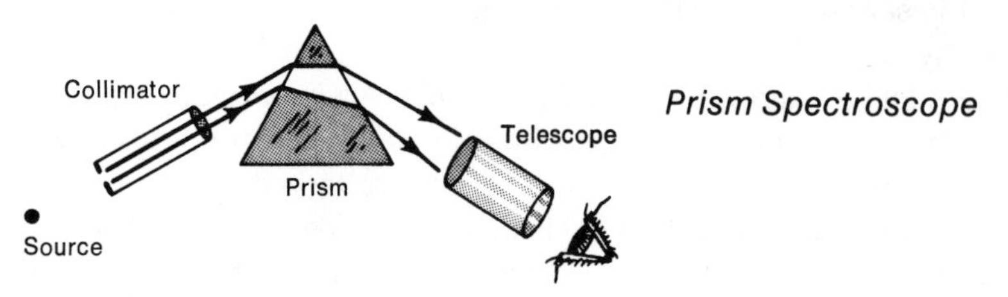


*Prism Spectroscope*

rectangular slit before it reaches the prism which disperses the light. The rectangular slit is part of a tube known as the collimator tube. A telescope is then usually used to look at the dispersed light. A diffraction grating may be used instead of the prism. The grating will be described later.

**Continuous Spectrum** When white light was used above with the triangular prism, a continuous spectrum was produced: there was no gap as we went from one extreme, the red, to the other extreme, the violet. A continuous spectrum may be obtained from a glowing solid or liquid, or a glowing gas under high pressure.

**Bright Line Spectrum** A bright line spectrum is obtained from a glowing gas under low or moderate pressure. Lines of light appear with gaps in between where there is no light. These lines of light can be thought of as images of the illuminated slit. Different chemical elements in gaseous form produce different spectra. The spectrum is characteristic of the element and, since they are different for the different elements, can be used to identify small quantities of the element. The line spectrum has its origin in the atoms of the element. This will be discussed later.

20 ** **Band Spectrum** Under certain conditions of moderate temperatures, a substance emits a band spectrum, a spectrum in which bands of light appear with regions of no light in between. Careful analysis shows that these bands are actually fine lines of light very close together. The band spectra have their origin in combination of atoms (molecules) rather than in the individual atoms.

21 **Dark Line Spectrum** When Newton observed the sun's spectrum, it seemed to be continuous. Careful examination in 1802 with better instruments available then, showed that there are many dark lines in the sun's spectrum. A dark line spectrum is a continuous spectrum which is interrupted by thin dark lines. It is an absorption spectrum. The dark lines result from the absorption of the energy of the missing wavelengths by gases that are between the hot source and the spectroscope.

When a gas is cool, it can absorb some of the wavelengths which it emits when hot. The dark lines in the sun's spectrum are known as *Fraunhofer lines*. They are produced mostly by gases in the sun's atmosphere, which absorb some of the energy emitted by the hot core. Study of these Fraunhofer lines has shown that many of the elements on earth are also present in the sun's atmosphere. The gas helium was identified in the sun before it was found on earth.

22 **Electromagnetic Spectrum** We have already mentioned that Maxwell's work in 1864 indicated that light is an electromagnetic wave. Light is a small portion of the electromagnetic spectrum which includes, in decreasing wavelength: radio waves, infrared, visible light, ultraviolet, x-rays, gamma rays, and secondary cosmic rays. Some of these divisions are not well defined; for example, x-rays and gamma rays are similar.

Many of the beautiful color effects produced with light are the result of the interference of light. This will be discussed in the next chapter.

# Questions and Problems

1. Where should a point source of light be placed with respect to a lens so that a beam parallel to the principal axis will be obtained? What kind of lens should be used?
2. A convex lens has a focal length of 15 cm. An object is placed 10 cm from the lens. *a.* Calculate the location of the image. *b.* Calculate the size of the image. *c.* Make a ray diagram for this situation. (Assume object size is 4 cm.)
3. Repeat question 2 for an object distance of 20 cm.
4. Repeat question 2 for an object distance of 30 cm.
5. Repeat question 2 for an object distance of 40 cm.

6. Repeat question 2 for an object distance of 10 cm.
7. Repeat question 2 for an object distance of 5 cm.
8. A concave lens has a focal length of 10 cm. Make a ray diagram to show the formation of the image if the object distance is *a.* 5 cm *b.* 15 cm *c.* 10 cm *d.* 20 cm
9. For the lens in question 8, calculate the image distance if the object distance is *a.* 5 cm *b.* 15 cm *c.* 10 cm *d.* 20 cm
10. For the lens in question 8, calculate the size of the image, if the object is 4 cm tall, and the object distance is *a.* 5 cm *b.* 15 cm *c.* 10 cm *d.* 20 cm
11. What is meant by chromatic aberration? How is it corrected?
12. Draw a diagram to show how white light is dispersed by a triangular glass prism.
13. What are the characteristics of the image produced by the objective of an astronomical telescope?
14. Why is the focal length of a lens different for red light than for blue light?
15. An object 6 cm from a convex lens produces an image 4 cm on the other side of the lens. Calculate the focal length of the lens. Is the lens converging or diverging?
16. Two converging lenses are placed 20 cm apart on a meter stick so that light from a candle flame placed 40 cm from the lens nearer to it passes through both lenses. The candle is mounted on the meter stick and each lens has a focal length of 10 cm. Where should a screen be placed to get a sharp image of the flame?

## Test Questions

*(Unless stated otherwise, in the following questions lenses are assumed to be made of glass and are surrounded by air.*

1. Convex lenses always 1. have two spherical surfaces 2. are thinner in the middle than at the edges 3. are thicker in the middle than at the edges.
2. An object is more than one focal length away from a converging lens. As the distance of the object from the lens increases, the distance of the image from the lens 1. increases 2. decreases 3. remains the same.
3. In question 2, the size of the image 1. increases 2. decreases 3. remains the same.
4. When a person uses a convex lens as a simple magnifying glass, the object must be placed at a distance from the lens of 1. a little less than one focal length 2. a little more than one focal length 3. a little less than two focal lengths 4. a little more than two focal lengths.
5. The image produced by a concave lens is 1. always virtual and enlarged 2. always virtual and reduced in size 3. always real 4. sometimes real, sometimes virtual.
6. Real images formed by single convex lenses are always 1. on the same side of the lens as the object 2. inverted 3. erect 4. smaller than the object.

7. A virtual image is formed by 1. a slide projector 2. the ordinary camera 3. a simple magnifier 4. a motion picture projector.
8. A convex lens produces a virtual image; the distance of the object from the lens 1. must be less than on focal length 2. may be more than two focal lengths 3. may be between one and two focal lengths 4. must be two focal lengths.
9. The distance of an image from a convex lens is the same as that of the object. The distance of the object from the lens 1. must be less than one focal length 2. may be more than two focal lengths 3. may be between one and two focal lengths 4. must be two focal lengths.
10. An object is placed 12 cm from a convex lens whose focal length is 10 cm. The image must be 1. virtual and enlarged 2. virtual and reduced in size 3. real and reduced in size 4. real and enlarged.
11. An object is placed 12 cm from a concave lens whose focal length is 10 cm. The image must be 1. virtual and enlarged 2. virtual and reduced in size 3. real and reduced in size 4. real and enlarged.
12. An object is placed 25 cm from a convex lens whose focal length is 10 cm. The image distance is ........... cm.
13. If the object in question 12 is 4 cm tall, the height of the image is ........... cm.
14. A camera is used to photograph a distant object. If the focal length of the lens is $f$, the distance of the film from the lens should be 1. slightly more than $2f$ 2. slightly less than $2f$ 3. slightly more than $f$ 4. slightly less than $f$.
15. In monochromatic yellow light, a blue cloth will appear 1. yellow 2. blue 3. green 4. black.
16. As compared with the wavelength of green light, the wavelength of yellow light is 1. greater 2. smaller 3. the same.
17. In sound, frequency determines pitch. In light, frequency determines ............
18. A beam of monochromatic light cannot be 1. reflected 2. dispersed 3. refracted 4. absorbed.
19. Grass appears green because it ........... green light.
20. The pupil in the eye corresponds to the ........... in the camera.

## Topic Questions

1.1 What are some instruments which use lenses?

2.1 What is another name for a converging lens?

2.2 What is the principal focus of a converging lens?

2.3 How does an increase in the index of refraction affect the focal length of a lens?

3.1 Under what conditions does a converging lens produce a real, enlarged image? What other characteristics will such an image have?

4.1 Under what conditions does a convex lens produce a virtual image?

4.2 If an image is formed on a screen, what kind of an image is it?

5.1 What is the lens equation?

6.1 Where must an object be placed with respect to a concave mirror, if a real, enlarged image is to be produced?

7.1 What are the characteristics of the image produced by a concave lens?

8.1 Under what conditions does a converging lens produce an image which is the same size as the object? What other characteristics will such an image have?

9.1 What parts of the eye correspond to the following parts of a camera? *a.* diaphragm *b.* diaphragm opening or aperture *c.* film *d.* lens

10.1 If an *f*-number setting is changed from *f*/8 to *f*/4, it is possible to take pictures in dimmer light. Why is this so?

11.1 What kind of lenses are needed to help farsighted people?

12.1 How is a convex lens used as a magnifying glass?

13.1 How are the objective and eyepiece of a microscope arranged?

14.1 How are the objective and eyepiece of a refracting telescope arranged?

15.1 Dispersion of white light from an incandescent filament produces a spectrum. What are the colors of this spectrum, arranged in order of decreasing wavelengths?

15.2 What is speed of red light compared to that of blue light in *a.* vacuum *b.* glass?

16.1 If a red object is exposed in succession to red, green, and blue light, which colors will it reflect? What color will the object appear to be when exposed to each of these lights?

17.1 What is the minimum number of monochromatic colors that can be used to get white light? What name is given to such colors?

18.1 Why do many cameras have an achromat?

19.1 What light sources produce continuous spectra?

19.2 Why can a line spectrum but not a continuous spectrum be used to identify an element?

20.1 What is the source of band spectra?

21.1 *a.* What are Fraunhofer lines? *b.* How are they produced?

22.1 What are some portions of the electromagnetic spectrum other than visible light?

# 11 Interference, Diffraction, Polarization

## 1 Introduction

We have already studied reflection and refraction of light. Explanation of these phenomena did not give a clear-cut advantage to proponents of either the wave or the particle theory of light. Observation of interference and diffraction of light gave a definite advantage to those who believed in the wave theory of light. Observation of the polarization of light showed that the wave has to be transverse. We shall now look at these phenomena in more detail.

## 2 Interference of Light

**Coherent Sources** In the chapter on Wave Phenomena and Sound we noted that when two waves go through the same portion of the medium at the same time, interference occurs. The two waves are superposed. They may reinforce each other, giving constructive interference; or they may annul each other, giving destructive interference. This is readily observed in sound, where we can easily hear beats. It is also readily observed in standing waves in a string, or in water waves.

In order to observe interference we need two *coherent sources*, sources that produce waves with a constant phase relation. For example, two waves that are always in phase will always reinforce each other. If two waves are of opposite phase, that is, one wave is 180° out of step with the other, they will tend to annul each other. We have no trouble maintaining this constant phase relationship with mechanical vibrations. If we strike a tuning fork, both prongs will keep vibrating at the same frequency and in step. If we hold the vibrating fork vertically and rotate it around the stem as an axis, we can observe the interference produced by the waves produced by the two prongs. The two prongs are coherent sources.

Until recently it has been impossible to get two independent light sources which are coherent. The reason for this is that most sources of light emit it in short bursts of unpredictable duration. This would be like a tuning fork momentarily stopping every few cycles. Two light sources might be out of phase during one burst but not during the next one. Annulment would be masked by the rapid change in phase relationship. It is now possible to get two lasers to be coherent. For most purposes, if we want interference of light, we take light from one source and have it travel by two different paths to the same place. This will be illustrated below. (Lasers will be discussed in a later chapter.)

*Interference of light by a thin film*
*Gegbold—Heraeas*

3
*

**Interference with Thin Films** The beautiful colors of soap bubbles and of thin films of oil on a city pavement or highway are due to interference of light and give further evidence that light behaves like a wave. Let us assume that we have monochromatic light incident on a thin film. The film may be oil, soap solution, or even air trapped between two flat glass plates. Some of the light is reflected at the top surface of the film and is represented by ray 1. Most of the light goes through the film; some of this is reflected at the bottom surface of the film and comes out again. This is represented by ray 2. If we look down on the film, rays 1 and 2 may

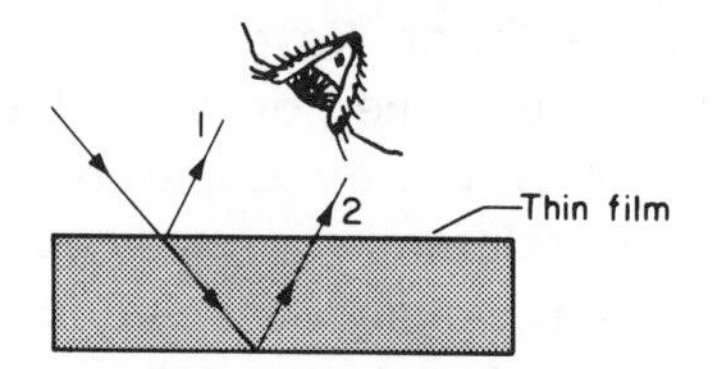


enter our eye and be focused on our retina. Coherent light is coming to our eye by two different paths. Ray 2 had to go back and forth through the film. The two rays (or light waves) arriving simultaneously on our retina, will interfere with each other.

Ray 2 may be delayed a whole number of wavelengths by having to go through the film. If so, it will arrive at the eye in phase with ray 1, and the two waves will reinforce each other. The observer looking at the film in that direction will see a bright spot.

Ray 2 may be delayed an odd number of half-wavelengths. $\lambda/2$, $3\lambda/2$, $5\lambda/2$, etc. If so, it will arrive out of phase with ray 1, and the two waves will annul each other. The observer looking at the film in that direction will see a dark spot. In general, such patterns consist of bright fringes alternating with dark fringes.

(A complicated effect observed with all waves and pulses under certain conditions is phase reversal on reflection. This means that pulses and waves are reflected as though they were delayed one-half a wavelength when they are reflected from a medium with a greater index of refraction than the one they travel in. This change of phase on reflection will be ignored here. The phase difference in rays 1 and 2 will be assumed to be due only to the difference in path.)

If the film is 1/4 wavelength thick, and if the incident ray is almost perpendicular to the surface, the path difference for ray 2 is 1/2 wavelength. The two waves cancel each other. This idea is applied in the coating of expensive photographic lenses to minimize undesirable reflections.

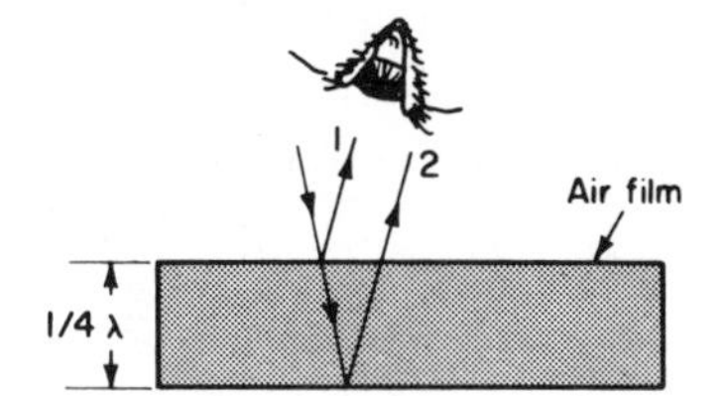

*Rays 1 and 2 Interfere*

In ordinary daylight the film is illuminated by many different wavelengths. The thickness of the film may be just right in one direction to cause annulment of one color. All the other wavelengths will combine to give a colored spot. In another direction the thickness may be just right for the cancellation of a different wavelength, and the complement of the canceled color will be seen. As a result, when viewing a soap bubble or an oil film we see the full colors of the rainbow.

4 **Interference with the Double Slit** In 1801 Thomas Young performed a famous experiment which suggested the interference of light. Its essential features are similar to the set-up described: Light from a small source $F$ illuminates a barrier which has two narrow slit openings, $S_1$ and $S_2$, very close to each other. The light which gets through these two slits falls on a screen. The screen may be thought of as the retina of our eye, because we usually look at the light source through the two slits. Assume that the light source is monochromatic. What we see on the screen is surprising. We see a whole series of bright lines alternating with dark

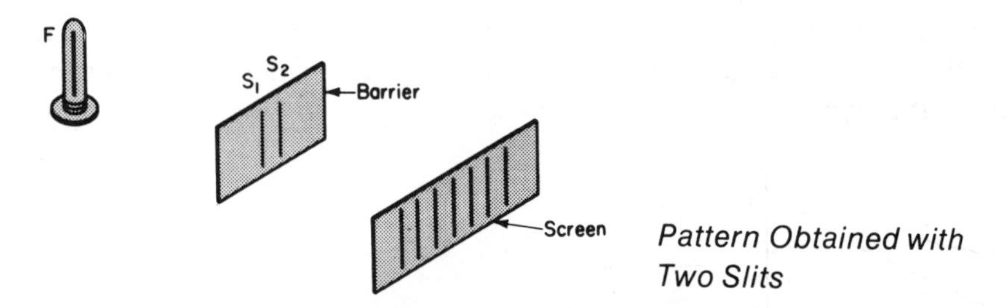

*Pattern Obtained with Two Slits*

regions. The technique we used for drawing ray diagrams with lenses and mirrors is inadequate. On that basis we might have expected to see two bright lines, one for each slit which lets light through.

We apply *Huygens' Principle* and the wave theory of light to explain what we actually see. Imagine that we are looking down at the above set-up. The light from the source spreads out and the wavefronts are represented by circles with source $F$ at the center. The two slits are equidistant from the source. Each wavefront reaches $S_1$ and $S_2$ at the same time. According to *Huygens' Principle*, we can think of the portions of the wavefront at each slit as acting like independent sources of light, producing their own wave fronts. These are represented as two sets of circles with $S_2$ being the center of the top set, $S_1$ the center of the bottom set. Since the light originally came from a single source, $F$, the waves produced by $S_1$ and $S_2$ are coherent.

The waves overlap on the screen and interfere with each other. In some places

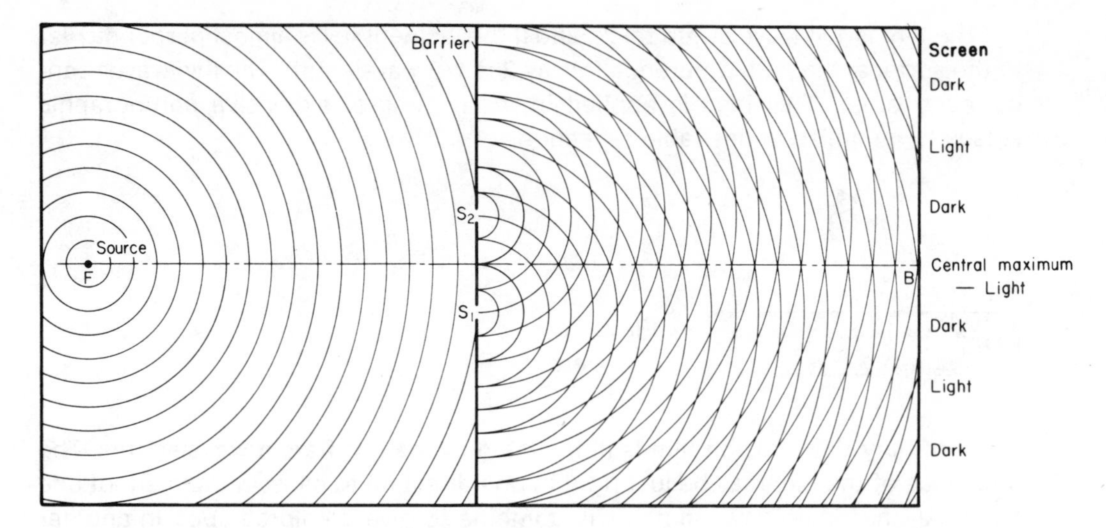


the waves arrive in phase, reinforce each other, and give a bright region or line. One such place is $B$, which is equidistant from $S_1$ and $S_2$; at this place we have the central maximum, or the central bright line. There are other places, on either side of the central maximum, where we get bright lines, places where the two waves are in phase and reinforce each other. The waves are in phase, but one is a whole number of wavelengths behind the other. In between the bright lines we have dark regions. The two waves cancel each other, producing nodes, where they are out of step by 1/2 of a wavelength, 3/2 wavelengths, or any other odd number of half-wavelengths.

* Let us see if we can calculate the distance between bright lines or fringes. Let us look at the right half of the above diagram in a slightly different way. Let us think of the first bright line on one side of the central maximum and represent it as point $P$. It is bright because the wave from $S_2$, traveling in the direction $S_2P$, arrives in phase with the wave from $S_1$ which travels in the direction $S_1P$. The length of path $S_1P$ is greater than $S_2P$; how can the waves arrive in phase? Only if the difference in the length of the two paths is exactly one wavelength; for the second bright

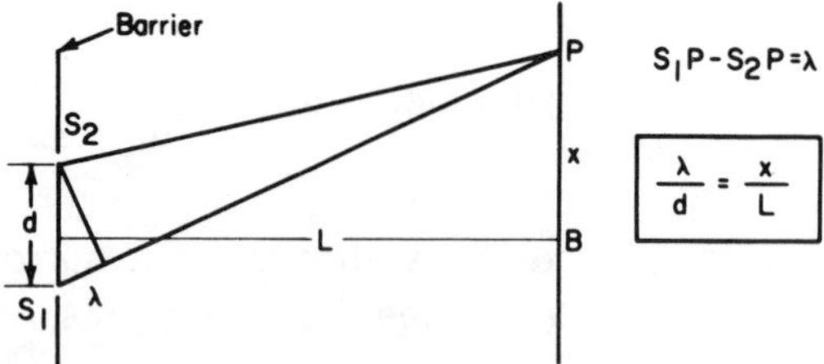


*Double Slit—Location of Bright Fringes*

fringe, the difference in path is two wavelengths; for the third bright fringe, the difference in path is three wavelengths, etc.

If $d$ is the distance between the two slits, $L$ the distance between the barrier and the screen, and $x$ the distance between the central maximum and the first bright fringe, it can be shown by selecting two triangles and making some approximations, that

$$\frac{\lambda}{d} = \frac{x}{L}$$

It can also be shown that the distance between any two adjacent bright fringes has the same value of $x$.

Complete annulment occurs midway between the bright fringes; the distance between nodes is also $x$.

Let us look at the important relationship shown above to see what predictions we can make. Since

$$\frac{\lambda}{d} = \frac{x}{L},$$

1. If the wavelength of light is increased ($\lambda$), the distance between fringes increases. If we use red light instead of blue light, the distance between successive red lines on the screen is greater than between successive blue lines.
2. If the distance between the two slits ($d$) is decreased, the distance between the bright fringes increases.
3. If white light is used, the central maximum (at $B$) is white, since all the wavelengths are reinforced there. On either side of the central maximum we get bright fringes consisting of all the colors of the spectrum, with violet closest to the central maximum.
4. If the distance between the screen and the slits ($L$) is increased, the distance between the bright fringes increases.

**EXAMPLE:** Calculate the distance between adjacent bright fringes if the light used has a wavelength of 5460 Angstroms and the distance between the two slits is $1.0 \times 10^{-3}$ meter. The distance between the screen and the slits is 1.0 meter.

$$\frac{\lambda}{d} = \frac{x}{L};$$

$$\frac{5460 \times 10^{-10}\text{ meter}}{1.0 \times 10^{-3}\text{ meter}} = \frac{x}{1.0\text{ meter}};$$

$$x = 5460 \times 10^{-7}\text{ m, or approx. } 0.00055\text{ m.}$$

**5 Diffraction Grating** If instead of having only two parallel slits, as above, we have many such slits close together, we have a diffraction grating. It is possible now to have cheap gratings with thousands of slits per inch. The full analysis of how it works is complicated, but in some respects it is similar to the double slit. However, the more slits we have, the brighter is the pattern. For approximate calculations, the above formula (for the double slit) may be used. It provides a convenient method for measuring the wavelength of light. Instead of the triangular glass prism the grating may be used in the spectroscope.

As with the double slit, the pattern obtained on the screen with a grating has a central maximum which is the same color as the light used. On either side of the central maximum we get bright fringes; if white light is used, the bright fringes are continuous spectra. The spectrum closest to the central maximum is known as the first order spectrum. Notice there is a first order spectrum on either side of the central maximum. If the source of light is a luminous gas, each "bright fringe" is the characteristic spectrum of the element.

# 6 Diffraction

**Wave Characteristic** As was pointed out earlier, when we deal with a wave, we expect to find *diffraction*: the bending of a wave around obstacles. We know that this happens in the case of sound because, for example, we can hear a speaker even if our direct view of him is blocked by a crowd of people. In the case of a water wave we can see the wave going around objects on the surface of the water, so that after a short distance the wave looks again as though there had been nothing in the way.

If light is a wave, why do objects cast sharp shadows? Why doesn't the light bend around the object and illuminate the region behind it? Theory shows that the smaller the wavelength, the smaller the diffraction effects. We already know that the wavelength of light is only a small fraction of a millimeter, while that of radio and sound waves is many centimeters, (and even meters for radio waves). It is because the wavelength of light is so small, that shadows are so sharp, that a beam of light tends to travel along straight lines.

However, diffraction of light does occur; because its wavelength is so small, we must look very carefully. For example, if we photograph the shadow of a small ball, using a suitable light source, we find a small illuminated spot in the center of the shadow.

7
*
**Single Slit Diffraction** Suppose we shine a beam of monochromatic light on an opaque barrier which has a narrow rectangular opening cut into it, a single narrow slit. Let the light which goes through the slit fall on a screen held parallel to the barrier. On the basis of what we did with ray diagrams, we might expect that the

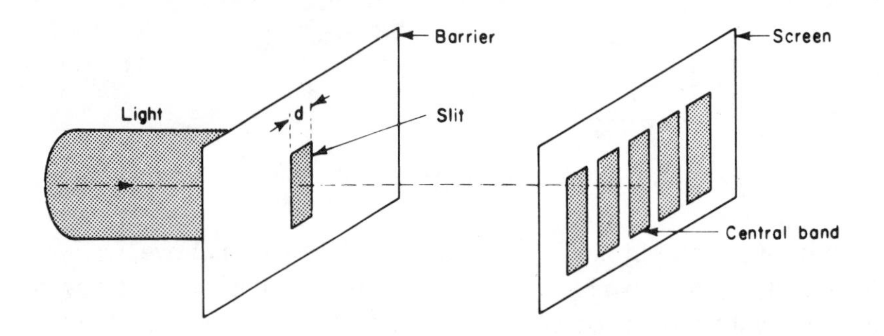


part of the beam which gets through the slit will have the same cross section as the slit, and that the illuminated part of the screen will look exactly like the slit in shape and size.

If the slit is narrow and we look carefully, what we actually see on the screen is many rectangular, parallel bands of light, (*images of the slit*). Each band may be a little wider than the slit. The central band is very bright; to either side are bands of decreasing brightness. (The central band's intensity is about 20 times as great as the first bright band to either side.) Below is a rough graph of the intensity of the observed pattern plotted against distance along the screen.

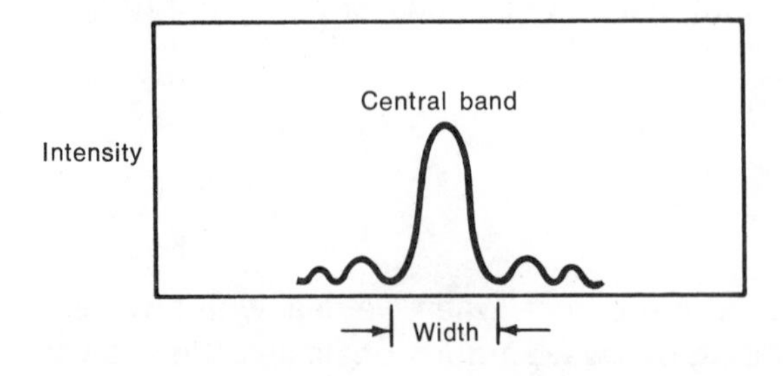


*Intensity of Single Slit Diffraction Pattern*

In the diffraction pattern, the central band is also called the central bright line or the central maximum. Careful analysis shows that

1. the width of the central maximum varies inversely as the width of the slit. That is, if we make the slit narrower, the central bright line becomes broader; or, in other words, there is more and more diffraction, more and more bending of the light around the edges of the slit.
2. the width of the central maximum varies directly as the wavelength. If we use red light, we get a wider or broader central line than if we use blue light.
3. The distance between successive dark bands is approximately the same as the distance between the first dark band and the central bright maximum.

If the opening is large compared with the wavelength, light practically travels along straight lines; diffraction is negligible.

8 ** **The Wave Theory and Single Slit Diffraction** As in the case of the double slit interference pattern, we can apply Huygens Principle and the wave theory of light to explain the single slit diffraction pattern. Let us view the barrier and screen edge-on. Assume the slit has a width *d*, and that the barrier is a distance *L* from the screen. Think of a wave front arriving at the slit opening. According to *Huygens' Principle*, every point on the wavefront can be thought of as sending out its own little waves. Two such points are shown at *A* and *B*. Think of the waves from *A* and *B* going towards *P*, a point on the screen. If *P* is to be a dark spot on the screen, the two waves traveling along *AP* and *BP* must annul each other; and, similarly, the

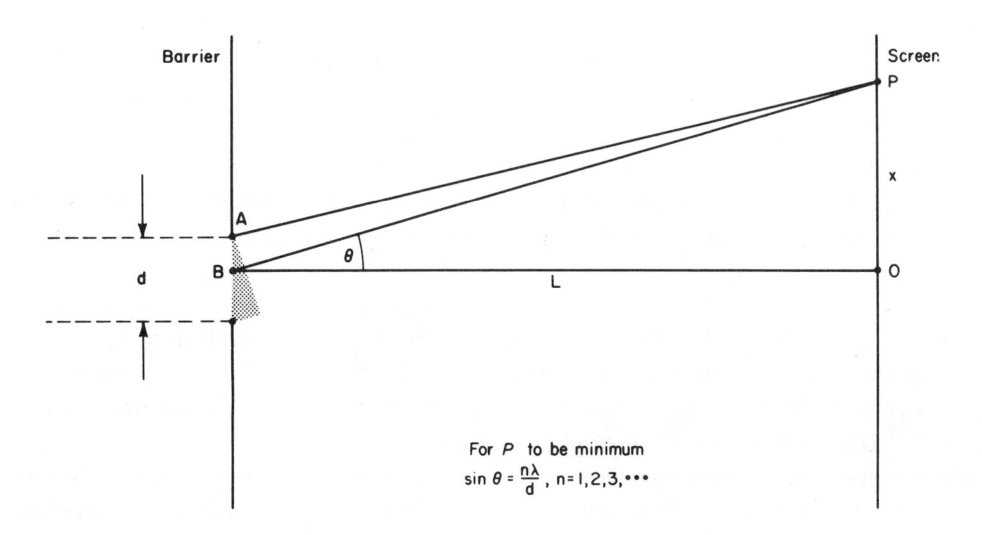


waves from the two points below *A* and *B* on the wave front at the slit, etc. Now path *BP* is larger than path *AP*. If this difference in path length is one-half wavelength, the two waves will annul each other.

With a little geometry it can be shown, that for a dark point on the screen, $\sin\theta = n\lambda/d$, where the letters have the values shown in the diagram, and $\lambda$ is the wavelength of the light used. When $n = 1$, *P* gives us the point of minimum intensity for the central line. In other words, the distance *OP* gives half of the width of the central band. By looking at the relation ($\sin\theta = \lambda/d$), we can see that if the wavelength $\lambda$ increases, $\sin\theta$, and therefore the width of the central band, must increase. Also, if the width of the slit $d$ decreases, $\sin \theta$, and therefore the width of the central band must increase.

9
* **Diffraction and Lens Defects** We have already indicated, that even a perfectly made spherical lens suffers from chromatic aberration. Spherical lenses have other defects even if monochromatic light is used. One of these is due to diffraction. As with a single slit, the light going through a lens is diffracted. The amount of diffraction is usually negligible, but one effect is that parallel light entering the lens is not brought to a point focus, but to a disk focus. Around this bright diffraction disk we may be able to see bright diffraction rings. The fact that we get a disk diffraction pattern instead of a point focus with a lens, sets a practical limit to the magnification we can get with a lens.

10
* **Resolving Power** The ability of a lens or optical instrument to distinguish between two point sources is its *resolving power*. Resolution or resolving power is increased by increasing the diameter of the lens; it works similarly for spherical mirrors. Suppose, for example, we look through a telescope at two stars that are close to each other in the sky. If the objective lens has a small diameter, the diffrac-

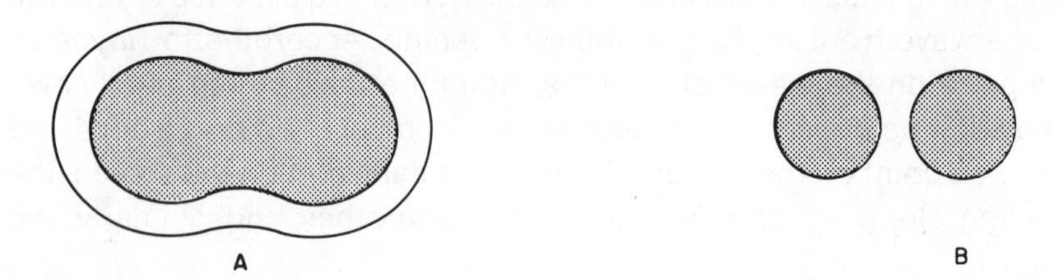


tion pattern of each star will be large, and the two disks will overlap as in diagram *A*. The resolution is poor: it is hard to tell whether we are looking at one star or two. If we try to use greater magnification we will not improve resolution, because we also magnify the size of the disk. If, instead, we use an objective lens with a larger diameter, we decrease the amount of diffraction, we decrease the size of the diffraction disk, and we increase the resolving power. The effect of using a larger diameter lens is shown in diagram *B*.

11
* **Doppler Effect with Light** When we studied sound, we noticed that the pitch we hear depends on the frequency of the vibration of the source. The pitch is affected by the relative motion between the source and the observer. A similar phenomenon is observed in the case of electromagnetic waves.

If the distance between the source and the observer is decreasing, the observed frequency is greater than that of the source. In the case of light, frequency determines the color of light we see. If the frequency is greater than the source frequency (wavelength is smaller), the light is more violet than it would have been if there were no relative motion.

If the distance between the source and the observer is increasing, the observed frequency is lower than that of the source. In the case of light, this means that the observed light would be shifted towards the red.

There are some interesting applications of the Doppler Effect with electromagnetic waves. Radar may be used to measure the velocity of approach of airplanes, automobiles, and similar objects. At a radar installation a transmitter sends out an electromagnetic wave in a narrow beam. The beam is reflected back by many objects, especially metallic ones. The reflected wave is picked up by a radar receiver. If the object is stationary, the reflected wave has the same frequency as that sent out by the transmitter. If the object's distance from the radar is increasing, the reflected frequency is less than the radar frequency. If the object's distance is decreasing, the reflected frequency is greater. The change in frequency depends on the velocity of approach. The greater the velocity, the greater the change in frequency.

The famous *Red Shift* is explained by the *Doppler Principle* as being due to an expanding universe. What does that mean? The light from the sun, stars, and other heavenly objects has been analyzed with a spectroscope. Present are the characteristic spectra of many elements we know on earth. By comparing the spectrum of an element on earth with the corresponding spectrum from stars, it has been found that the spectrum from some stars is shifted towards the blue end of the spectrum; this means that these stars must be approaching the earth. The spectrum from some other stars has been found to be shifted towards the red end of the spectrum. This means that these stars are moving away from the earth. There are some very distant objects in the sky known as nebulae (singular is nebula). The spectra from these nebulae are shifted towards the red. The further the nebulae are, the greater the shift. This has been interpreted to mean that the distant parts of the universe are moving further and further away from us: the universe is expanding.

12 **Polarization of Light** We have been saying that light behaves like a transverse wave. What is the evidence? We have already shown how light behaves like some kind of wave. We explained interference and diffraction of light by making use of wave theory. We can readily see how two waves can get to the same place at the same time and annul each other. We cannot explain this annulment if we think of the energy as arriving as streams of particles. However, this did not tell us whether the wave is longitudinal or transverse. We observed interference and diffraction with sound, and decided that the sound wave is longitudinal.

An effect observed with light and not with sound is polarization. *Polarized light* is light whose direction of vibration has been restricted in some way. The direction of a longitudinal wave cannot be restricted; either it vibrates in the direction in which the energy is traveling (the direction of propagation), or it doesn't vibrate at all and we have no wave. On the other hand, in a transverse wave there is an infinite number of ways in which we can have vibrations at right angles to the direction of propagation of the wave.

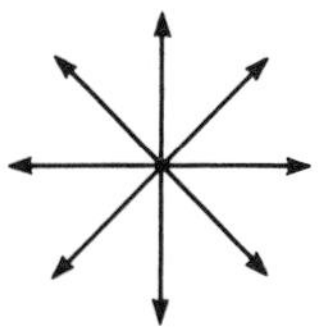

In the diagram, imagine a ray coming out of the paper. Vibration can be towards the top and bottom edges of the paper, towards the right and left, and in all other directions in between. We think of ordinary natural light, such as we get from our tungsten filament bulbs, as a transverse wave with all directions of vibrations present. Some crystals, natural and synthetic, allow only that light to go through whose vibration is in the direction of the axis of the crystal. Such light is then polarized:

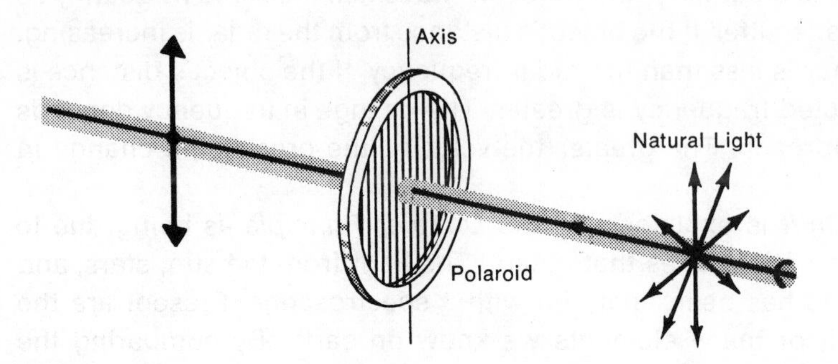


*Polaroid* is a synthetic material that does this. It consists of plastic sheets with special crystals embedded in them. If we let ordinary light shine through one polaroid disk, the light is polarized but the human eye can't tell the difference. If we allow this polarized light to shine through a second polaroid disk kept parallel to the first one, the amount of light transmitted by the second disk depends on how we rotate the second disk with respect to the first one. If we rotate the second disk so that its axis is at right angles to the axis of the first disk, practically no light is transmitted. The maximum amount of light is transmitted when the two axes are parallel to each other.

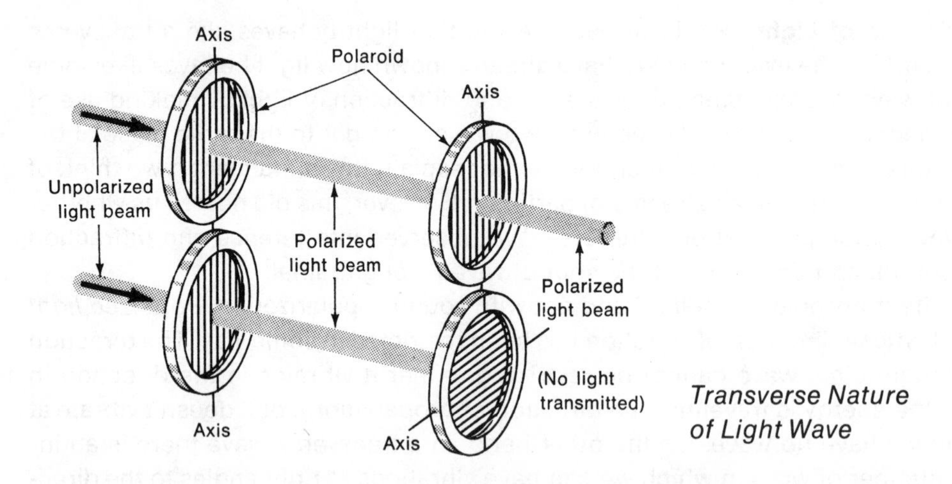


*Transverse Nature of Light Wave*

According to *Maxwell's Theory*, light is a transverse wave in which electric and magnetic fields vibrate or fluctuate at right angles to the direction of propagation. Electric and magnetic fields will be discussed further in later chapters.

## Questions and Problems

1. In *Young's Double Slit* experiment, how were two coherent beams of light obtained?
2. In an interference pattern obtained in a ripple tank by two probes dipping into water at the same frequency, what motion, if any, does a piece of cork on a nodal line undergo?

3. In *Young's Double Slit* experiment using monochromatic light, what is the effect on the spacing between two adjacent bright lines in the pattern on the screen, of tripling the *a.* distance between slits *b.* distance between the screen and the double slit? *c.* What is the effect of using blue light rather than yellow light?
4. The light from two automobile headlights overlap. Should an interference pattern be observable? Explain.
5. A diffraction grating is used to get the spectrum of white light. *a.* Describe the color of the central bright region on the screen. *b.* Give the order of colors in the first order spectrum starting with the color closest to the central bright region. *c.* Compare this with the order of colors in the spectrum of white light obtained with a glass prism.
6. In diffraction with a single slit, what change is observed in the appearance of the diffraction pattern when the slit width is decreased?
7. Two parallel slits, 0.3 mm apart, are illuminated by light whose wavelength is $5 \times 10^{-7}$ m. A viewing screen is 2.0 meters away from the slits. What is the distance between two adjacent bright lines on the screen?
8. In question 7, light whose wavelength is $6.0 \times 10^{-7}$ meter is substituted for the original light. What is the new spacing between two adjacent bright lines?
9. A double slit, with a spacing of 0.20 mm, is illuminated with light whose wavelength is 5000 Angstrom units. An interference pattern appears on a screen 2.0 meters away. How far from the central bright image is the next bright line?
10. Monochromatic light passes through two slits which are $2.0 \times 10^{-4}$ m apart. An interference pattern is obtained on a screen 140 cm away. The first order bright line is 0.18 cm away from the middle of the central bright region. Calculate the wavelength of the light.
11. In problem 9, what is the answer if the light used has a wavelength of 7000 Angstrom units instead of the original light?
12. Calculate the frequency of light whose wavelength is 7000 Angstrom units.
13. Calculate the frequency of light whose wavelength is *a.* 6000 Angstrom units *b.* $5.5 \times 10^{-5}$ cm.
14. Light whose wavelength is $4.5 \times 10^{-5}$ cm passes through a single slit and falls on a screen 180 cm away. What is the distance between two adjacent bright lines in the diffraction pattern, if the width of the slit is *a.* 0.020 cm *b.* 0.030 cm?
15. Calculate the wavelength of light in water, if the wavelength of this light in air is 6600 Angstrom units.

## Test Questions

1. The continuous spectrum seen when looking at an oil film on street pavements is due primarily to 1. pigment in the oil 2. prism effect for the different wavelengths 3. interference 4. diffraction.
2. A beam of monochromatic light illuminates a narrow slit. The light which passes through the slit is allowed to fall on a screen which is two meters away from the slit. If the intensity of the light on the screen is

plotted against distance along the screen, the graph will be most like

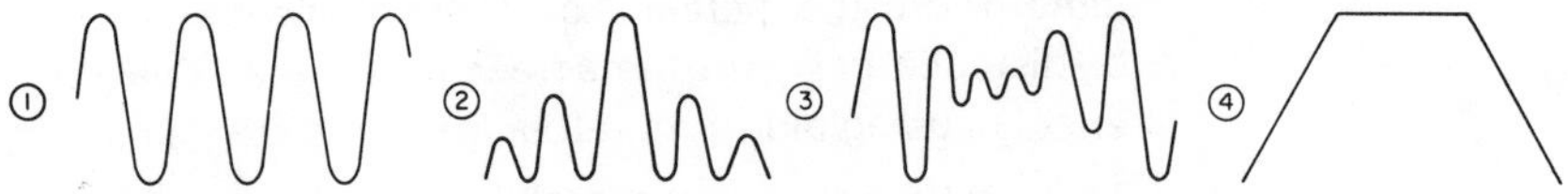

① ② ③ ④

3. In order to observe interference of light, the two light sources used must be 1. independent 2. coherent 3. dispersed 4. polarized.

4. A mixture of monochromatic red and monochromatic green light is used to illuminate a narrow slit. In the diffraction pattern obtained on the screen, the distance between two adjacent red lines, as compared with the distance between two adjacent green lines is 1. greater 2. the same 3. smaller.

5. George and Jonathan look at the same white hot wire through a narrow slit. George's slit is narrower than Jonathan's. The distance between two adjacent red lines seen by George as compared with that seen by Jonathan is 1. greater 2. the same 3. smaller.

6. The pattern seen when white light is viewed through two adjacent narrow slits is evidence that light 1. behaves like a wave 2. behaves like a stream of particles 3. can be polarized 4. is electromagnetic in nature.

7. A continuous spectrum may be seen when looking at white light through each of the following *except* 1. a diffraction grating 2. two polaroid sheets 3. a narrow single slit 4. two narrow slits near each other.

8. Newton's rings are produced by 1. a pebble thrown obliquely into water 2. a lighted cigarette in pure oxygen 3. diffraction 4. interference.

9. The fact that light usually appears to travel along straight lines is best explained by 1. recognizing that light is a stream of particles 2. showing that light can be polarized 3. recognizing that light has a rather short wavelength 4. noting that the speed of light has a maximum value.

10. Evidence for the wave nature of light can not be obtained from the phenomenon of 1. reflection 2. refraction 3. interference 4. diffraction.

11. Two light waves traveling by different paths reach a point on the screen simultaneously. One has an amplitude $a$, the other has amplitude $b$. The resulting amplitude is 1. $a+b$ 2. $a-b$ 3. $1/2(a+b)$ 4. indeterminate on the basis of the given information.

12. An interference pattern has a black line. This results from two waves getting to the screen at the same time which are 1. in phase 2. one-half wavelength out of step 3. one wavelength out of step 4. two wavelengths out of step.

13. Which of the following cannot be polarized? 1. blue light 2. infrared radiation 3. ultraviolet radiation 4. sound.

14. In the double slit experiment using monochromatic red light, if the distance between slits is halved, the distance between two adjacent red lines in the interference pattern is 1. halved 2. doubled 3. quadrupled.

15. In the single slit diffraction pattern using monochromatic red light, if the width of the slit is doubled, the width of the central maximum 1. increases 2. decreases 3. remains the same.
16. As the wavelength of the light used in a microscope increases, the resolving power of the microscope 1. increases 2. decreases 3. remains the same.
17. As the distance between two narrow slits decreases, the distance between bright lines in the interference pattern produced with the slits 1. increases 2. decreases 3. remains the same.
18. When the wavelength used to illuminate two narrow slits changes from red to blue, the distance between bright lines in the interference pattern produced with the slits 1. decreases 2. remains the same 3. increases.
19. If the distance between two slits and the screen is doubled, the distance between bright lines in the interference pattern produced with the slits 1. is halved 2. is doubled 3. is quadrupled 4. remains the same.
20. As a plane taking off from the earth increases its altitude at a greater and greater speed, the frequency of its radio signal as observed by stationary observers on earth 1. seems to increase 2. seems to decrease 3. seems to remain constant.

## Topic Questions

1.1 Which phenomena supported the wave theory of light rather than the particle theory?

2.1 What is meant by coherent sources?

2.2 How do we usually get two coherent beams of light?

3.1 If interference is produced as a result of the reflection of light from a thin film, what is the minimum path difference for the two beams so that there will be annulment? From what surface are these beams reflected?

4.1 When we observe a double-slit interference pattern with monochromatic light, we see some bright lines. How do the distances from the two slits to the bright lines compare?

5.1 What is one advantage in using a diffraction grating rather than a double slit?

6.1 Why is it so difficult to observe the bending of light around obstacles?

7.1 In the diffraction pattern obtained with a single slit, how does the width of the central maximum depend on *a.* the width of the slit *b.* the wavelength?

9.1 Light going through a lens is diffracted. What effect does this have on the sharpness of focus?

10.1 How can the resolving power of a telescope be increased?

11.1 What are some applications of the Doppler Effect?

12.1 What is a polarized wave?

# 12 Static Electricity

## 1 Introduction

**Structure of Matter** Electricity, like nuclear energy, can be used to hurt or to help man. In lightning it terrifies and destroys. In electric motors it is a great boon to modern civilization.

From earlier education you remember some of the facts about the structure of matter. All matter is made of atoms. There are more than 100 different kinds of atoms. All atoms except ordinary hydrogen atoms are made of protons, neutrons, and electrons; ordinary hydrogen atoms do not contain neutrons. Neutrons are electrically neutral, protons are positive, electrons negative. Neutrons are slightly heavier than protons, and almost 2000 times as heavy as electrons. The protons and neutrons are in the nucleus of the atom, electrons are in special arrangements on the outside of the nucleus. A neutral atom has the same number of electrons as protons.

How do we know all this? To save time we shall make use of some of these facts as we go along, but the next few chapters will show how this knowledge was gained, and also how this and other knowledge has been useful.

2 **Early Discoveries** The Greek philosopher Thales (600 B.C.) discovered that if substances like amber are rubbed with a piece of cloth, they can pick up little shreds of cloth and other small pieces of matter. The word *electricity* comes from the ancient Greek word for amber – elektron. The English scientist and physician, William Gilbert (about 1600) showed that many substances can be *electrified* by rubbing, that is they get the ability to attract small bits of matter. The French investigator Dufay (about 1700) showed that there are two kinds of electricity, that opposite kinds of electricity attract each other and like kinds repel each other. In 1752 the American Benjamin Franklin performed his dangerous kite experiment and showed that lightning is a grandiose example of electricity. He also suggested the terms positive and negative electricity which we still use.

A rubber rod rubbed with a piece of fur or wool becomes electrified negatively. Another way of expressing this is to say that the rod becomes *charged* negatively, or acquires a negative charge. We can think of electrons leaving the fur and going to the rubber rod. The fur does become positively charged.

A smooth glass rod rubbed with silk becomes positively charged. The silk gets some electrons from the glass rod, leaving the glass rod positive and making the silk negative. Lucite may be used instead of glass, polystyrene instead of rubber.

We find that two rubber rods repel each other after they have been rubbed with fur. The two rods became negatively charged; negative charge repels negative charge.

We find that two glass rods repel each other after they have been rubbed with silk, but the charged glass rod attracts a charged rubber rod. This shows that the charge on the glass rod is different from the charge on the rubber rod. The charge on the glass rod is called positive. Positive charge repels positive charge but attracts a negative charge.

In summary, like charges repel, unlike charges attract. This is sometimes called the *Law of Charges*.

3 **Coulomb's Law** About 1780 the Frenchman Coulomb performed some quantitative experiments and found that the electrical force of attraction and repulsion between charges follows laws similar to the gravitational force of attraction between masses. Before we look at his findings we need to define some terms. The charge on the electron is the smallest negative charge we have been able to find in nature. The most careful measurements made so far show that the charge on the proton is equal and opposite to that of the electron. (Some recent theories indicate that there should be a charge smaller than the electron. It has been called the *quark*, but has not been found yet.)

The charge of an electron is sometimes referred to as the elementary negative charge; the charge of a proton as the elementary positive charge. Charges can exist only as integral multiples of these elementary charges.

The charge of an electron is extremely small. For practical purposes, especially involving electric currents, we need a larger unit. In the *MKS* system, the unit of electrical charge is the *Coulomb*. The *Physics Reference Table* shows that one coulomb is the charge of $6.25 \times 10^{18}$ electrons; or, that the charge of one electron is about $1.60 \times 10^{-19}$ coulomb. A *micro-coulomb* is one-millionth of a coulomb.

Ordinary matter contains a vast number of electrons and protons, and also neutrons. It is electrically neutral. When this matter gains or loses electrons it is said to be electrically charged. When we speak about the *charge* on the object, we mean this excess or deficiency of electrons.

Coulomb made measurements on small charged spheres. His findings apply to the force of attraction and repulsion if the charges are concentrated at points or small spheres. The findings are expressed in *Coulomb's Law*: The force is proportional to the product of the two charges and varies inversely as the square of the distance between them.

**EXAMPLE:** Suppose the force between two point charges is *N* newtons. If the distance between them is doubled, what happens to the force? The force becomes less, since the distance between the charges becomes greater. The force becomes one-fourth as great, since the square of two is four.

4 **Coulomb's Law** in Equation Form
★

$$F = \frac{kq_1q_2}{d^2},$$

where $F$ is the force acting on either charge, in newtons

$q_1$ and $q_2$ are the two charges, in coulombs

$d$ is the distance between the charges (between the centers of the spheres), in meters

*k* is the proportionality constant appropriate for the units; the *Physics Reference Table* gives the value in the *MKS* system:
$9 \times 10^9$ $Nt\text{-}m^2/coul^2$. This is the value in vacuum or air.

**EXAMPLE:** Two charges are 6.0 cm apart in air. The charges are 20, and 12 microcoulombs, respectively. What is the force between them?

$$F = \frac{(9.0 \times 10^9)(20 \times 10^{-6})(12 \times 10^{-6})}{(0.060)^2}$$
$$= 6.0 \times 10^2 \text{ newtons}$$

## 5 Methods of Charging, and Electrostatic Instruments

**Charging by Contact** If two neutral objects made of different material are brought into good contact, as by rubbing, one object usually takes electrons away from the other. The one that gains electrons becomes negative, the one that loses electrons becomes positive. We saw an example of this when we rubbed a rubber rod with fur. In a solid, the protons usually do not move readily.

A neutral object usually acquires the same kind of charge as that of the charged object it touches. (Notice that we have been discussing a subject usually referred to as *static electricity*—electricity at rest—but this—includes these quick transfers of charges.) In these experiments with static electricity, we often use a pith ball suspended by means of a silk thread. If we hold a charged rubber rod next to a pith ball, we notice that the pith ball is first attracted to the rod, but after touching it, is repelled. On touching the rod, the pith ball acquires the same charge as the rod, and then is repelled because like charges repel.

Notice that the silk is an *insulator*: it does not let electric charges move through it readily. Other insulators are rubber, glass, dry wood, dry air. Metals are good *conductors*: they let electrons move through them readily. Other reasonably good conductors are water solutions of salts, acids, and substances called bases.

**6 Conservation of Charge** When a neutral rubber rod is rubbed with a neutral piece of fur, the negative charge produced on the rubber rod is numerically the same as the positive charge produced on the fur. (The rod gained as many electrons as the fur lost.) The net charge after rubbing the two neutral objects together is still zero, as it was before rubbing. If a charged rubber rod is brought in contact with a neutral pith ball, the charge gained by the pith ball is exactly the same as the charge lost by the rod. Again, the net charge after contact is the same as it was before contact. This illustrates the *Principle of Conservation of Charge*: In a given system, the net charge remains constant.

**FOR EXAMPLE:** We find that under special conditions a neutron splits into a proton and an electron. The charge of a neutron is zero. Since a proton and electron have equal but opposite charges, their combined net charge is also zero.

**7 Charging by Induction** In charging by induction, a charged object is held near (but not touching) a neutral conductor while the latter is grounded. The latter thus acquires a charge opposite to that of the original charged object. The ground connection is removed before the charged object is removed. *Grounding* may be accomplished by connecting a wire between the object to be grounded and a water pipe. This provides a good conducting path between the object and the moist ground. In practice, the object is effectively grounded if a person standing on an ordinary floor touches the object.

We shall show how an electroscope may be charged by induction. The *electroscope* is an instrument used to determine the presence and sign (positive or negative) of small electrical charges. In one type of electroscope, the indication is given by gold leaves which deflect more or less. The leaves hang from the lower end of a

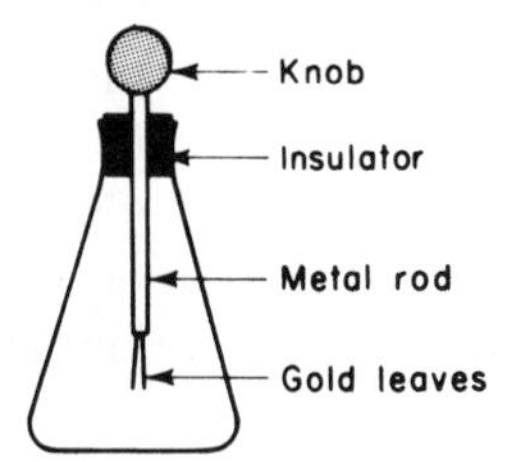


*The Gold Leaf Electroscope*

vertical metal rod. At the other end of the rod is a metal knob. To begin with, we make sure that the electroscope is neutral. We can do this by momentarily touching the knob, thus grounding the electroscope. If it was negative, grounding it allowed the excess electrons to go to ground. If it was positive, grounding it allowed enough electrons to come up from ground to neutralize the positive charge. Remember, positive and negative charges attract each other, but only the electrons normally move through the solid metal.

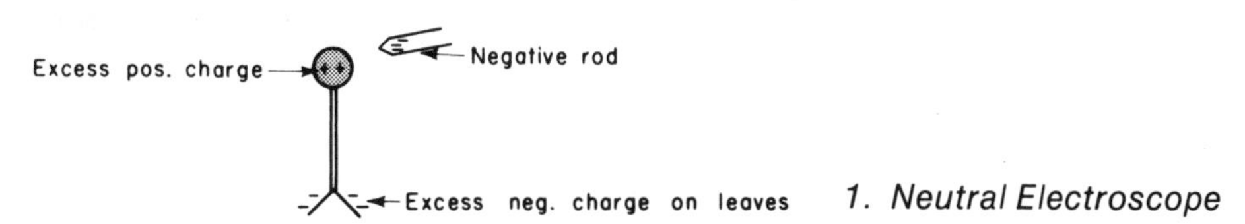


*1. Neutral Electroscope*

To charge the electroscope positively by induction, several steps are required. First we bring a negatively charged rod near the knob of the electroscope. Since like charges repel, electrons from the knob will be repelled to the leaves. Both leaves become negatively charged and repel each other. The knob has lost electrons to the leaves and is now positive. The electroscope as a whole is still neutral, since it has neither gained nor lost electrons. Some of its electrons have merely been distributed non-uniformly. The knob (not the electroscope) has been charged by induction, charged as a result of redistribution of its charges brought about (without contact) by the proximity of another charged object. To charge the elec-

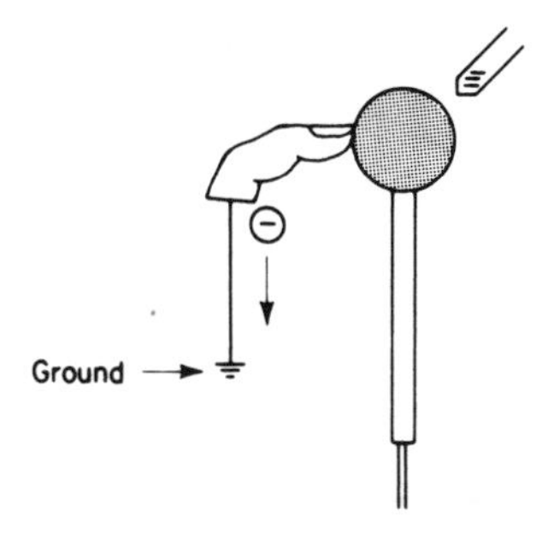


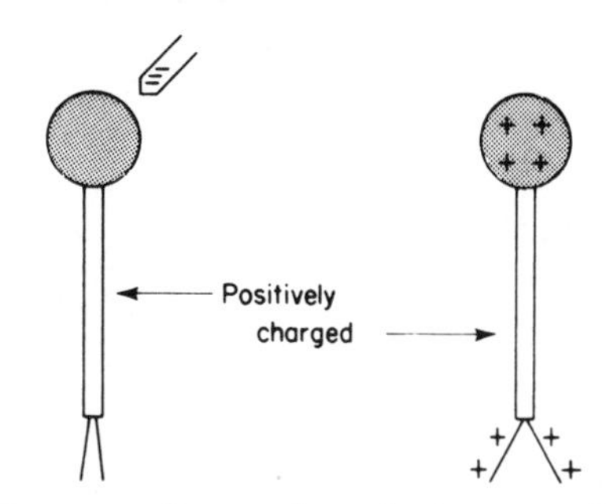


*2. Ground Electroscope* *3. Remove Ground* *4. Remove Rod*

troscope, we momentarily ground the knob by touching it. This provides a new path for electrons. The negative rod repels electrons from the electroscope to ground. After removing ground from the knob, we take the rod away. This allows the charges on the electroscope to redistribute themselves. The whole electroscope has lost electrons, and therefore it is positively charged. Both leaves are positive and repel each other. Note that the electroscope acquired a charge opposite to that of the rod. If we want to charge an object negatively by induction, we must use a positive rod.

8 **Electric Field** We have seen that a negatively charged rod repels another negatively charged rod. They exert an electric force on each other without touching. In order to describe this force more exactly, physicists use the concept of an electric field. An *electric field* is said to exist wherever an electric force acts on an electric charge. For example, we know from the above illustration that there is an electric field around a charged rod. In fact there is an electric field around every charged object.

**Electric Field Intensity** We speak about the strength or intensity of the electric field. The *electric field intensity* at a certain point is the force per unit positive charge placed there. The positive charge we place at the point is sometimes called the test charge. We may use a positively charged pith ball as the test charge. If it is pulled to the right, we know that there is an electric field to the right. The field exerting the force on the pith ball may be produced by a negative charge to the right of it, or by a positive charge to the left of it. The test charge does not give us the cause of the field, it gives us the direction of the field. Or, to put it a little differently, the direction of the field at a given point is the same as the direction of the force on a positive charge placed at that point, but opposite to that on a negative charge placed there.

Note that the field intensity is a vector quantity: it has magnitude and direction. Physicists arbitrarily define the direction of the field as the direction of the force on a positive charge. If we use a negative test charge, the force on it is directed opposite to the field.

**Field around a Point Charge or Metal Sphere** Since electric field intensity is a vector quantity, we can represent the field by vectors. Since we already know that like charges repel, we know that a positive test charge placed anywhere near a positive point charge, or a positively charged sphere, will have a force acting on it, always away from the charge. The field is represented by vectors drawn away from the point; in the case of the positive sphere, by vectors drawn along the direction of the radius away from the surface of the sphere. In other words, the electric field outside a charged sphere is radial. If the point charge or metallic sphere is nega-

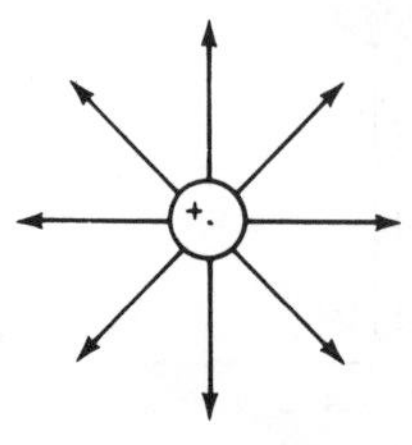


Positive Sphere

Negative Sphere

*Field near charged Metal Sphere*

tive, the positive test charge is attracted, and the direction of the field is towards the point charge and towards the center of the sphere.

*The electric field intensity due to a point charge* is proportional to the charge and varies inversely as the square of the distance from the charge. The intensity of the electric field outside a charged metallic sphere is just what it would be if the charge were concentrated at the center of the sphere. The field inside a charged metallic sphere is zero. The latter fact is applied to the electroscope: when not used, a metal can is placed over the knob as a shield against external electric fields. The field inside the can is practically zero.

9 **Electric and Gravitational Fields Compared** We introduced the concept of electric field to describe the force between two charged rods which are not in contact. We came across a similar force before. The earth pulls on the apples in the apple tree even before they touch the ground. The earth also pulls on the sun, moon, and satellites. The gravitational force is the name we use for this force produced by the mass of the objects. We say every mass is surrounded by a gravitational field, the earth's pull is produced by its gravitational field. The strength or intensity of the gravitational field is defined as the force it exerts per unit mass.

**FOR EXAMPLE:** If the earth's pull on a 5-kilogram object is 50 newtons, the earth's gravitational field near the object is (50/5) newton per kilogram, or 10 *Nt*/kg. Its direction is towards the earth, because the gravitational force is always towards the objects producing it.

*Coulomb's Law* tells us that the electric force due to a point charge varies inversely as the square of the distance away from it. Therefore, the intensity of the electric field due to a point charge varies inversely as the square of the distance away from the charge. If we double the distance away from the charge, the field intensity becomes one-fourth as great. Similarly, *Newton's Law of Gravitation* tells us that the gravitational force produced by a point mass varies inversely as the square of the distance from it. Therefore, the gravitational field intensity produced by a point mass varies inversely as the square of the distance from it. In the case of a point outside of a sphere we make the measurements from the center of the sphere. The field then, too, varies inversely as the square of the distance.

**FOR EXAMPLE:** 4000 miles from the surface of the earth the gravitational field is one-fourth as much as on the surface, because we are twice as far from the center.

The gravitational force between two rods is extremely small. By comparison, the electric force between two charged rods is quite large, and we can neglect the gravitational force.

10 * **Formulas for Field Intensity** We have stated that the definition of electric field intensity is the force per unit positive charge placed in the field. Therefore,

$E = F/q$ where, $F$ is the force exerted on positive charge $q$
$E$ is the electric field intensity.
In the *MKS* system, $F$ is in newtons, $q$ in coulombs, and $E$ is then in newtons per coulomb.

**EXAMPLE:** A charge of $40. \times 10^{-6}$ coulomb is placed in an electric field whose intensity is $20. \times 10^{3}$ newton/coulomb. What is the force acting on the charge?

$$20 \times 10^3\ Nt/\text{coul} = F/(40 \times 10^{-6}\ \text{coul})$$
$$F = 800 \times 10^{-3}\ Nt$$
$$= 0.80\ Nt.$$

This is the magnitude of the force. We do not have enough information to give the direction.

There is an interesting relationship for gravitational field intensity. This was defined as the force per unit mass placed in the field: $F/m$. Now recall *Newton's Second Law of Motion*, which states that a net force acting on a mass produces an acceleration in the direction of the force. This was expressed as: $F = ma$. If the object of mass $m$ is released above the earth, the earth's gravitational force produces the acceleration $g$. Then, $F = mg$; or, rewriting this, $g = F/m$. In other words, gravitational field intensity (or field strength) at any point is the same thing as the acceleration due to gravity at that point. Therefore, its units in the *MKS* system are newtons per kilogram, or meters per second$^2$.

11 **Shape of Electric Field** We have seen that we can represent the electric field intensity by vectors which show the direction and strength of the electric field. Sometimes we represent the field by drawing lines to show the direction of the field; the length of the line does not depend on the strength. These lines are sometimes called electric lines of force. Of course, the lines themselves do not exist; they are drawn to represent the field. The stronger the field, the more the lines of force drawn in that area. This will be further illustrated by a few examples.

**Field Around a Point Charge** The field intensity varies inversely as the square of the distance, and is directed away from a positive charge, and towards a negative charge. Note that the lines of force are straight, and that they spread out more and

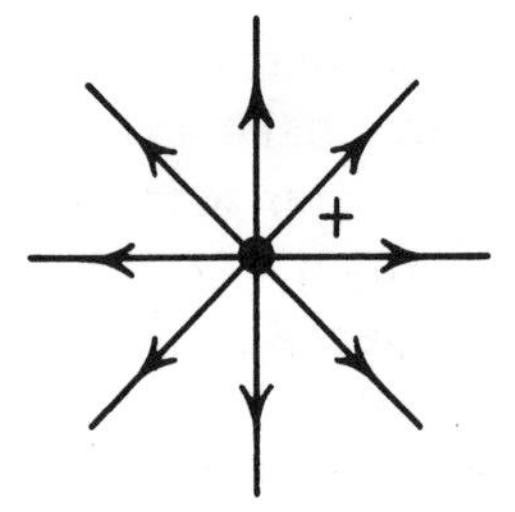


*Positive Point Charge*

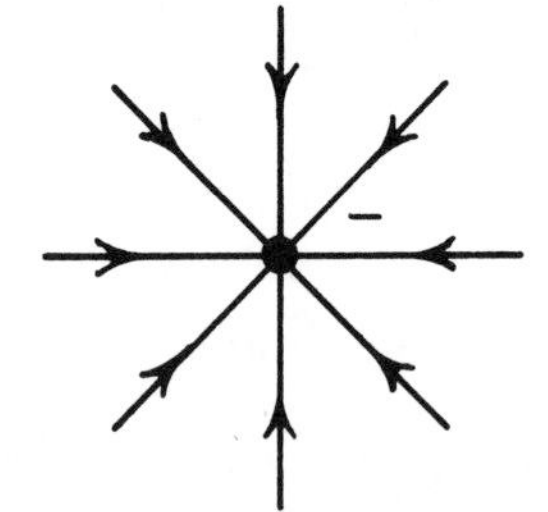


*Negative Point Charge*

more and more as we go away from the point charge. This indicates that the field is weaker as we go further away from the charge. Note the similarity with the field due to a charged sphere, which we described earlier.

12 **Field Around Charged Metal Sphere** The field is radial, and was described earlier. Remember that outside the sphere, the field is the same as though the charge were at the center of the sphere. There is no field inside the charged metallic sphere. In a charged metallic sphere, the excess charge is on the outside surface of the sphere and is uniformly distributed.

13 **Field Around Uniformly Charged Metal Rod** The lines of force are perpendicular to the surface. The field varies inversely with the distance from the axis of the rod, that is the line going down the center of the rod. Note that the field decreases as

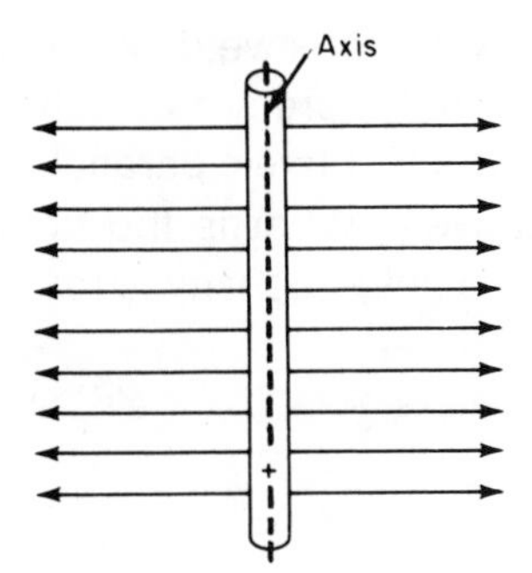


*Field Around Positively Charged Metal Rod*

we go further away from the rod, but not inversely as the square of the distance. This does not show up in a two-dimensional drawing.

14 **Field Near Pointed Conductor** When a metal sphere is charged, the charges distribute themselves uniformly on the surface of the sphere. When the metal object is not uniform, the charges are more concentrated on the more pointed portions of the surface. This is shown on the pear-shaped object. The charge is greatest on the most pointed portion. The electric field is greatest near the most pointed portion. This is often demonstrated with the electric whirl. This consists of a wire bent as shown in the diagram; the wire is pivoted at its center of gravity so that it is free to rotate in a horizontal plane. When the whirl is charged strongly, as by being connected to an electrostatic machine, it always rotates in the direction shown, no matter whether it gets charged positively or negatively. The direc-

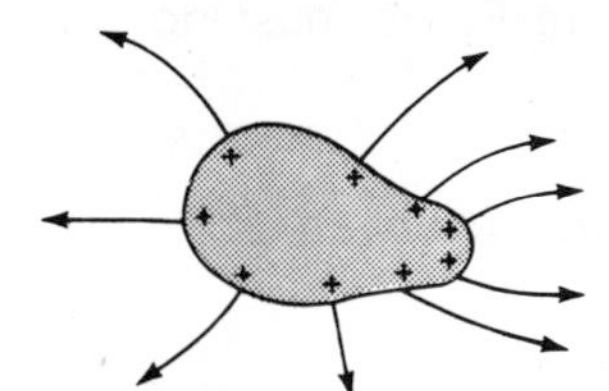

*Field is Greatest Near the Most Pointed Portion*

tion of rotation is away from its points. The air always contains some ions. An *ion* is a charged atom or groups of atoms. Near the whirl's points, the electric field is very strong. Therefore, the force on the ions near the points is very large. According to Newton's Third Law of Motion, the force on the whirl is equal and opposite to the force on the ion. When the whirl is positively charged, its points are strongly repelled by positively charged ions in the air. When the whirl is negatively charged,

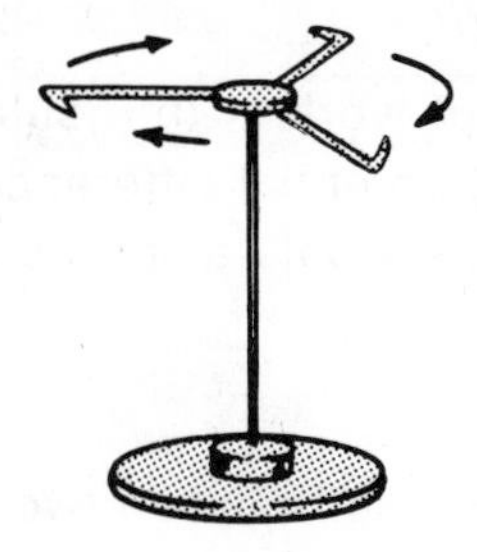

it is repelled by negative ions in the air. Why isn't this effect neutralized by ions in the air of opposite charge? Opposite charges are attracted to the points, and by successive collisions overcome the friction and inertia of the wheel. This produces rotation in the same direction as before.

**15 Field Between Oppositely Charged Plates** In the above situations we described the field due to one charged object with the assumption that there are no other charged objects around. Now let us consider two large metallic plates, parallel to each other, one charged positively and the other negatively. What is the field between two such plates placed close to each other? Since field intensity is force

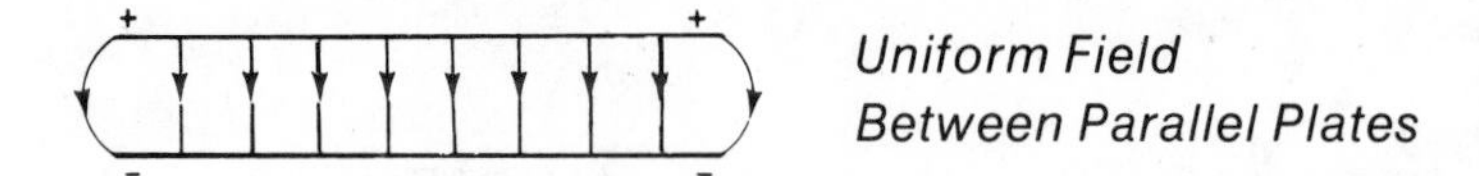

*Uniform Field Between Parallel Plates*

per unit positive charge, we consider what would happen to a positive charge placed between the two plates. The positive charge is repelled by the positive plate and attracted by the negative plate. Therefore the field is directed from the positive to the negative plate. The field is uniform except near the edges. This means that the force on a given charge is the same no matter where we put it between the two plates. The field is perpendicular to the plates.

The arrangement of two such plates is known as a capacitor, and will be discussed later.

**16 Electric Potential and Potential Difference** One of the most important concepts in electricity is the one of potential difference. Suppose we imagine an electric field represented by the lines drawn below. We need not be concerned about the cause of the field. Let $A$ and $B$ be two points in the field, as shown. If we put a positive charge $q$ at point $A$, the field will exert a force $F$ on the charge in the direction of the field. If we want to move the charge from $A$ to $B$, we must do work.

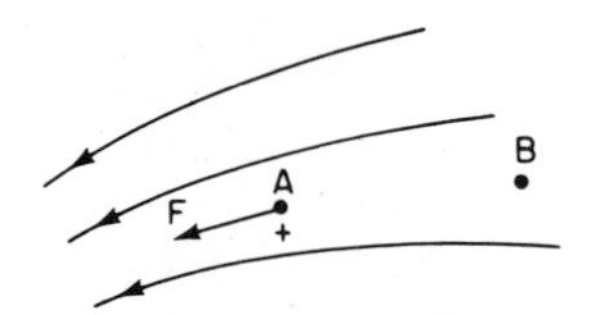


*Work is Required to Move a Charge from A to B*

(Remember, work done equals the product of the force and distance moved in the direction of the force.)

We define the *potential difference*, $V$, between two points in an electric field as the work per unit charge required to move a charge between the points.

$$\text{potential difference} = \frac{\text{work}}{\text{charge}};$$

$$V = W/q.$$

In the *MKS* system, the unit of work is the joule, and the unit of charge is the coulomb. Therefore the unit of potential difference is joule per coulomb. It is such an important unit, we give it a special name: the *volt*.

$$1 \text{ volt} = 1 \text{ joule/coulomb}.$$

**EXAMPLE:** The potential difference between two points is 120 volts. What is the work required to move a charge of $6.0 \times 10^{-4}$ coulomb between these points?

$$120 \text{ volts} = \text{work}/6.0 \times 10^{-4} \text{ coul}$$

$$\text{work} = 720 \times 10^{-4} \text{ joule}$$

$$= 0.072 \text{ joule}.$$

**Electric Potential** Electric potential is really the same as electric potential difference, except that we arbitrarily select a special point as reference. In some practical work, the special reference "point" is the ground. In our work, the special point is any point at infinity. We then define the *electric potential* at any point in an electric field as the work required to bring a unit positive charge from infinity to that point.

**FOR EXAMPLE:** If we have to do work of 0.80 joule to move a charge of $2.0 \times 10^{-4}$ coulomb from infinity to point *C*, the potential of point *C* is 0.80 divided by $2.0 \times 10^{-4}$, or $4.0 \times 10^{3}$ volts.

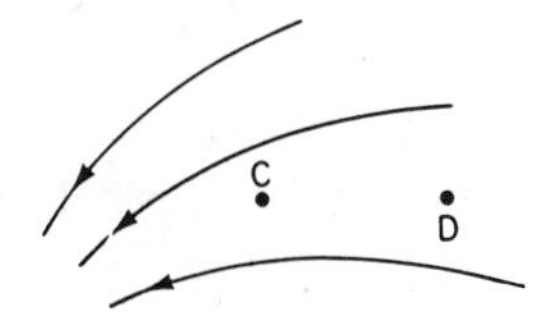


*Potential of D is Higher than C*

Notice that the potential of point *D* is higher than of point *C*, because additional work has to be done to move a positive charge from point *C* to point *D*. Also note that, in this case, when we move the charge from *C* to *D*, the charge will have more potential energy at *D* than at *C*. This is due to the fact that we did work in moving the charge from *C* to *D*. If we let go of the charge after getting it to *D*, the field will move the charge. The potential energy is then converted to kinetic energy. You may notice that this is similar to lifting an object against the earth's gravitational field, except that the earth's gravitational field is always vertical. Also, all objects released in the earth's gravitational field, move in the direction of the field, namely towards the earth. In the case of the electric field, if we release a positive charge, it moves in the direction of the field. If we release a negative charge, it moves in the opposite direction. We arbitrarily select the positive charge for reference.

17 * **Capacitance** Earlier we described the electric field between two charged plates. We implied that the space between the plates is vacuum or air, and that the plates are metallic. Any arrangement of two conductors separated by an insulator is known as a *capacitor*. The insulator is also referred to as the *dielectric*. It may be air, vacuum, mica, ceramic, paper, glass, etc. A capacitor's ability to store a charge is known as its *capacitance*. When we tune a radio, we usually vary the capacitance of an air capacitor by changing the area of plates used to store a charge. For a capacitor, *the capacitance is directly proportional to the area of the plates, and is inversely proportional to the distance between the plates.*

**FOR EXAMPLE:** If we double the distance between the plates of a capacitor, the capacitance is halved. The unit of capacitance is the *farad*. Practical capacitors have a capacitance of only a small fraction of a farad. A capacitance of 10 microfarads is quite large.

18 ** **A Formula for Capacitance** When a capacitor is charged, one plate positive and the other negative, there is a difference of potential between the plates. The negative charge is numerically the same as the positive charge. If we increase the charge on a given capacitor, the potential difference also increases. The ratio between the charge on one of the plates and the potential difference remains constant; this ratio is the capacitance.

$$\text{capacitance} = \frac{\text{charge}}{\text{potential difference}},$$

When the charge is in coulombs,
and the potential difference in volts,
the capacitance is in farads.

Note that one farad is the same thing, therefore, as one coulomb/volt.

**19 Electron Volt (ev)** The electron volt is a small unit of energy. It is especially useful when dealing with small bits of matter, such as atoms and electrons. According to the *Physics Reference Table*, 1 electron volt equals $1.60 \times 10^{-19}$ joule, a very small fraction of a joule.

Let us see why it has this value. We have defined the potential difference between two points as the work required to move a unit charge between the points.

Potential difference = work/charge.

Therefore,

work = charge × potential difference.

When we move a coulomb of charge through a potential difference of 1 volt, the work we do is 1 coulomb-volt, which is called a joule. When we move an electron through a potential difference of 1 volt, the work we do is: charge of 1 electron × 1 volt. We may call this product 1 electron volt. This helps to remind us what we mean by it: an *electron volt* is the energy required to move an electron through a potential difference of 1 volt.

To find out how much energy this is, expressed in joules, we must substitute for the electron its equivalent charge in coulombs. The *Reference Table* gives the charge of 1 electron as $1.60 \times 10^{-19}$ coulomb. Therefore,

$$\begin{aligned} \text{work} &= 1.60 \times 10^{-19} \text{ coulomb} \times 1 \text{ volt} \\ &= 1.60 \times 10^{-19} \text{ joule.} \end{aligned}$$

Some atom smashers can accelerate particles to energies of several billion electron volts.

**20 * Electric Field Between Charged Plates** Let us again consider two parallel plates, oppositely charged. We have already said that the electric field intensity is uniform in this case. It turns out that the electric field intensity is equal to the ratio between the potential difference of the two plates and the distance between the plates:

$E = V/d$, where $E$ is the field intensity,
$V$ is the difference in potential (in volts)
$d$ is the distance between the two plates (in meters).

Note that in the *MKS* system, from this expression, the field intensity unit is volts per meter. From the definition of electric field intensity, force per unit charge, we get the unit: newton per coulomb. It follows that

1 volt/meter = 1 newton/coulomb.

It is important to understand how the above formula can be derived. Suppose we have a positive charge $q$ between two charged plates, as shown. Assume that the distance between the plates is $d$, and the difference in potential between the plates is $V$. From the definition, field intensity equals force per unit charge,

$$E = F/q,$$

we get that the force on the charge is $Eq$. The work required to move the charge from the negative plate to the positive plate is the product of this constant force and the distance:

$$W = Fd = Eqd.$$

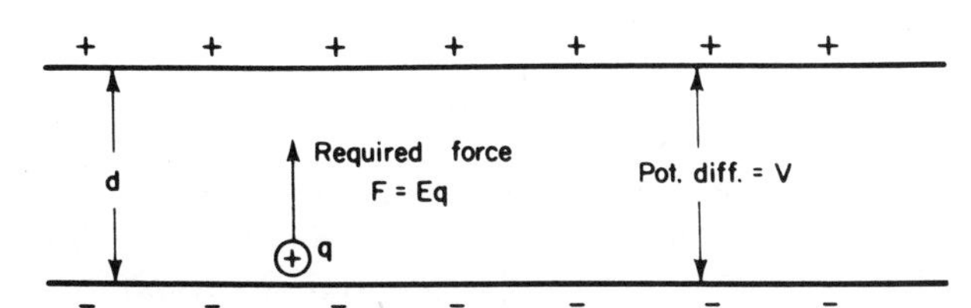


*Work Required to Move Charge between the Two Plates* $= Eqd$.

We can also calculate the work required by using the definition of potential difference: work per unit charge.

$$V = W/q.$$

Therefore the work required to move the charge is

$$W = Vq.$$

Now we set the two expressions for work equal to each other:

$$Eqd = Vq.$$

This gives us the expression we wanted:

$$E = V/d.$$

21 **Millikan's Oil Drop Experiment** We have described electric charges in terms of electrons, as though electric charge come, in little bundles which can't be divided indefinitely. How do we know this? How do we know what the charge of one of these bundles is?

In 1913 Robert Millikan, an American physicist performed an ingenious experiment which answered these questions. We don't need to review the details of his experiment here, but the basic ideas are simple. He injected oil drops into the space between two horizontal, parallel plates. The oil drops fell toward the bottom plate because of the pull of gravity. The pull of gravity is equal to the weight of the drop: $F = mg$, where $m$ is the mass of the drop, and $g$ is the acceleration due to gravity. If the drop is charged negatively, and there is a potential difference between the plates with the top plate positive and the bottom negative, there will be an electric force acting upward on the drop, unlike charges attract, etc. The electric force will depend on the potential difference. The force is increased by increasing the potential difference. At a certain value of potential difference, the upward force will just equal the weight of the drop, which will then be stationary or move with constant velocity.

As will be shown later, when this happens, the charge on the drop can be calculated by using the expression

$$q = mgd/V.$$

The four quantities on the right can be calculated or measured.

$m$ is the mass of the drop
$g$ is the acceleration due to gravity
$d$ is the distance between the plates
$V$ is the potential difference; it can be measured with a *voltmeter*.

Millikan found that the charge is always an integral multiple of a small quantity. We sometimes say that electricity is granular rather than continuous. The smallest quantity of negative charge is the charge of the electron. Millikan was able to get various integral multiples of this charge on the oil drops by shining X-rays on the drops.

* We can derive the above expression for charge by using the relationships we mentioned earlier. The electric force on the drop is equal to the product of the electric field between the plates and the charge on the drop: $F = Eq$. The field tween the parallel plates is $V/d$. Therefore, $F = Vq/d$. Set this equal to the weight of the drop:

$$(Vq/d) = mg$$
$$\therefore q = mgd/V.$$

# Questions and Problems

1. With the aid of diagrams explain how an electroscope may be charged by induction *a.* negatively *b.* positively. Discuss motion of charges, if any.
2. When a flame of a match is held near the knob of a charged electroscope, the electroscope discharges. Explain, assuming the electroscope is charged *a.* negatively *b.* positively.
3. A negatively charged rod attracts either end of a compass needle near which it is held. Explain.
4. Two point charges are a distance $d$ apart. What happens to the electrostatic force between them if the distance is *a.* halved *b.* doubled *c.* tripled.
5. Two point charges are a distance $d$ apart. What happens to the electrostatic force between them if the charge on *a.* one of them is doubled *b.* on both is doubled *c.* on both is tripled.
6. Two equal positive point charges are a distance $d$ apart. What is the electric field intensity midway between them?
7. A positive test charge is midway between two oppositely charged particles. The force on the test charge due to one of the charged particles is $F$. What is the resultant force on the test charge if the charge on the second particle is *a.* numerically the same as that on the first particle *b.* numerically twice as large as on the first one *c.* numerically three times as large as on the first one.
8. Sketch the electric field between *a.* two positive charged particles *b.* two negatively charged particles *c.* two oppositely charged particles *d.* two oppositely charged parallel plates.
9. Two equal positive charges of $4.0 \times 10^{-5}$ coulomb are 2.0 meters apart. Calculate the electrostatic force between them.

10. Two charges, one $4.0 \times 10^{-5}$ coulomb, and the other $5.0 \times 10^{-5}$ coulomb, are 1.5 meter apart. Calculate the force on the positive charge. How does this compare with the force on the negative charge?

11. ★ Calculate the force on a 40 microcoulomb positive charge exerted by a negative 60 microcoulomb charge 50 cm away.

12. ★ Calculate the force on a 250 microcoulomb charge exerted by a 300 microcoulomb charge placed 0.50 meter away. Assume both charges are negative.

13. ★ Using the information in the *Reference Tables*, calculate the electrostatic force between a proton and an electron which are $10^{-10}$ meter apart.

14. ★ A positive test charge of 250 microcoulombs experiences a force of 0.35 newton at a certain point in an electric field. *a.* Calculate the electric field intensity at that point. *b.* What would be the force at this point on a test charge twice as great?

15. ★ The field intensity at a certain point is 12,000 newtons/coulomb. Calculate the force on a charge in this field, if the charge is *a.* 1 microcoulomb *b.* 3 microcoulombs *c.* $4.5 \times 10^{-6}$ coulomb.

16. ★ Two oppositely charged parallel plates, 10 cm apart, have a difference of potential between them of 35 volts. What is the electric field intensity between them?

17. How much work is done in moving a charge of 400 microcoulombs from the negative plate in problem 16, to the positive plate?

18. ★ A positively charged oil drop weighing $10^{-13}$ newton, is kept at rest between two oppositely charged horizontal plates. The plates are 1 cm apart and have a difference of potential between them of 200 vols. Calculate *a.* the electric force on the oil drop *b.* the electric field between the plates *c.* the charge on the oil drop.

19. A hollow metal sphere is electrically neutral. As it gets charged positively, what happens to the electric field intensity at the center?

# Test Questions

1. A rubber rod rubbed with a piece of fur acquires 1. north poles 2. south poles 3. negative charge 4. positive charge.

2. A negatively charged object is one that has an excess of 1. north poles 2. south poles 3. negative charge 4. positive charge.

3. A charged rod is brought near the knob of a negatively charged electroscope; it is observed that the leaves of the electroscope diverge more. This indicates that the rod is 1. charged negatively 2. charged positively 3. not charged and not magnetized 4. magnetized but not charged.

4. The fundamental particle in an atom which is readily transferred from one object to another is the 1. proton 2. electron 3. neutron 4. positron.

5. An electroscope is readily discharged by placing a burning match near its knob. This can be best explained by referring to 1. neutrons 2. protons 3. positrons 4. ions.

6. The work needed to move an electric charge between two points is measured in units known as 1. coulombs 2. watts 3. joules 4. amperes.

7. Two point charges repel each other with a force of $9.0 \times 10^{-5}$ newtons when they are 1.0 meter apart. We can be sure that both charges are 1. negative 2. positive 3. unlike 4. like.

8. If the distance between the two charges in question 7 is increased to 3.0 meters, the force between the charges becomes, in newtons, 1. $1.0 \times 10^{-5}$ 2. $27 \times 10^{-5}$ 3. $3.0 \times 10^{-5}$ 4. $4.5 \times 10^{-5}$.

9. Two point charges are a distance of $x$ meters apart and exert a force of $F$ on each other. If the distance between them is doubled, the force becomes 1. $F/4$ 2. $F/2$ 3. $F/2x$ 4. $2F$.

10. Most metals are good conductors of electricity because they have 1. a large number of molecules 2. a large number of free electrons 3. a shiny surface 4. a low temperature coefficient.

11. As the electric field intensity at a certain point decreases, the force per unit charge at the point 1. increases 2. decreases 3. remains the same.

12. When two materials such as glass and silk, are rubbed together, the net charge on one material, after rubbing, is numerically, 1. less than that of the other 2. greater than that of the other 3. the same as the other.

13. Six joules of work are required to transfer 12 coulombs between two points. The potential difference between the two points is, in volts, 1. 0.5 2. 2 3. 12 4. 72.

14. A capacitor is a device used to 1. store sparks 2. store an electric charge 3. store radio programs 4. generate electrons.

15. If the distance between two plates of a capacitor is doubled, its capacitance is 1. doubled 2. quadrupled 3. halved 4. reduced to 1/4.

16. If the charge on a metal sphere is doubled, the intensity of the electric field inside the sphere 1. is doubled 2. is quadrupled 3. decreases 4. remains the same.

17. A positive test charge is on the line connecting two small equally charged spheres, one of which has the charge $+Q$, the other $-Q$. The test charge is at a point one-third of the distance between them and closer to the positive charge. If the negative charge exerts a force $F$ on the test charge, what is the force exerted by the positive charge? 1. $F$ 2. $2F$ 3. $3F$ 4. $4F$.

18. The electric field intensity between two oppositely charged parallel metal plates 1. varies inversely as the distance 2. varies inversely as the square of the distance 3. varies as the square of the distance 4. is uniform. (In this question, distance is measured from the positive plate.)

19. The force between two small charges is 48 newtons. If the distance between the charges is halved, and each charge is doubled, the force will be, in newtons, 1. 12 2. 24 3. 48 4. $48 \times 16$.

20. The *MKS* unit of electric potential is the 1. volt 2. voltmeter 3. microvolt 4. coulomb.

# Topic Questions

1.1 Which three particles are present in almost all atoms?

2.1 What property do electrified objects have?

2.2 What is another term that means the same thing as electrified?

2.3 What is the *Law of Charges*?

3.1 What is an elementary charge? How does it compare with the coulomb?

3.2 What is meant by the charge on an object?

3.3 What is *Coulomb's Law*?

5.1 What kind of charge is acquired by a neutral object when it touches a positive object?

5.2 What are some good insulators in your home?

6.1 What is meant by the statement that electric charge is conserved in a system?

7.1 What kind of rod must be used if we want to charge an electroscope positively by induction?

7.2 How can we ground an electroscope?

8.1 What is meant by electric field intensity?

8.2 What is arbitrarily taken as the direction of an electric field?

8.3 What is the direction of the electric field outside a positive sphere?

8.4 How can we get a region in which the electric field is zero?

9.1 In what respects are electric and gravitational fields similar? How do they differ?

10.1 In the *MKS* system, what are two units that may be used for the gravitational field?

11.1 What are the characteristics of the electric field around a point charge?

12.1 What are the characteristics of the electric field around a charged metal sphere?

13.1 What are the characteristics of the electric field around a pear-shaped conductor?

14.1 What are two characteristics of the electric field between two parallel plates which are oppositely charged?

15.1 *a.* What is the definition of potential difference? *b.* To what is one volt equivalent?

# 13 Electric Circuits

## 1 Introduction

**Electric Current** Up to now we have considered static electricity—situations in which the electric charges are essentially at rest. Practical applications of electricity depend primarily on charges in motion. The electric *current* is the rate of flow of electric charges. The *ampere* (amp) is a unit of electric current. When 1 coulomb passes through a conductor in one second, we say we have one ampere:

1 ampere = 1 coulomb/second.

In general, if we want to know the rate of flow, we divide the charge ($q$) by the time it takes the charge to move through the conductor. This rate of flow is the current $I$:

$$I = q/t.$$

**EXAMPLE:** 10. coulombs go through a conductor in 5.0 seconds. What is the current?

$I = q/t$; $I = 10$ coul/5.0 sec; $I = 2.0$ amp

The *ammeter* is an instrument for measuring electric current.

**2 Conductivity in Solids** Since we are going to study the electric current, we need to know a little more about substances which allow charges to move, the conductors; and substances which do not allow charges to move, the insulators. We may use an arrangement such as shown below:

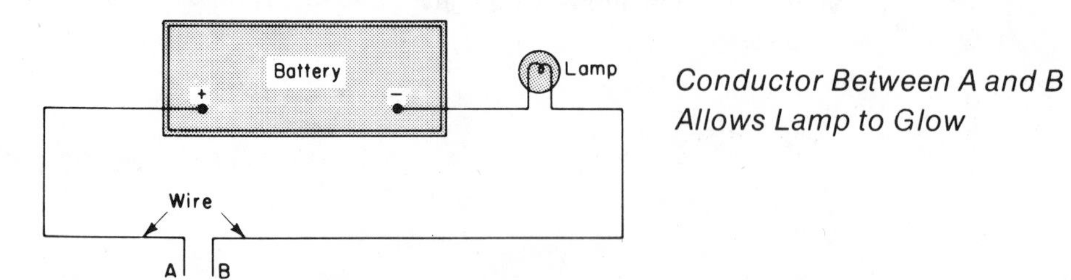


*Conductor Between A and B Allows Lamp to Glow*

Instead of the lamp an ammeter may be used. If wires *A* and *B* are allowed to touch, the lamp glows, indicating that there is current in it. If we separate *A* and *B*, the lamp goes out, indicating that the air between them does not allow the electric charges to move. Air is an insulator, normally. By connecting different substances between *A* and *B*, we can find out to what degree they allow charges to move.

At this time, we do not need to know the details of operation of the *battery*. We only need to think of it as a device in which chemical energy is used. As a result of the chemical action in the battery, electrons are pumped on to a terminal which becomes negative, and are taken away from another terminal which becomes positive. Because the terminals are oppositely charged, there is a potential difference between them. In a flashlight cell this potential difference is about 1.5 volt. The electrons on the negative terminal repel each other. If we connect a conductor between the negative terminal and the positive terminal, the electrons can move over to the positive terminal; but remember, the chemical action in the battery continues to pump electrons on to the negative terminal.

We find that, in general, metals are good conductors of electricity. Silver is the best; copper and aluminum are also good. Metals are good conductors because they have many electrons that are relatively free to move; electrons which are not held firmly by the atoms. Non-metals are poor conductors; they have few free electrons. Very poor conductors are known as non-conductors or insulators.

The electric current in solids consists of moving electrons.

**Conductivity of Liquids** Pure water is a poor conductor. A water solution of sugar is a poor conductor. Water solutions of acids and salts conduct electricity. Acids, salts, and some other chemical compounds dissociate into positively and negatively charged particles called *ions*. In such solutions, the ions are free to move; the electric current in them consists of moving positive and negative ions.

**Conductivity in Gases** We have already seen that normally the air is an insulator. This is generally true of gases. However, gases may be ionized and then will conduct electric current. For example, if we hold a lighted match near the knob of a charged electroscope, the electroscope rapidly discharges. Also, neon tubes and fluorescent lamps depend for their operation on conduction through ionized gases.

The earth is surrounded by belts of ionized gases. Gases may be ionized by strong electric fields, by cosmic rays, x-rays, and collision of atoms. Heat increases the violence of these collisions.

The electric current in gases may consist of motion of electrons as well as positive and negative ions.

## 3 Ohm's Law

As we saw above, some substances are better conductors than others. We can express this differently by saying that some substances offer more resistance to electric current than others. We shall now look at this notion of electrical resistance in more detail.

If we connect a coil of wire to a battery, we know from the above discussion that there will be a current in the wire. If we insert an ammeter, we can measure the current, $I$. If we connect a voltmeter to the ends of the wire, we find that there is a potential difference, $V$, between the ends. If we increase the current by using a different battery, we find that the potential difference between the ends also increases. Ohm found that for metallic conductors, the ratio between potential difference and current remains nearly constant. We define this ratio as the resistance $R$ of the conductor.

$R = V/I.$

This expression is known as *Ohm's Law*. It is assumed that the temperature of the conductor is not changed. The resistance is in ohms when the potential difference is in volts and the current is in amperes. The *ohm* is a unit of resistance. Of course, Ohm's Law can also be written in the form

$V = IR.$

**EXAMPLE:** The current through a lamp is 2.0 amperes when the potential difference between its terminals is 120 volts. What is the resistance of the lamp?

$R = V/I$
$= (120 \text{ volts})/2.0 \text{ amp}$
$= 60. \text{ ohms}.$

**Note:** Instead of saying potential difference between the terminals, we also say *potential difference across* the device. Instead of using the expression potential difference, we may use the word *voltage*.

## 4 Other Electrical Units

Prefixes are often used with the basic electrical units to express larger or smaller quantities. This will be illustrated for a few of the important ones.

| Prefix | Meaning | Example |
|---|---|---|
| meg(a)- | $10^6$ | 1 megohm = $10^6$ ohms = 1 million ohms |
| kil(o)- | $10^3$ | 1 kilovolt = $10^3$ volts = 1 thousand volts |
| milli- | $10^{-3}$ | 1 milliamp = 0.001 amp |
| micro- | $10^{-6}$ | 1 microamp = $10^{-6}$ amp |
| micromicro- (or pico-) | $10^{-12}$ | 1 micromicrofarad = $10^{-12}$ farad (or 1 picofarad) |

## 5 Factors Affecting Resistance

These are temperature, length, cross-section, and materials.

**Effect of Temperature** The resistance of most metallic conductors increases with increasing temperature. The resistance of non-metals and solutions usually decreases with increasing temperature. (At temperatures near absolute zero, some materials have no measurable resistance. This phenomenon is known as superconductivity, and the materials are then said to be superconductors.)

**Effect of Length** The resistance of a conductor is proportional to its length. If two conductors differ only in length, the longer conductor has the greater resistance. For example, if we cut wire from a given spool, one foot of the wire will have one-fourth the resistance of a wire four feet long.

**Effect of Cross-section** The resistance of the conductor varies inversely with the cross-section area. The thicker the wire, the less its resistance. For example, if the cross-section area is twice as great, the resistance is one-half. (Remember, most wires have a circular cross-section, and the area of a circle is $\pi r^2$.)

**Effect of Material** If wires have the same length, cross-section area, and temperature, the wire made of silver will have the least resistance. Copper wires also have a low resistance. By comparison, wires made of tungsten and nichrome have a high resistance.

**Combined Effect** The resistance of a conductor at constant temperature is proportional to its length and is inversely proportional to its cross-section area.

** This can be expressed by a formula:

$$R = \frac{kL}{A},$$ where $k$ is the resistivity of the material.

# 6 The Series Circuit

A circuit consists of a source of potential difference and one or more conductors connected to this source. Some of the conductors are selected to have a desired resistance. Such devices are called *resistors*. Some circuits include a switch. We *open the circuit* by *opening the switch*: we set the switch in a position so that there is no conducting path. When we *close the switch*, we provide a conducting path. We then have a closed circuit.

The source of potential difference most often is a battery or a generator. The principle of operation of the generator will be described later; the battery has already been explained. In either case, think of an internal action which pumps electrons on to one terminal thus making it negative, and simultaneously taking electrons off the other terminal.

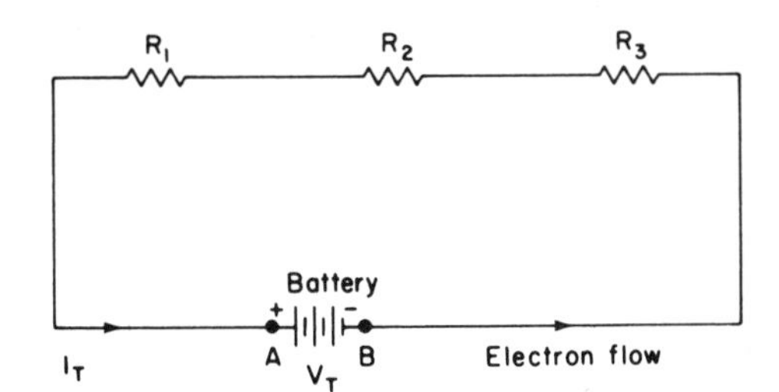


*The Current is the Same Throughout the Circuit*

The diagram represents a series circuit. Three resistors are connected in series with the battery. The terminals of the battery are $A$ and $B$. The straight lines represent conductors of negligible resistance. Notice that in the series circuit there is only one path for the current. The current consists of electron flow from terminal $B$ through resistors $R_3$, $R_2$, $R_1$, and then to terminal $A$. Inside the battery, the current consists of the motion of ions. Electrons are attracted to the positive terminal at the same time that electrons are repelled from the negative terminal. (Some books follow an old convention in which the direction of the current is shown opposite to the direction of electron flow.)

Since there is only one current path in the series circuit, *the current through the battery, $I_T$, is the same as the current through each resistor*:

$$I_T = I_1 = I_2 = I_3,$$ where $I_1$ is the current through $R_1$, etc.

*Ohm's Law* tells us that there is a potential difference across each resistor since there is current through it. For each resistor this potential difference is equal to the product of the current and the resistance through which it goes:

$$V_1 = I_1 R_1$$
$$V_2 = I_2 R_2$$
$$V_3 = I_3 R_3.$$

Now, potential difference is work per unit charge. In other words, the above three equations remind us that work must be done to push electrons through the resistors. Where does the energy come from to do this work? The energy is supplied by the battery, which also has a potential difference across it, $V_T$. The battery supplies to each charge the energy which is needed to push it through the circuit:

$$V_T = V_1 + V_2 + V_3$$

Substituting in this equation, and remembering that all the currents in the series circuit are the same, we get:

$$V_T = IR_1 + IR_2 + IR_3$$
$$= I_T(R_1 + R_2 + R_3)$$
$$\therefore V_T = I_T R_T$$

where $R_T = R_1 + R_2 + R_3.$

In words, *the resistance of a series circuit is the sum of the individual resistances.* This is reasonable, because connecting wires in series is similar to using a longer wire.

**Other Terms** The potential difference across a resistor is sometimes called a *potential drop*, a *voltage drop*, or an *IR-drop*.

**EXAMPLE:** Two resistors, one of 20. ohms and the other of 30. ohms, are connected in series to a battery of 100 volts. Calculate: *a.* The combined resistance of the resistors *b.* the current through the 20-ohm resistor *c.* the voltage across the 30-ohm resistor.

*a.* $R_T = R_1 + R_2$

$= (20 \text{ ohms} + 30 \text{ ohms}) = 50 \text{ ohms}.$

*b.* $V_T = I_T R_T$

$100 \text{ volts} = I_T (50 \text{ ohms})$

$I_T = 2.0 \text{ amp} = I_1$

*c.* $V_2 = I_2 R_2$

$= (2.0 \times 30) = 60 \text{ volts}.$

Note that in a series circuit, only a portion of the battery voltage appears across each resistor. *The applied voltage, or battery voltage, is equal to the sum of the voltages across the individual parts of the circuit.*

7 * **Internal Resistance** Batteries and generators also offer resistance to the flow of charges through them. This is known as their internal resistance. In the above illustrations, we assumed that the internal resistance was negligible. If it is not negligible, the internal resistance can be treated as an external resistance in series with the rest of the circuit.

However, when the internal resistance is not negligible, the voltage at the terminals of the battery is not always the same, but depends on the current through it. The voltage between the terminals of a battery or generator when there is no current through it is known as its electromotive force or *emf*. (The term electromotive force is poor, because it refers to work per unit charge, not force. When the term was invented, physicists were still confused about some of the concepts. To avoid confusion, we usually just say the three letters, *emf*.) When a resistor is connected to the battery, the terminal voltage is less than the *emf* if the internal resistance is not negligible.

**EXAMPLE:** A battery having an emf of 40. volts and internal resistance of 2.0 ohms is connected to a resistance of 18. ohms. Calculate the current and voltage of the resistor. For many circuit problems, it is desirable to have a diagram. Note how the

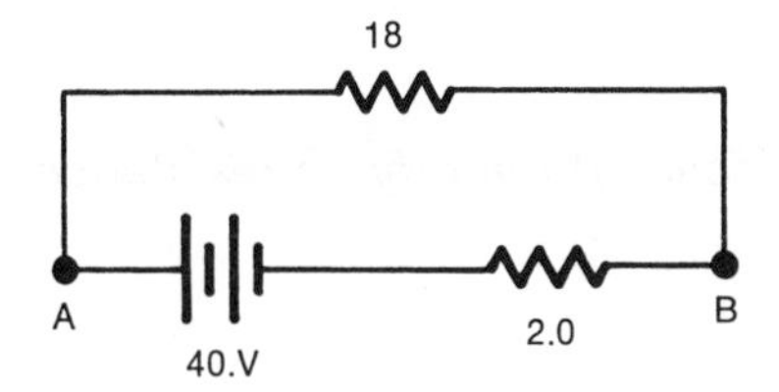


internal resistance of 2.0 ohms is shown in series with the rest of the circuit. *A* and *B* are the terminals of the battery.

$$R_T = R_1 + R_2$$
$$R_T = (18 + 2) \text{ ohms} = 20 \text{ ohms}$$
$$I_T = I_1 = I_2 = V_T/R_T$$
$$= 40/20 = 2.0 \text{ amp}$$
$$V_1 = I_1R_1 = 2.0 \text{ amp} \times 18. \text{ ohms} = 36. \text{ volts.}$$

Note that the voltage across the resistor connected to the terminals of the battery is only 36 volts, not 40. volts. The reason for this is that there is an internal voltage drop across the internal resistance of the battery. This drop is (2.0 amp × 2.0 ohms), or 4.0 volts. Therefore, the voltage available at the terminals of the battery is (40. − 4.0) or 36. volts.

# 8 Parallel Circuit

A *parallel circuit* is one in which there is more than one path for current from the battery or other source. The circuit shows three resistors connected of the terminals of the battery, which is assumed to have negligible internal resistance. The current from the battery, $I_T$, has three paths, one through each of the resistors. The three resistors are in parallel with each other and the battery.

Since each resistor is connected to the terminals of the battery, *the voltage across each resistor is the same as the voltage between the terminals of the battery.*

$$V_T = V_1 = V_2 = V_3$$

Since there is no place for charges to disappear, the current which leaves the battery has to be equal to the current which returns to it, and also, *the current which leaves the battery has to be equal to the sum of the branch currents.*

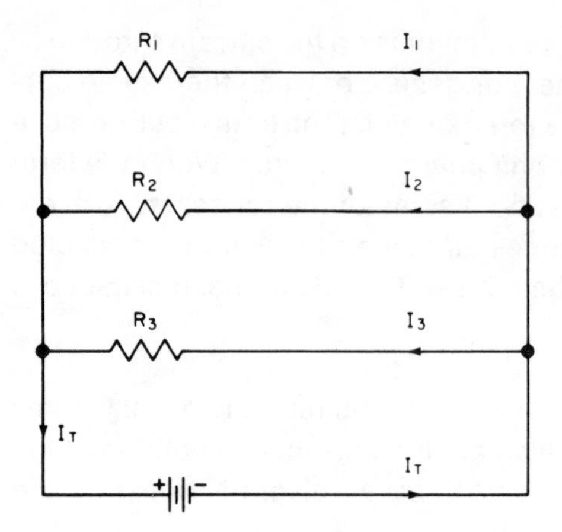


$$I_T = I_1 + I_2 + I_3$$

It can also be shown that *the sum of the reciprocals of the branch resistances equals the reciprocal of the combined resistance.*

$$\frac{1}{R_1} + \frac{1}{R_2} + \frac{1}{R_3} = \frac{1}{R_T}$$

It is usually desirable to use a special form of this relationship:

1. When all resistances in parallel are equal, their combined resistance is equal to one resistance divided by the number of equal resistances:

$$R_T = R/n.$$

**EXAMPLE:** 10 resistors of 60. ohms each are connected in parallel. What is their combined (or effective) resistance?

$$R_T = 60.\ \text{ohms}/10$$
$$= 6.0\ \text{ohms}.$$

2. When only two resistors are connected in parallel, their combined resistance is equal to their product divided by their sum:

$$R_T = \frac{R_1 R_2}{R_1 + R_2}.$$

**EXAMPLE:** What is the combined resistance of a 40. and 60. ohm resistor connected in parallel?

$$R_T = \frac{(40 \times 60)}{40 + 60}$$
$$= 24.\ \text{ohms}.$$

**EXAMPLE:** Calculate the total current, total resistance, and the current through each resistor. This is a parallel circuit.

$$R_T = (R_1 R_2)/(R_1 + R_2)$$
$$= (20 \times 30)/(20 + 30)$$
$$= 12\ \text{ohms}.$$

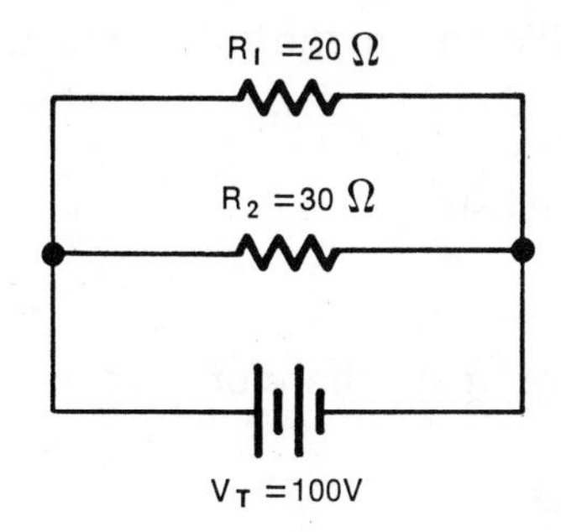


$I_T = V_T/R_T$
$= (100 \text{ volts})/12 \text{ ohms}$
$= 8\frac{1}{3} \text{ amp.}$

$I_1 = V_1/R_1 = (100 \text{ volts})/20 \text{ ohms} = 5 \text{ amp.}$

$I_2 = V_2/R_2 = (100 \text{ volts})/30 \text{ ohms} = 3\frac{1}{3} \text{ amp.}$

Check: $I_T = I_1 + I_2 = (5 + 3\frac{1}{3}) \text{ amp} = 8\frac{1}{3} \text{ amp.}$

9 **Electrical Power and Energy** Power is the rate of using or supplying energy. By using this definition, and the definitions of potential difference and current, it can be shown that the power used by a device is equal to the product of its voltage and the current through it:

$P = VI$

where $P$ is the power in watts,
$V$ is the potential difference across the device, in volts,
$I$ is the current through it, in amperes.

If we apply *Ohm's Law* to this device, which has resistance $R$, in ohms, then we can derive two additional useful formulas for power used by the device:

$P = I^2R$
and
$P = V^2/R.$

**EXAMPLE:** A 60-watt bulb is connected to its rated voltage of 120 volts. What is the current through it, and what is its resistance when hot?

$P = VI;$

$60 \text{ watts} = (120 \text{ volts})I;$

$I = 60/120 = 1/2 \text{ amp.}$

$R = V/I$
$= 120 \text{ volts}/(1/2) \text{ amp}$
$= 240 \text{ ohms.}$

**Notes:** 1. The rated voltage is usually stamped on a device, and is the voltage at which the bulb will use the 60 watts. At a lower voltage, it will use less power.

2. The calculation shown gives the resistance when the lamp is hot.

When it is cold, the resistance of the tungsten filament in the bulb is much less.

**Energy** Since power is the rate of using or supplying energy,

$$P = \text{energy/time},$$

energy = power × time. The energy used by a resistor may, therefore, be expressed in a unit of power multiplied by a unit of time:

1 joule = 1 watt-second.

If the 60-watt bulb in the above example is used for 4 hours, the energy it uses is: 60 watts × 4 hours or 240 watt-hours, or 0.240 kilowatt-hours.

kilowatt-hours = watts × hours/1000

Also, since energy = power × time, we may write these formulas for the electrical situation:

$$\text{energy} = VIt, \quad \text{and} \quad \text{energy} = I^2Rt;$$

if $V$ is in volts, $I$ is in amperes, $R$ in ohms, and $t$ in seconds, the energy will be in watt-seconds or joules.

**Heat** Current through a resistor produces heat. Heat is energy, and the amount of heat produced is given by any of the energy formulas given above. If we want to express the heat produced in kilocalories, we must use the proper conversion factor:

$$\text{kilocalories} = 2.4 \times 10^{-4}\, I^2Rt$$

or,

$$\text{kilocalories} = 2.4 \times 10^{-4}\, VIt.$$

**Cost** Our electric bill depends on how much energy we use. The rate may be 8 cents per kilowatt-hour. If we use 40 kilowatt-hours, we pay $3.20.

**EXAMPLE:** What is the cost of operating an air conditioner during the month of June, if it is used 4 hours every day on 120 volts, and draws a current of 7.5 amperes? The rate is 8 cents per kilowatt-hour.

$$P = VI$$
$$= (120 \text{ volts})(7.5 \text{ amp})$$
$$= 900 \text{ watts}.$$

kilowatt-hours = watts × hours/1000
= (900 watts)(120 hours)/1000
= 108.

cost = 108 kilowatt-hours × 8 cents/kilowatt-hour
= $8.64.

**Energy Supplied by Source** In an electric circuit, as in other situations, energy is conserved. The power, or energy, supplied by the source, is equal to the power, or energy, used by the devices in the circuit. The power supplied by the source can also be calculated by using one of the formulas given above:

$P = VI$. where $V$ is the *emf* of the source, and
$I$ is the current through the source.

**10 Heat In Series and Parallel Circuits** If a 30- and a 60-ohm resistor are connected in a circuit, which resistor will produce more heat? It depends on the circuit.

In a series circuit, the larger resistance will produce the greater amount of heat.

In a parallel circuit, the smaller resistance will produce the greater amount of heat.

This can be illustrated by an *example.* Assume the two resistors are connected in series to 180 volts. The total resistance is (30 + 60) ohms, or 90 ohms. The current through each is (180 volts)/90 ohms, or 2 amps. The power used ($I^2R$) for the 30-ohm resistor is ($2^2 \times 30$) or 120 watts, and for the 60-ohm resistor is ($2^2 \times 60$) or 240 watts. The 60-ohm resistor uses more power, and therefore produces more heat.

In the parallel circuit, each resistor has a potential difference of 180 volts. The power used ($V^2/R$) for the 30-ohm resistor is ($180^2/30$) or 1080 watts. For the 60-ohm resistor, it is ($180^2/60$) or 540 watts. This time the 30-ohm resistor uses more power.

## 11 Series-Parallel Circuit

**

Many practical circuits involve a few components in parallel, with the combination connected in series to a source of power. The solution is indicated by an example.

**EXAMPLE:** The 20-ohm and 30-ohm resistors are in parallel. The combination is in series with a battery and a three-ohm resistor. (The 3 ohms might include the in-

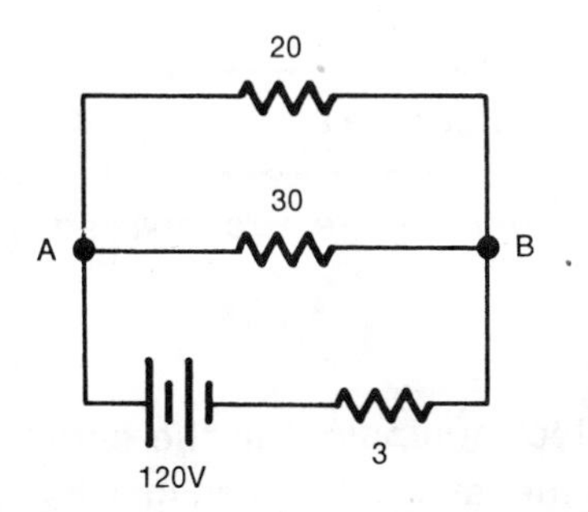


ternal resistance of the battery.) The combined resistance of the two resistors in parallel:

$$R = \frac{20 \times 30}{20 + 30} = 12 \text{ ohms.}$$

This gives the equivalent circuit drawn below the original circuit. The new circuit is a series circuit. Its resistance is (12 + 3) ohms or 15 ohms.

$$I_T = \frac{V_T}{R_T} = \frac{120 \text{ volts}}{15 \text{ ohms}} = 8.0 \text{ amp.}$$

This is the current through the 3-ohm and the 12-ohm resistance. The voltage between points $A$ and $B = IR = (8 \times 12) = 96$ volts. This is also the voltage across the 20-ohm resistor, as well as the 30-ohm resistor.

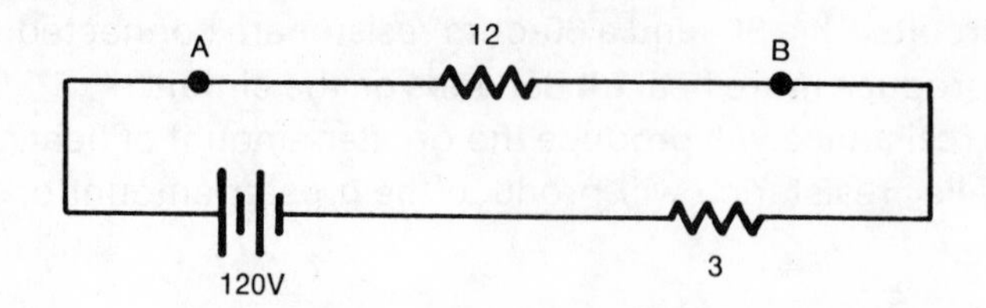


## 12 Summary of Formulas

$I_T = V_T/R_T;\ R_T = V_T/I_T;\ V_T = I_T R_T$

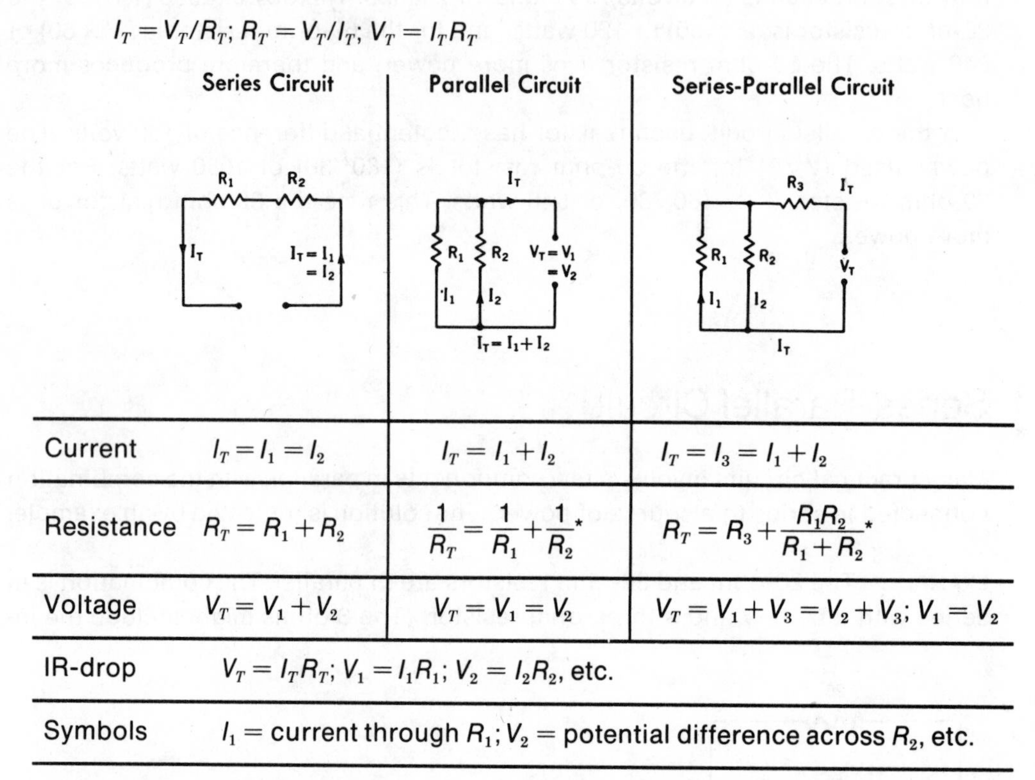


| | Series Circuit | Parallel Circuit | Series-Parallel Circuit |
|---|---|---|---|
| Current | $I_T = I_1 = I_2$ | $I_T = I_1 + I_2$ | $I_T = I_3 = I_1 + I_2$ |
| Resistance | $R_T = R_1 + R_2$ | $\frac{1}{R_T} = \frac{1}{R_1} + \frac{1}{R_2}$* | $R_T = R_3 + \frac{R_1 R_2}{R_1 + R_2}$* |
| Voltage | $V_T = V_1 + V_2$ | $V_T = V_1 = V_2$ | $V_T = V_1 + V_3 = V_2 + V_3;\ V_1 = V_2$ |
| IR-drop | $V_T = I_T R_T;\ V_1 = I_1 R_1;\ V_2 = I_2 R_2$, etc. | | |
| Symbols | $I_1$ = current through $R_1$; $V_2$ = potential difference across $R_2$, etc. | | |

*If only two resistors are connected in parallel, the combined resistance is equal to their product divided by their sum: $R_T = \frac{R_1 R_2}{R_1 + R_2}$.

**13 Fuses** Fuses or circuit breakers are used to protect equipment in the circuit. Fuses melt when the current through them exceeds the rated value. You may have several fuses in your house. Each fuse protects a different circuit. If too large a fuse is used, the wires in the wall may get too hot and start a fire. The circuit shows a fuse to protect the whole circuit ($F_1$), and a second fuse to protect the ammeter in one of the branches ($F_2$). If $F_1$ "blows" (burns out), the whole circuit loses power. If $F_2$ blows, there is still power for $R_1$.

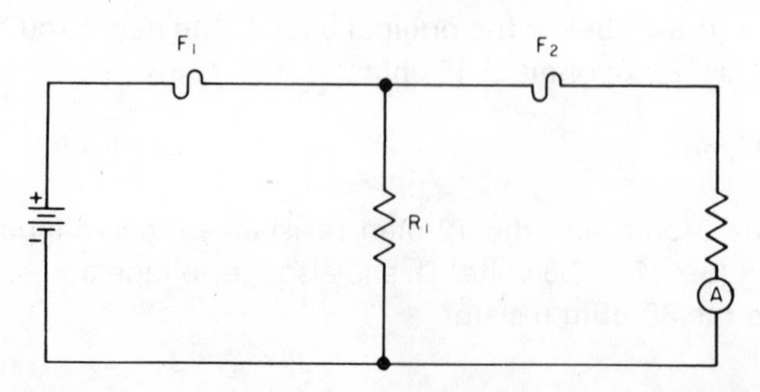

# Questions and Problems

*(Assume that power sources have negligible internal resistance unless otherwise stated.)*

1. A resistor of 60 ohms is connected to a battery supplying 120 volts. Calculate *a.* the current in the resistor *b.* the power used by it *c.* the power supplied by the battery.

2. Two resistors are connected in series to a generator with an emf of 100 volts. The resistance of one resistor is 20 ohms, that of the other is 30 ohms. Calculate *a.* the current in each *b.* the power supplied by the generator *c.* the heat developed by the 30-ohm resistor in 2 seconds.

3. Two resistors are connected to a battery supplying 45 volts. The potential difference across each resistor is 45 volts, and each resistor has a resistance of 15 ohms. *a.* How are the resistors connected? *b.* What is the current in each resistor? *c.* How much power is used by each resistor? *d.* What is the combined resistance of the two resistors? *e.* What is the current in the battery?

4. Three resistors are connected in series to a generator supplying 150 volts. Their resistance is 20, 25, and 30 ohms, respectively. Calculate *a.* their combined resistance *b.* the current through the 30-ohm resistor *c.* the voltage across the 30 ohm resistor *d.* the power used by the 30-ohm resistor *e.* the power used by the whole circuit.

5. Three equal resistors, each 45 ohms, are connected in parallel to a source of 5.0 volts. Calculate *a.* their combined resistance *b.* the current through each *c.* the current from the source *d.* the power supplied by the source *e.* the heat developed by each resistor in 2 minutes.

6. What is the cost of operating a 600-watt toaster for one minute at a cost of 5 cents per kilowatt-hour?

7. How much does it cost in a 30-day month to operate 10 500-watt lamps 4 hours each day, if the cost per kilowatt-hour is 4 cents?

8. A 100 watt lamp operates on its rated voltage of 120 volts. Calculate *a.* the current in the lamp *b.* the resistance of the lamp under these conditions *c.* the energy supplied to the lamp in 5 seconds *d.* the heat produced by this lamp in 10 seconds, if all the energy supplied to it is converted to heat; express your answer in calories and kilocalories.

9. A 50-ohm and a 60-ohm resistor are connected in series to a source of 55 volts. Calculate *a.* the work required to push a coulomb of charge through the 50-ohm resistor *b.* the work required to push an electron through the same resistor.

10. A 40-ohm resistor is connected to a source of 100 volts while it (the resistor) is kept submerged in 250 grams of water. *a.* How long will it take, at the minimum, for the temperature of the water to rise from 20°C to 50°C? *b.* Why might it actually take longer?

11. ** Two 60-ohm resistors in parallel are connected in series with a 40-ohm resistor and a source of 35 volts. Calculate *a.* the resistance of the combination *b.* the current through the 40-ohm resistor *c.* the voltage across the 40-ohm resistor *d.* the voltage across each of the other resistors *e.* the current through each 60-ohm resistor.

12. When a resistor is connected to a 30-volt battery, the current in it is 5

amperes. *a.* Calculate the resistance of the resistor. *b.* If this resistor is to be used with a sixty-volt battery instead, what resistance must be in series with it, so that the current in it will still be only 5 amperes?

13. * A battery has an emf of 6.0 volts and an internal resistance of 0.2 ohm. If a resistance of 1.8 ohm is connected to the battery, what will be the potential difference between the terminals of the battery?

## Test Questions

1. Solids which are good conductors of electricity have many loosely bound 1. protons 2. electrons 3. atoms 4. neutrons.
2. A unit of electric current is the 1. volt 2. watt 3. ampere 4. coulomb.
3. A 30-ohm and a 60-ohm resistor are connected in series. The current in the 60-ohm resistor is how many times that in the 30-ohm resistor? 1. 1 2. 2 3. 1/2 4. 4.
4. A man has two nichrome wires, each 6 feet long. One wire has a cross-section area of 4 units, the other a cross-section area of 8 units. The resistance of the first wire is ........... times that of the second. 1. 1/2 2. 2 3. 1/4 4. 4.
5. Two copper wires have the same cross-section but one wire is three times as long as the first. The ratio of the resistance of the long wire to that of the shorter one is 1. 1/3 2. 1/9 3. 3 4. 9.
6. The combined resistance of the two resistors in question 3 is, in ohms 1. 20 2. 90 3. 30 4. 1800.
7. Two wires are made of nichrome. Wire *A* is twice as long as wire *B* and has a cross-section area twice as great. The ratio of the resistance of wire *A* to that of wire *B* is 1. 1 2. 2 3. 1/2 4. 4.
8. The current in a 60-ohm resistor connected to a battery of negligible resistance having an emf of 30 volts is ............
9. The voltage across a resistor of 20 ohms, with a current in it of 3.5 amperes, is ............
10. A broiler operates on 120 volts and uses 960 watts. The current in the broiler is ........... amperes.
11. A 20-ohm and a 40 ohm resistor are connected in series. The ratio of the power used by the 40-ohm resistor to that used by the 20-ohm resistor is 1. 1 2. 2 3. 1/2 4. 4.
12. A 20-ohm and a 40-ohm resistor are connected in parallel. The current in the 40-ohm resistor is how many times that of the 20-ohm resistor? 1. 1 2. 2 3. 4 4. 1/2.
13. For the two resistors in question 12, the power used by the 40-ohm resistor is how many times that of the 20-ohm resistor? 1. 1/2 2. 2 3. 1/4 4. 4.
14. For the resistors in question 12, the rate at which heat is developed in the 40-ohm resistor is how many times that of the 20-ohm resistor? 1. 1/2 2. 2 3. 1/4 4. 4.
15. If the potential difference across a resistor is doubled, the current in it is 1. unchanged 2. doubled 3. halved 4. quadrupled.

16. If the potential difference across a resistor is doubled, the power used by it is 1. unchanged 2. doubled 3. halved 4. quadrupled.

17. Several electrical appliances are arranged in order to decreasing wattage rating. Assuming that they are rated at 120 volts, as the wattage rating decreases, the resistance of the appliances 1. increases 2. decreases 3. remains the same.

18. As more appliances are added in parallel to a battery of negligible internal resistance, the voltage across each appliance 1. increases 2. decreases 3. remains the same.

19. A 100-watt heater operates on 120 volts. The current in the heater is ............ amperes.

20. Of the following, the meter which reads energy used, is the 1. ammeter 2. voltmeter 3. ohmmeter 4. kilowatt-hour meter.

## Topic Questions

1.1 What is one ampere?

2.1 What does the electric current consist of in *a.* metals *b.* liquids *c.* gases?

3.1 *a.* What is *Ohm's Law*? *b.* What is one ohm?

4.1 What is a milliampere?

5.1 What is the effect of each of the following on the resistance of a metallic conductor: *a.* doubling the length *b.* doubling the cross section area *c.* increasing the temperature?

6.1 What is meant by a closed circuit?

6.2 What is the potential difference across one of the resistors in a series circuit?

7.1 When is the terminal voltage of a battery less than its emf ?

8.1 What is the combined resistance of resistors connected in parallel?

9.1 How is the energy used by a device related to its power rating?

10.1 If two unequal resistors are connected in parallel, which resistor develops heat at the greater rate?

13.1 Why is it usually not desirable to replace a burned-out fuse with a good one which has a larger rating?

# 14 Magnetism and Electromagnetism

## 1 Introduction

**2 Some Basic Terms** We have already described the battery as a source of electric power. Much more important is the dynamo, or electric generator. In order to understand its operation, and that of the electric motor, we need to learn something about magnetism. You may already know some facts from other science courses.

A *magnet* attracts iron and steel; when freely suspended it assumes a definite position (because of the earth's magnetism). A *magnetic* substance is one that can be attracted by a magnet. The magnetic materials include iron, nickel, cobalt, and alloys of iron, such as alnico.

*Non-magnetic* substances are only feebly affected by a very strong magnet; e.g. glass, wool, brass, wood. *Paramagnetic* substances are attracted feebly, *diamagnetic* ones repelled feebly. The effect of paramagnetic and diamagnetic substances is so feeble that we usually ignore it.

*Ferromagnetic* substances like iron are attracted strongly by a magnet. A *magnetized* substance is a magnetic substance which has been made into a magnet. A magnetic *pole* is a region of a magnet where its strength is relatively great; every magnet has at least two poles. The *North pole* (N-pole, or north-seeking pole) of a suspended magnet points towards the earth's magnetic pole in the northern hemisphere. A compass needle is a magnet. The end that points in the general direction of north is a North pole, the other end is a South pole.

**3 Law of Magnets** Like poles repel, unlike poles attract. A North pole repels another North pole, but attracts a South pole. If the poles are concentrated at points, the force between two poles is proportional to the product of their strengths and varies inversely with the square of the distance between them. (Note similarity of Coulomb's law for electric charges.)

**4 Magnetic Field** So far we have mentioned three kinds of forces between matter: gravitational force, electrostatic force, and now the magnetic force. We shall show later that the magnetic force is due to electric charges in motion. As in the case of the first two types of force, so in the case of the magnetic force: it is convenient to explain some of the effects in terms of a field. The *magnetic field* is the region around a magnet (or moving charge) where its influence can be detected by another magnet or by a moving electric charge. The *direction* of the field is the direction in which the N-pole of a compass needle would point if placed in the field. (Also see paragraph 13 of this chapter.)

**Magnetic Flux** The magnetic field is mapped or represented by magnetic lines of force, also called magnetic flux lines. As in the case of the electric field, the lines of force are imaginary. Their purpose is to help us visualize some of the properties of a particular field. The number of lines drawn (or imagined) through an area perpendicular to the magnetic field is proportional to the intensity or strength of the field. We shall describe this later under the heading of flux density. In the meantime, it is sufficient to think of the strength of a magnet in terms of its ability to pick up nails; the stronger the magnet, the more nails it can pick up. The field of the stronger magnet is represented by more flux lines than that of the weaker magnet.

A representation of the magnetic field around a magnet can be obtained by sprinkling iron filings on a horizontal surface in the field. The little iron filings are magnetized by the field, and then behave like compass needles, aligning themselves in the direction of the field.

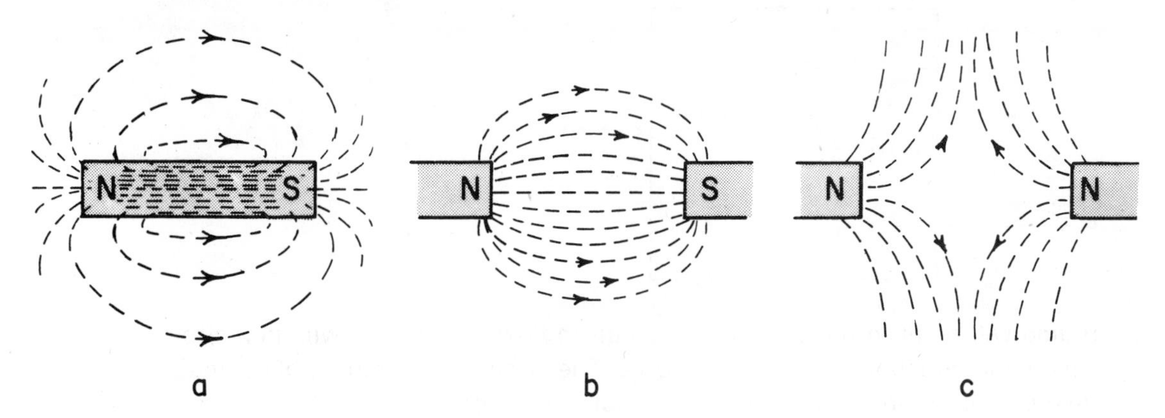


**Properties of the Magnetic Lines of Force**

1. Outside the magnet, their direction is from the N-pole to an S-pole.
2. They never cross each other.
3. They are most crowded at the poles.
4. The greater the strength of the field, the greater the crowding of the lines.
5. They are closed curves.
6. The magnetic field is distorted by the presence of magnetic substances in the field. *Permeability* is the ability of a substance to concentrate the lines of force when placed in a magnetic field. Iron is more permeable than nickel. A permeable substance is magnetized when placed in a magnetic field. The more permeable the substance, the more readily it gets magnetized.

5 ** **Terrestrial Magnetism** In a rough way the earth acts as if it had a huge bar magnet inside it with the S-seeking pole in the northern hemisphere. (However, this pole is called the North Magnetic Pole.) Remember that it attracts the N-pole of the compass needle, and that the earth's magnetic poles do not coincide with the geographic poles, which are the points where the earth's axis of rotation intersects the surface of the earth.

The angle between the direction of the compass needle at a particular location and the direction of geographic north (or true north) is the *angle of declination* or compass variation of that location.

The *dipping needle* is a compass needle mounted on a horizontal axis. It measures the *angle of dip* or angle of inclination at a given place. At the magnetic poles the angle of dip is 90°.

## 6 Electromagnetism

In 1819 Oersted discovered that there is a magnetic field around a wire carrying current. The stronger the current, the stronger the magnetic field.

The magnetic field around a *long straight wire* carrying current is represented by lines of force which, in a plane perpendicular to the wire, are concentric circles around the conductor. To determine the direction of these magnetic lines of force, we may use the *Left Hand Rule*: Grasp the wire with the left hand so that the

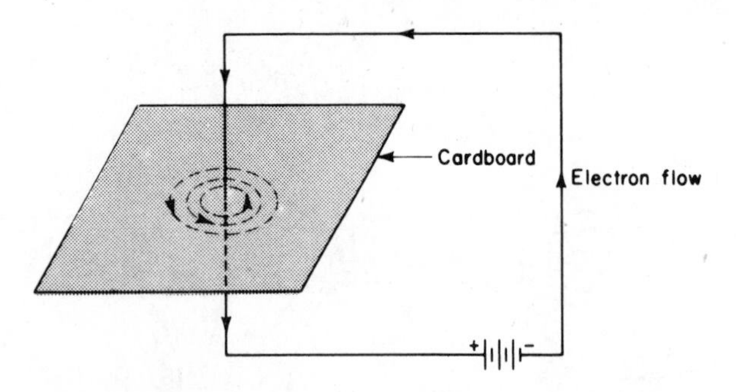


thumb will point in the direction of *electron flow*; the fingers will then circle the wire in the direction of the lines of force. The strength or intensity of the magnetic field is proportional to the current. No poles are produced.

## 7 **Circular Loop**

The magnetic field around a *circular loop* carrying current is such that the faces show polarity. We may apply the same left hand rule as above to see this. Notice that both the left half and the right half of the loop produce lines of force which have the same direction near the center of the loop. In this region the

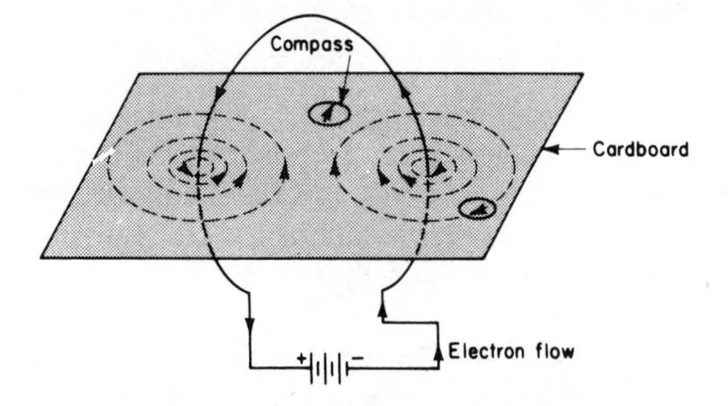


two fields add up to give a relatively strong field. The face of the loop from which the lines are directed is the north pole: in this diagram, it is the face away from us.

## 8 **Magnetic Field Produced by a Solenoid**

A solenoid is a long spiral coil carrying current. It produces a magnetic field similar to that of a bar magnet, and has an N-pole at one end, and an S-pole at the other end. The location of the poles can be found by the use of the Left Hand Rule for Solenoids and Electromagnets: grasp the coil with the left hand so that the *fingers* will circle the coil in the direction of *electron flow*. The extended thumb will then point to the N-pole and in the direction of the lines of force.

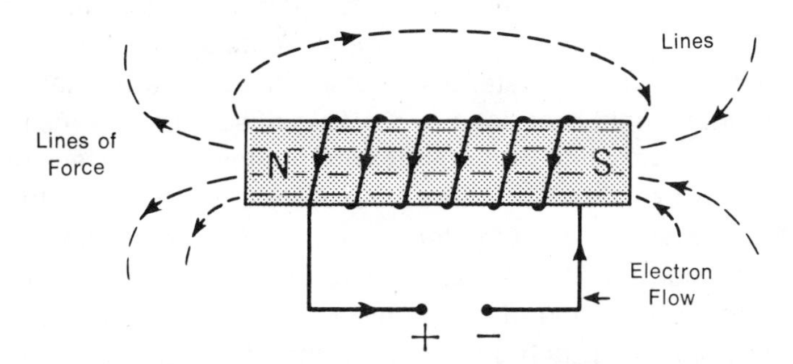


Note the properties of the lines of force of the solenoid. They are nearly parallel inside the coil. They are continuous. Outside the solenoid they are directed away from the North pole and towards the South pole.

The *strength* of a solenoid is proportional to the number of turns in the coil and the current. The effect of these two is combined in the concept of *ampere-turns*: it is the product of the number of turns and the number of amperes of current through the wire.

**FOR EXAMPLE:** A solenoid with 50 turns and 2 amperes going through it, has 100 ampere-turns. If the current is increased to 4 amperes, it has 200 ampere-turns, and the strength of its magnetic field is doubled.

An *electromagnet* is a solenoid with a permeable core such as soft iron. The *strength* of an electromagnet depends on its ampere-turns and the permeability of the core. Over a wide range, the strength of an electromagnet is proportional to the number of ampere-turns. This is affected by the nature of the core. Iron is said to have a high permeability: it helps to give a strong electromagnet. Air, wood, paper, etc. have a low permeability; their presence does not affect the strength of the solenoid.

The electromagnet is a temporary magnet. Its strength becomes practically zero when the current is turned off. The electromagnet has many practical applications. It is used in devices such as the electric motor, meter, generator, bell, telegraph, and telephone.

9 **Electromagnetism Due to Moving Charge** Considering what you already know about the electric current, it is probably no surprise to you that any moving electric charge produces a magnetic field; that, in fact, the magnetic field around a current-carrying wire is due to a charge moving in the wire. This was not at all obvious 100 years ago. This was first shown experimentally in 1876 by H. A. Rowland. He used a hard-rubber disk with metallic pieces inserted near its rim. He charged the metallic pieces electrically, and then set the disk into rapid rotation. A magnetic field resulted which was the same as that produced by a current in a circular loop. When the disk was stationary, there was no magnetic field. Of course, there was still an electric field.

10 **Theory of Magnetism** We have already reviewed some aspects of our present theory of the atom. Every atom has electrons moving around its nucleus. Each

orbiting electron produces a magnetic field. In addition, we believe that each electron spins around its own axis, similar to the way the earth rotates around its axis at the same time as it revolves around the sun. The spin of the electron also produces a magnetic field. The direction of spin of some electrons in an atom is opposite to that of other electrons and the magnetic fields cancel each other.

In *ferromagnetic* materials (magnetic materials like iron), each atom has a residual or uncancelled magnetic field produced by the spin of some electrons in the same direction. If we sprinkle very fine iron powder on the polished surface of a magnetic material, and then examine it under a microscope, we see that the powder forms definite patterns which are affected by the degree to which the material is magnetized. This, and other evidence, leads to the following theory of magnetism for magnetic materials. Atoms of magnetic materials are grouped in microscopic clusters called *domains*. (Each domain consists of about $10^{15}$ atoms and is about 0.001 inch long.) Within a domain, the magnetic fields of the atoms are in the same direction and add up so that each domain is like a little bar magnet. However, in a given piece of magnetic material the domains are normally arranged in random fashion, and they cancel each other's magnetic field. If the magnetic material is placed in an external magnetic field, some domains grow at the expense of others, and the direction of magnetization of the domains tends to rotate in the direction of the external magnetic field. If all the domains are magnetized in the same direction, the material is fully magnetized. If some of these effects on the domains persist after the external magnetic field is removed, we have produced a *permanent* magnet. (A material which becomes magnetized by being placed near another magnet is said to be magnetized by *induction*.)

**11 Force on a Current-Carrying Conductor** When electric charges move across a magnetic field, a force acts on the moving charges which is not present when the charges are stationary. In the diagram, imagine the vertical magnetic field between the North and South poles of a horseshoe magnet. A wire is suspended at right angles to this field. When the switch is closed permitting electrons to move through the circuit, the wire moves. If the direction of the current is reversed, the wire moves in the opposite direction. If the current is increased, the force increases. If we suspend the wire so that the current is parallel to the magnetic field, the magnetic field does not exert a force on the wire.

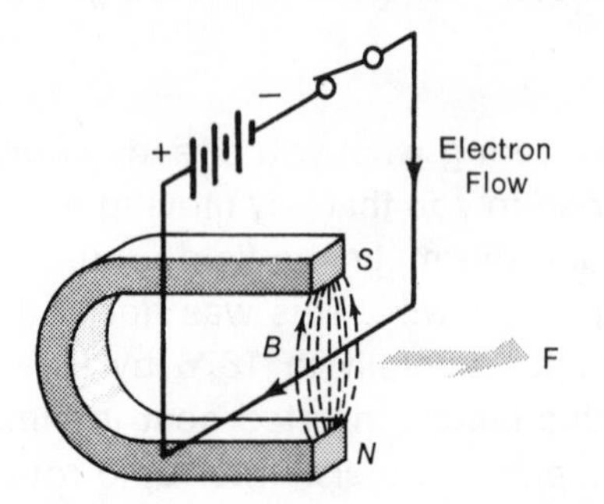


*The Force Is At Right Angles to Both Current And Field*

Further experimentation shows that the force on the current-carrying wire is greatest when the wire is perpendicular to the magnetic field. Strangely enough, the force is then perpendicular to both the direction of the current and the direction of the magnetic field. The magnitude of the force is proportional to the current and also to the strength of the magnetic field.

We can determine the direction of the force by thinking of an interaction between two magnetic fields: the magnetic field produced by the current and the magnetic field produced by the horseshoe magnet, the so-called external field. Imagine that the circle with the dot in the center represents a wire perpendicular to the paper with the electron current coming out towards the reader. The magnetic field produced by this current is represented by the dotted circles. We get the

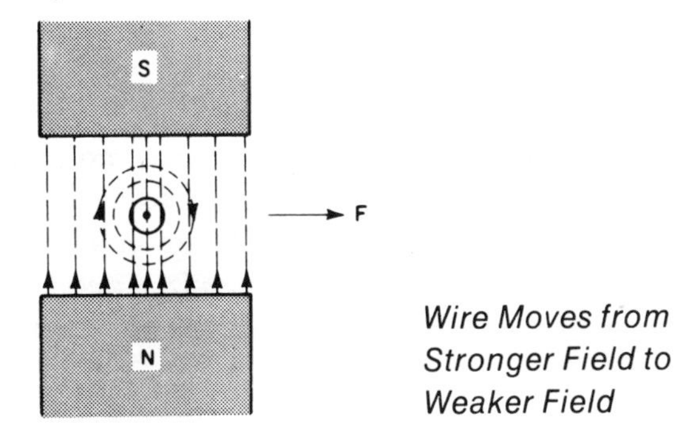

*Wire Moves from Stronger Field to Weaker Field*

direction of this magnetic field by using our Left Hand Rule for a wire. The external magnetic field is shown by straight lines from the N-pole to the S-pole. In this case, the two magnetic fields are in the same direction on the left side of the wire and reinforce each other. On the right side of this wire, they are in opposite directions and tend to cancel each other. (The actual field, which is a combination of the two fields, is not shown.) We think of the strong field on the left as pushing the wire towards the weaker field. Another rule has now been invented to fit these observed facts and which is quickly applied. It is called the *Open Left Hand Rule*: Hold the open left hand so that its fingers point in the direction of the magnetic field and the extended thumb in the direction of the electron motion. Then the palm faces in the direction of the force on the wire. (*Try this on the diagram on the previous page.*)

12 **Force Between Two Straight Parallel Conductors** If two straight parallel conductors carry current, a force is produced on both wires. If the current directions are the same, the force is one of attraction; the two wires tend to move towards each other. If the current directions are opposite, the force is one of repulsion.

We can again think of the force as being the result of the interaction of two magnetic fields. In the diagram, the two circles with dots in the center represent two wires with electron current directed towards the reader. The dotted curves represent the magnetic field produced by each wire. The direction of the magnetic

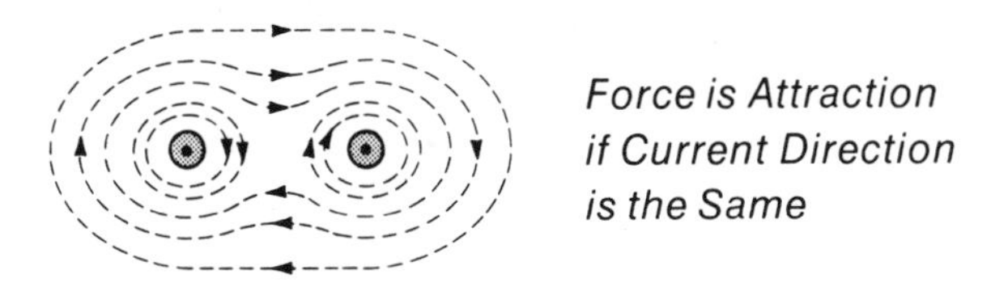
*Force is Attraction if Current Direction is the Same*

fields is given by the Left Hand Rule for a Wire. We can see that in the space between the wires, the two fields tend to cancel each other because they are in opposite directions. On the outside the fields reinforce each other. Again the wires tend to move from the stronger field to the weaker field.

The force increases if we increase either current. The force is proportional to the product of the two currents. If one current is tripled, the force becomes three times as great; if both currents are tripled, the force becomes nine times as great. The force is also inversely proportional to the distance between the wires

$$F \propto \frac{I_1 I_2}{d}$$

Note that this force does not vary inversely as the square of the distance. If the distance between the two wires is doubled, the force becomes one-half as great, not one-fourth.

13 **Force On a Moving Charge** We mentioned before that there is a force acting on a wire carrying current perpendicular to a magnetic field. This force is actually produced by the magnetic field on the moving charges. If the charges are in a wire, they transmit the force to the wire. If the moving charges are in an ionized gas, the force produced by the magnetic field changes the direction of motion of the ions.

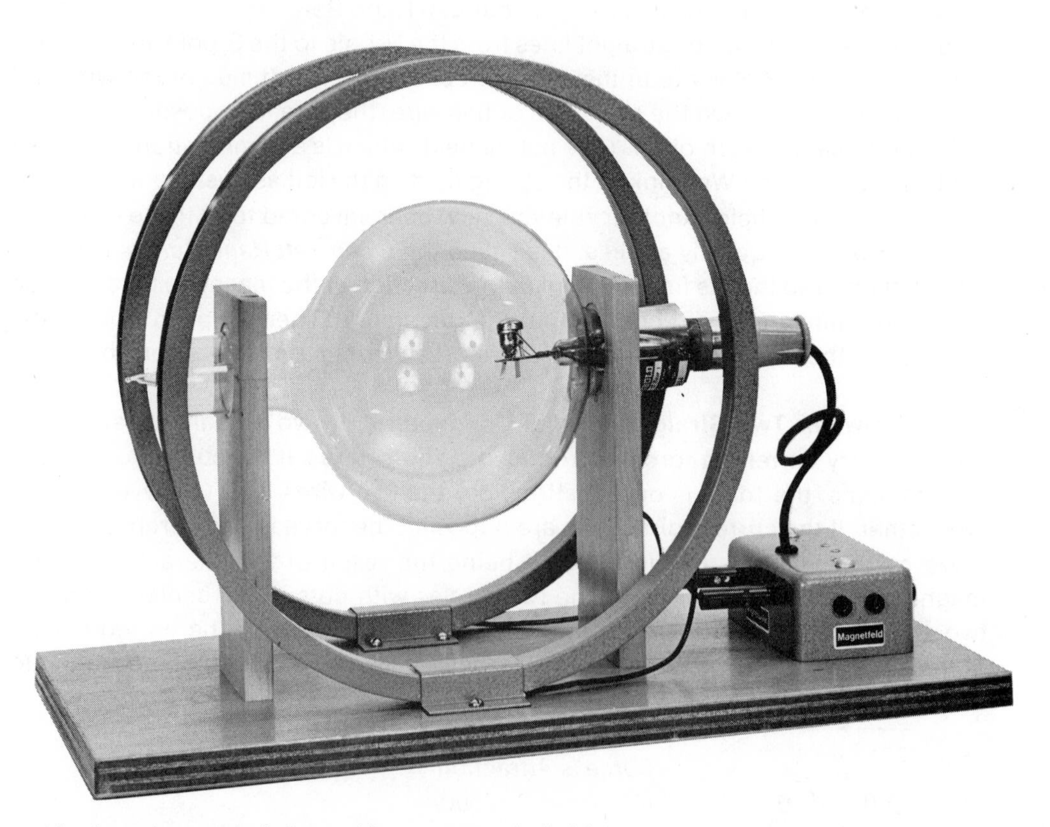


*Moving electrons deflected by a magnetic field*

The direction of the force on a charge moving in a magnetic field is perpendicular to the field and to the velocity. Since the force is perpendicular to the velocity, the speed of the charge does not change. If the direction of the charge originally was perpendicular to the magnetic field, the effect of the field is to make the charge move in a circular path. If the original direction was oblique to the magnetic field, the charge spirals along the magnetic field.

The force on a charge moving in a magnetic field is proportional to the magnitude of the charge, the strength of the magnetic field or its flux density, and the component of the charge's velocity perpendicular to the field.

**EXAMPLE:** If we double the speed with which we project an electron perpendicularly into a magnetic field, the force on the electron is doubled. This causes the electron to move in a circle of smaller radius.

* This effect is used to define magnetic flux density. We set the force equal to the product of the charge, velocity, and magnetic flux density, where the velocity and field are perpendicular to each other.

$F = qvB$, where $F$ is the force in newtons
$q$ is the charge in coulombs
$B$ is the magnetic flux density.

Several different units are used for $B$, the flux density. The above equation, and a similar one used for the force on a current-carrying wire, suggest the *newton per ampere-meter*. Since the flux is often represented graphically by flux lines, another unit used for $B$ is flux lines per meter$^2$, or *webers per meter*$^2$. Some people have suggested a new unit for $B$: the *Tesla*.

1 Tesla = 1 weber per meter$^2$.

** In the cgs system, the *gauss* is used as a unit for $B$, the flux density.

1 Tesla = $10^4$ gauss.

** When the magnetic field is at right angles to a *wire carrying current*, the force exerted on the wire by the field is equal to the product of the magnetic flux density, the current, and the length of wire in the field:

$F = BIL$.

14 **Force on Loop or Solenoid** We have shown that current through a loop or solenoid produces N- and S-poles. Therefore, if we place a current-carrying solenoid into a magnetic field, the solenoid will tend to turn the way a compass needle would, with the N-pole of the solenoid pointing in the direction of the magnetic field. This torque or turning effect of a magnetic field on a current-carrying solenoid is used in the design of many galvanometers and electric motors.

**The Galvanometer** A galvanometer measures the relative magnitude of small currents. The common ones are of the moving-coil type. This type has a coil which is free to rotate in a magnetic field provided by a permanent magnet. The current to be detected and measured goes through this coil, and produces a

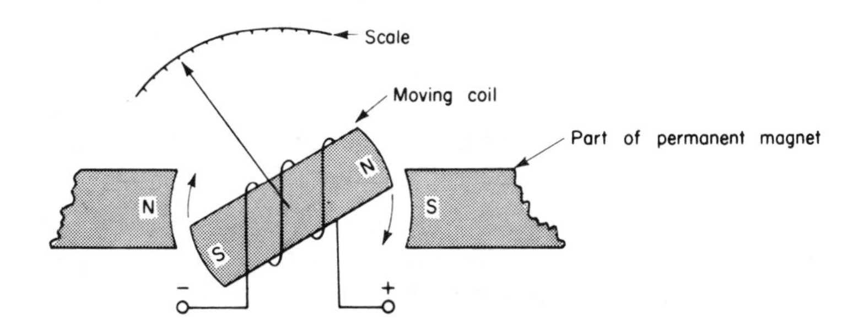

torque which tends to rotate the coil. A pointer attached to the coil rotates with the coil, and moves across a scale. Springs are used to keep the coil's rotation proportional to the current. When the current stops, the springs return the coil to the zero position. The electrical resistance of the galvanometer coil is usually small, about 50 ohms. The coil can usually carry only small currents, a few milliamperes. Galvanometers are modified for special purposes: an *ammeter* is used to measure currents, and a *voltmeter* is used to measure potential difference or voltage.

15
**

**The Ammeter** The ammeter is an instrument for measuring current; it is calibrated to give the actual magnitude of the current. The ammeter usually consists of a galvanometer with a low resistance in parallel with the moving-coil. This resistance is called a *shunt*. The ammeter itself is connected in series with the device whose current is to be measured. All the current to be measured enters the ammeter. Internally the current divides. A part of the current goes through the coil, the rest goes through the shunt.

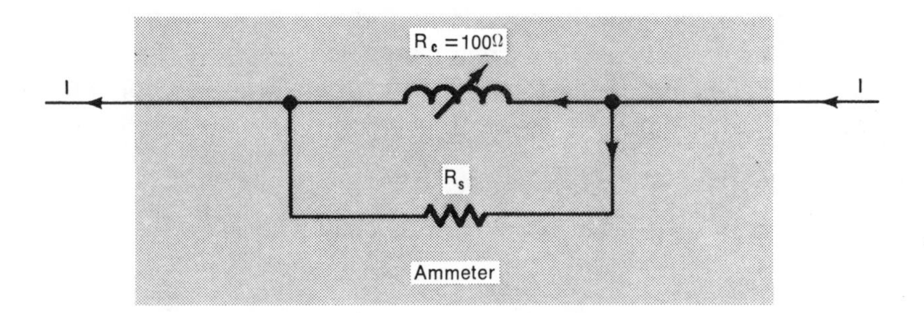


We can calculate the required resistance of the shunt ($R_s$), if we apply the rules for a parallel circuit. The shunt is in parallel with the coil ($R_c$), and therefore the voltage drops are equal:

$$I_s R_s = I_c R_c$$

If the coil has a resistance of 100 ohms and gives full scale deflection for 10 milliamperes, we can, for example, convert it to a 0 – 1 ampere meter, that is, a meter that gives full scale deflection for 1 ampere. Note, that at full scale deflection, the current through the coil is 10 milliamperes or 0.01 ampere. The other 0.99 ampere must then go through the shunt.

$$0.99 R_s = 0.01 \times 100$$

and $R_s$ is approximately 1 ohm.

16
**

**The Voltmeter** A voltmeter is an instrument calibrated to measure the potential difference connected to its terminals. If we want to measure the potential difference across a lamp, we connect a voltmeter in parallel with the lamp. A voltmeter usually consists of a galvanometer with a high resistance connected in series with it. This resistance is often called a *multiplier*.

We can calculate the value of the multiplier by applying the rules for a series circuit. Suppose we want to convert the above galvanometer to a voltmeter with full scale deflection at 100 volts. In other words, its range is to be 0 – 100 volts. At full scale deflection the current through both the coil and the multiplier ($R_m$) has to be 0.01 ampere. The $IR$-drop through both together is 100 volts.

$I_T R_T = 100$
$0.01\, R_T = 100$
and $R_T = 10{,}000$ ohms.

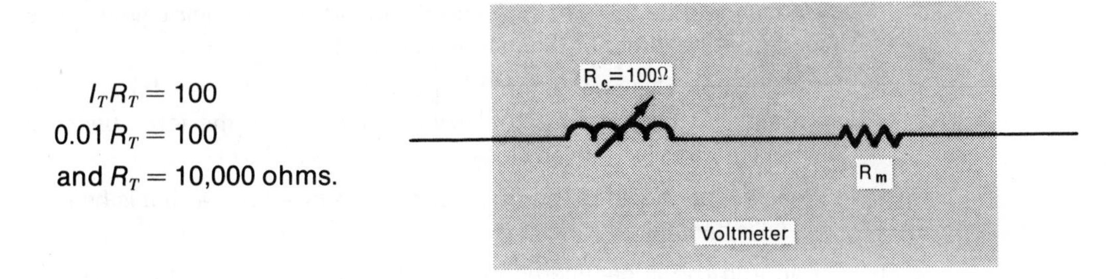


The coil supplies 100 ohms of this resistance. Therefore the required multiplier resistance is 9,900 ohms. The range of the voltmeter can be changed by changing multipliers.

## Questions and Problems

1. Use the concept of induced magnetism to explain why a magnet attracts an ordinary piece of iron.
2. By means of about 10 lines of force, sketch the magnetic field *a.* around a bar magnet *b.* between two south poles *c.* between two north poles *d.* between a north and a south pole
3. Use about 10 lines of force to represent the magnetic field around an electromagnet. Indicate the direction of the current.
4. Explain why breaking a bar magnet crosswise results in the production of new poles.
5. What should be the direction of current in two parallel wires so that they will tend to cancel the magnetic field produced between them? Explain.
6. Two parallel wires carry current. Name three different ways in which the electromagnetic force between them can be quadrupled.
7. A current-carrying wire is perpendicular to a magnetic field. *a.* If the current is doubled, what is the resulting change in the force on the wire? *b.* If the magnetic field strength only is doubled, what is the resulting change in the force on the wire?
8. In question 7, how can you determine the direction of the force on the wire?
9. In question 7, what would be the effect of doubling the current, if the wire were parallel to the magnetic field?
10. A beam of electrons is projected with a speed of $1.2 \times 10^7$ m/sec at right angles to a uniform magnetic field of $2.0 \times 10^{-2}$ Nt/amp.meter. What is the force exerted on the electron by the field?
11. If the beam in problem 10 consisted of protons moving with the same speed, *a.* what would be the force on each proton *b.* what would be the radius of the resulting path?
12. * A beam of electrons is projected horizontally into a magnetic field directed vertically downward. If the strength of the magnetic field is 1.2 weber/m$^2$, and the electrons move with a speed of $3.0 \times 10^6$ m/sec, what is the magnitude and direction of the resulting force on the electrons?

13. What should be the direction and magnitude of an electric field, if it is to cancel the effect of the magnetic field in problem 12?

14. Two long parallel conductors are 1.0 meter apart when the force due to currents in them is $1.4 \times 10^{-6}$ newton. What does the force become when the wires are moved closer to a distance of 0.5 m?

15. What will be the effect on the strength of a magnet, *a.* if it is heated, *b.* if it is hammered?

16. If an iron rod is hammered in a certain way, it is magnetized slightly. Explain this.

## Test Questions

1. A compass needle should be made of 1. steel 2. silver 3. soft iron 4. copper.

2. One pole of a magnet brought near a piece of iron attracts it. We can be sure that the iron 1. is a magnet 2. is magnetic 3. is electrified 4. has more than two poles.

3. The angle of dip at the North Magnetic Pole is, in degrees, 1. 0 2. 90 3. 60 4. 45.

4. An iron post stuck into the ground eventually becomes magnetized by the earth's magnetic field. It is said to be magnetized by 1. induction 2. reduction 3. conduction 4. production.

5. Lines of force produced by a magnet are directed away from the ........... pole.

6. When the distance between two magnetic poles is increased, the force between them 1. increases 2. decreases 3. remains the same.

7. The following electromagnets have the *same core*. Which one is the strongest? 1. 100 turns and 3 amperes 2. 200 turns and 2 amperes 3. 50 turns and 7 amperes.

8. Electrons are projected perpendicularly into a magnetic field. As the speed with which the electrons are projected increases, the force on the electron 1. increases 2. decreases 3. remains the same.

9. Electrons are projected perpendicularly into a magnetic field. As the flux density used increases, the force on the electron 1. increases 2. decreases 3. remains the same.

10. The magnetic force on a proton moving at right angles to a magnetic field, compared to an electron moving there at the same speed, is 1. the same 2. greater 3. less.

11. As the charge on particles projected at a certain speed into a magnetic field is increased, the force on the particles 1. decreases 2. increases 3. remains the same.

12. Magnetic lines of force are most readily concentrated in 1. aluminum 2. iron 3. silver 4. gold.

13. Permanent magnets are made of 1. aluminum 2. copper 3. alnico 4. rubber.

14. If the N-pole of a compass needle is repelled by the unmarked end of a magnet, that end of the magnet is a(n) ........... pole.

15. A compass needle is pivoted under a horizontal wire carrying electron current from east to west. The compass needle will point 1. east 2. west 3. north 4. south.

16. Objects lifted by an electromagnet become magnetized by 1. the current through the objects 2. induction 3. the earth 4. static charge.

17. Two parallel wires carry current in the same direction. The magnetic field which they produce midway between them is 1. zero 2. not necessarily zero, but less than it would be if the current in one of the wires were zero 3. greater than it would be if the current in one of them were zero.

18. Two parallel wires carry current. If the current in one of them is doubled, the force on the other wire is 1. doubled 2. quadrupled 3. not affected.

19. Two parallel wires carry current. If the current in both wires is doubled, the force on one of the wires is 1. doubled 2. quadrupled 3. multiplied by 8 4. multiplied by 16.

20. If an ammeter is to be used to measure the current in a device, it should be connected in ........... with the device.

21. Two long parallel wires are shown carrying currents in opposite direc-

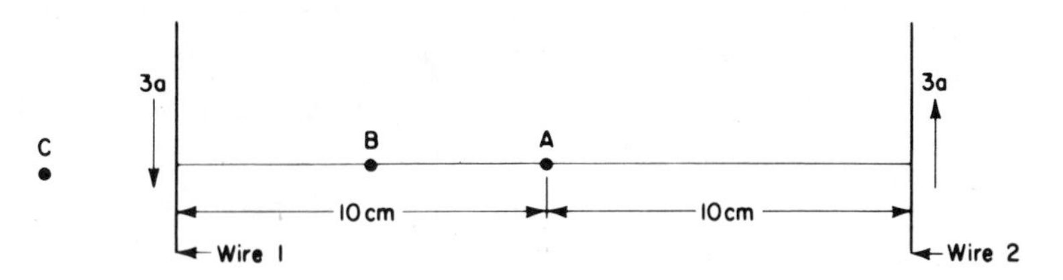


tions. The direction of the currents is that of electron flow. At point A the direction of the resulting magnetic field is 1. into the paper 2. out of the paper 3. towards the top of the paper 4. towards the bottom of the paper 5. none of the above.

22. Which of the above 5 choices is correct for point B in the diagram?

23. Which of the choices in question 21 are correct for point C in the diagram to the left of wire 1?

24. If the current in wire 2 increases a little, the intensity of the magnetic field at C 1. increases 2. decreases 3. remains the same.

25. If the current in wire 2 increases, the force on wire 1 1. increases 2. decreases 3. remains the same.

26. As a result of the currents in the two wires, the force on wire 1 as compared with the force on wire 2 is 1. greater 2. smaller 3. numerically the same 4. dependent on which wire has the greater current.

# Topic Questions

1.1 What is the difference between a magnetic substance and a magnetized substance?

2.1 How does the force between two magnetic poles depend on the distance between them?

3.1 Where can we find a magnetic field?

3.2 If we represent a magnetic field by lines of force, how do we differentiate between a strong and a weak field?

4.1 ** What is the location of the earth's *a.* geographic poles *b.* magnetic poles?

5.1 When a straight wire carries current, *a.* what is the shape of the magnetic field that is produced by the current; *b.* how can the direction of this magnetic field be determined?

6.1 If a circular wire loop carries current, why is there a relatively strong magnetic field in the center of the loop?

7.1 How can we find the North pole of a solenoid if we know the direction of electron flow in it?

7.2 What determines the strength of a solenoid?

8.1 Under what conditions does an electric charge produce a magnetic field?

9.1 Since a spinning electron is like a magnet, why aren't all atoms like magnets?

9.2 What are some important characteristics of magnetic domains?

10.1 A current-carrying wire is suspended in a magnetic field. *a.* What determines the magnitude of the resulting force on the wire? *b.* How can the direction of this force be predicted?

11.1 What factors affect the force between two parallel wires carrying current?

12.1 An electric charge moves in a magnetic field. What factors affect the resulting force?

13.1 Why does a solenoid tend to turn if it is placed in a magnetic field?

14.1 ** What is an ammeter?

15.1 ** What is a voltmeter?

# 15 Electromagnetic Induction

**1 Discovery** In the battery, chemical energy is converted to electrical energy. At the present time, the most important method of obtaining electrical energy is its conversion from mechanical energy. How is this done?

In 1819 Oersted discovered that an electric current produces a magnetic field in the space around it. Efforts started immediately to produce an electric current with a magnetic field. In 1831 this was finally achieved independently by Michael Faraday in England and by Joseph Henry in the United States.

*Michael Faraday*
*Niels Bohr Library A.I.P.*

If a wire moves so that it cuts across a magnetic field, an emf is induced in the wire. Actually it doesn't matter whether the wire or the magnetic field moves. An *induced emf* is produced in a wire whenever there is relative motion between the wire and a magnetic field. If the wire is part of a complete conducting path, the induced potential difference produces a current in this circuit in accordance with *Ohm's Law*. This current is sometimes referred to as *induced current* to stress its method of production. In its nature it is no different from other electric currents.

**2 Factors Affecting Magnitude of Induced EMF** The magnitude of the induced emf may be increased by increasing the length of wire moving across the magnetic field, by increasing the strength or intensity of the magnetic field, and by increasing the relative speed of motion between the wire and the magnetic field.

No emf is induced if the wire is moving parallel to the direction of the magnetic field. Maximum emf is induced when the wire moves at right angles to the direction of the magnetic field.

More precisely, the magnitude of the induced emf ($E$) is proportional to the flux density ($B$), the length of wire ($L$) in the magnetic field, and the speed ($v$) with which the wire moves perpendicularly to the magnetic field.

*$E = BLv$ where $E$ is in volts
$B$ is in webers per $m^2$
$L$ in meters
$v$ in meters per sec

* **EXAMPLE:** An emf of 0.003 volt is induced in a wire 0.6 meter long when it moves at right angles to a uniform magnetic field with a speed of 5 meters per sec. What is the flux density?

$$E = BLv;$$
$$0.003\,V = B \times 0.6\text{ m} \times 5\text{ m/sec}$$
$$B = (0.001\text{ weber/m}^2) = 1 \times 10^{-3}\text{ weber/m}^2$$

3 **Direction of the Induced Current** The direction of the induced current can be figured out by the use of *Lenz's Law*. This law is a special case of the law of Conservation of Energy.

We know from our study of electric circuits that energy is required to move the electric charges through the resistance of the circuit. When we use a battery to supply the emf of the circuit, the energy to move the charges, the energy for the current in the circuit, is supplied by a chemical change in the battery. When we have an induced current produced by moving a wire across a magnetic field, the energy for the current comes from the work done in moving the wire across the magnetic field.

Why must work be done to move a wire across a magnetic field? The induced current produces a magnetic field; this was Oersted's discovery. The Russian physicist Lenz discovered that this magnetic field interacts with the external magnetic field to hinder the motion. Therefore a force must be exerted to move the wire when a current is induced in it. The product of this force and the distance through which the wire is moved, is the work done on the wire and supplies the energy for the current in the circuit. Mechanical energy is converted to electrical energy.

**Lenz's Law** *Lenz's Law* may be stated as follows: the direction of the induced current is such as to produce a magnetic field which will oppose the motion that produced the current.

**EXAMPLE:** If a permanent magnet is moved away from a stationary coil, an emf is induced in the coil. In the diagram shown, the motion of the magnet is opposed if

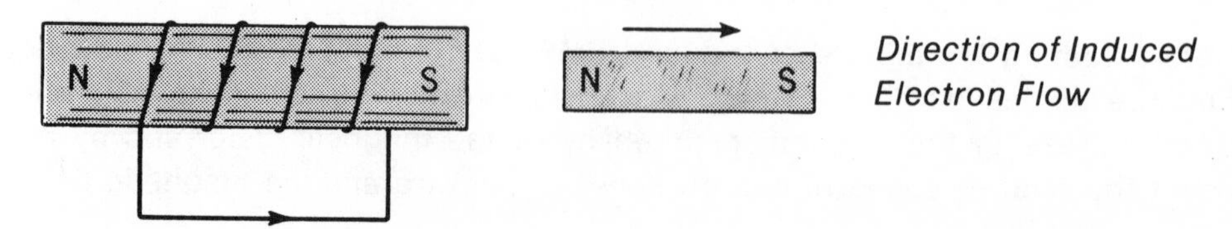

the induced current produces an S-pole to attract the retreating N-pole. This will happen if the induced current has the direction shown. We use our Left-hand rule for a coil to determine this direction: we grasp the coil with the left hand; when the outstretched thumb points in the direction of the N-pole, the encircling fingers point in the direction of electron flow.

If the permanent magnet is moved towards the coil, the direction of the induced current is reversed. The right end of the above coil has to become an N-pole to oppose the approach of the magnet.

If a single wire moves across a magnetic field, the direction of the induced current is indicated by a galvanometer connected to the ends of the wire. We can predict the direction of the current in the following way. In the diagram, suppose the wire is pulled downward across the magnetic field. The magnetic field is represented by lines of force going from the N-pole to the S-pole outside the magnet. The wire is imagined as cutting across these lines of force. With this picture in

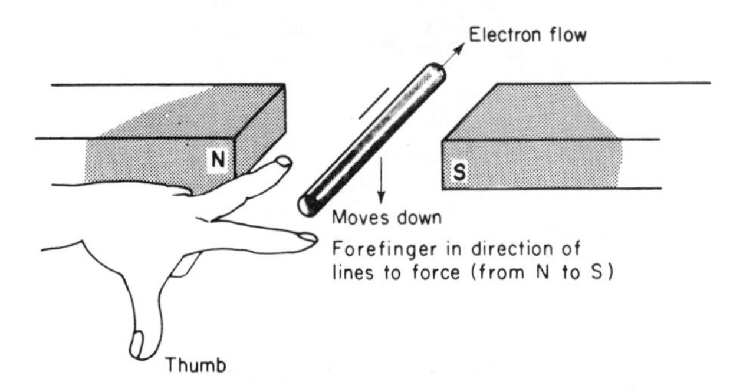


mind, we sometimes say that the induced emf is proportional to the speed with which the magnetic lines of force are *cut*. A special rule invented to predict the direction of the induced current is the Left-Hand Three Finger Rule: the thumb, forefinger, and center finger are kept at right angles to each other; point the Forefinger in the direction of the lines of Force, the thuMb in the direction of Motion of the wire, and the Centerfinger will then point in the direction of the electron Current. If the wire is moved upward, the direction of the current is reversed.

We may also apply here the *Open Left Hand Rule* described in section 11 of the previous chapter. In this case we interchange the function of the thumb and the palm. The fingers point in the direction of the lines of force, the *thumb* points in the direction in which the wire is moved, and *the palm faces in the direction of the induced current.*

4 **Cause of Induced EMF** We can explain the production of an induced emf by recalling something we stated in the previous chapter. When a charge moves at right angles to a magnetic field, the field exerts a force on the charge. Now, when we move the wire across a magnetic field, we are really moving the electrons and other particles in the wire across the magnetic field. The field will exert a force on these charges, and some of these electrons are free to move in the direction of the force. This results in the induced current, if the wire is part of a complete circuit. If the wire is not part of a complete circuit, excess electrons pile up at one end of the wire, making it negative. The other end will have a deficiency of electrons and be positive.

5 **Electric Generators** The electric generator or *dynamo* converts mechanical energy into electrical energy. Usually a coil is made to rotate in a magnetic field,

and an emf is induced in the coil. The source of the energy for turning the coil may be a waterfall, steam under pressure, etc.

In an actual generator the rotating coil has many turns of wire on a magnetic core. The rotating part is called an *armature*. Only one loop is shown in the simplified diagram. The stationary magnet provides a magnetic field which is cut by the rotating coil. This magnet is called a *field magnet*. Because the rotating coil cuts across magnetic lines of force, an emf is induced in it. This coil is called an *armature coil*. When it is part of a complete circuit, electric current will be present in the circuit, and electric energy can be taken from the armature coil.

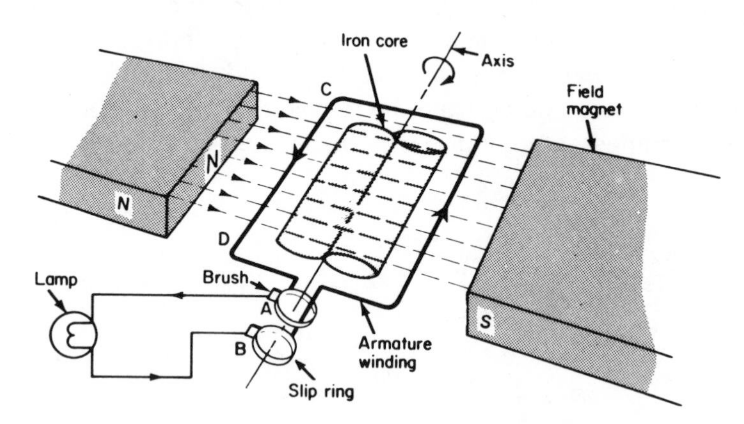


We can use the Three-finger Rule to determine the direction of the current in the coil. We note that at the instant shown, the wire on the left is moving up across the magnetic field while the wire on the right is moving down. Therefore the direction of the induced current in the two wires is opposite to each other, as shown. Looking down on the armature winding, the result is a counter-clockwise current in the loop.

In order that the wires from the rotating armature to the stationary lamp do not get entangled, we must use slip rings and brushes. The *slip* rings are on the same shaft as the armature and rotate with it. They are insulated from each other. The brushes are stationary and make wiping electrical contact with the slip rings. Current can therefore go from a rotating slip ring to a stationary lamp through a stationary brush. Notice that current through the lamp goes from negative brush A to positive brush B.

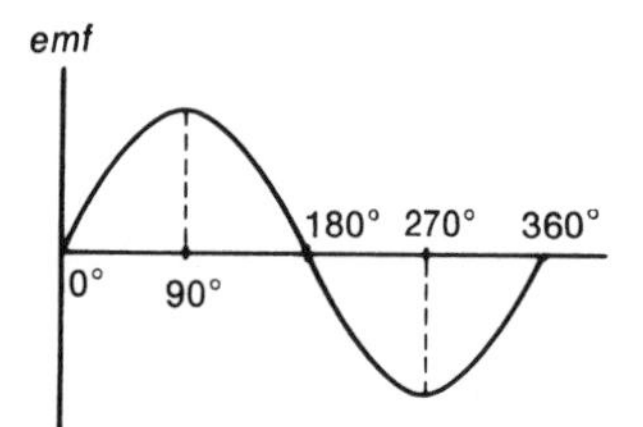


*Graph of Alternating EMF Induced in Armature*

However, the emf induced in the coil is not constant. It is maximum in the position shown. One-fourth of a rotation later, the emf will be zero, because the armature wires for an instant will not be cutting across the field, but will be moving parallel to it. Another fourth of a rotation later the wires will again be moving at right angles to the field, and the emf will be as large as before. But the wire which before was moving up across the field will now be on the right and moving down across the field. Therefore the direction of the induced current and emf reverse every half rotation, or revolution, as it is frequently called. This is shown in the graph below, where one complete rotation is represented by 360°. Current whose direction is constantly reversing is known as *alternating current*. In the generator described, using slip rings, the current in the external circuit (through the lamp), is also alternating.

One complete back-and-forth variation is known as a *cycle*. In the simple AC generator shown, there would be one cycle for each revolution of the armature. The above graph shows the one cycle of emf produced during this revolution. In the United States, alternating current is usually supplied to homes; its frequency is 60 cycles per second (60 cps).

Instead of using the three-finger rule to determine the direction of the current in the armature loop, we could have used *Lenz's Law*.

**FOR EXAMPLE:** Consider wire CD of the loop. The direction of the induced current in it should produce a magnetic field which opposes the upward motion of the wire. If the direction of the current is as shown in the diagram, from C to D, our left-hand rule tells us that the current produces a magnetic field which, above the wire, is in the same direction as the external magnetic field, and below the wire is in the opposite direction. Therefore the magnetic field above the wire is strengthened and tends to push the wire down as we try to push the wire up. This indicates that we have shown the correct direction of the induced current. You might check this with the wire on the right side of the loop. The *Conservation of Energy Principle*, of which *Lenz's Law* is a special case, indicates that as more current is drawn from the generator, more energy has to be used to turn the armature.

Sometimes the magnitude of the induced emf is described as being proportional to the rate of change in magnetic flux linking the loop.

**FOR EXAMPLE:** In the above diagram of the generator, the plane of the loop is shown parallel to the field. At that instant of its rotation, the emf induced in it is maximum, corresponding to the 90°-point on the graph. One-fourth of a rotation later, the plane of the loop is perpendicular to the field. The loop encircles or surrounds the maximum flux in this position; it is now linked with much flux, while before it was linked with no flux. At this instant, the induced emf is zero.

6 **Electromagnetic Radiation** We have seen that an electric charge is surrounded by an electric field. If the charge moves, we have a moving electric field. But a moving charge is an electric current which is surrounded by a magnetic field. In other words, a moving electric field produces a magnetic field.

Similarly, a moving magnetic field exerts a force on an electric charge just as a stationary electric field does. It is said that a moving magnetic field produces an electric field.

In 1864 the brilliant James C. Maxwell deduced with the aid of mathematics that whenever an electric charge is accelerated, a combination of electric and magnetic fields is generated which travels away from the charge with the speed of light. We can think of an interchange of energy between the electric and magnetic fields.

In 1888 Hertz verified Maxwell's predictions by discovering radio waves.

**FOR EXAMPLE:** If we connect a long wire to the negative terminal of an AC generator and an equally long wire to the positive terminal, electrons will oscillate back and forth along these wires. When a terminal is negative it repels electrons out towards the end of the wire; when it is positive it attracts electrons. The two wires become an antenna and radiate electromagnetic waves of the same frequency as the generator. To radiate effectively, the antenna must be about half of the wavelength radiated. This is not practical for 60 cycles per second, but is practical for radio, television, and radar. Light and X-rays are also electromagnetic radiation.

In the electromagnetic radiation the electric and magnetic fields are constantly changing in amplitude, and reverse direction at the same frequency as the oscillating charges in the antenna. This change in amplitude and direction of the fields is similar to the change in amplitude and frequency of a rope vibrating transversely. We think of an *electromagnetic wave* as a transverse wave in which the electric and magnetic fields do the vibrating: there is a wave-like change in the electric and magnetic fields.

If the motion of the electric field is restricted to one plane, the wave is said to be *plane polarized.* Light can be plane polarized by letting it pass through a polaroid filter. At the same time the motion of the magnetic field is restricted to a plane at right angles to the plane of the electric field. This can be represented by the next diagram.

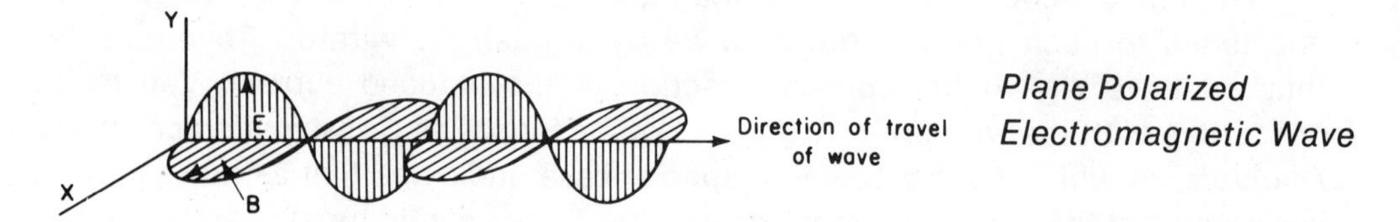

*Plane Polarized Electromagnetic Wave*

**7** ** **The Direct-current Generator** A DC or direct-current generator can be constructed with a slight modification of the above AC generator. We use the same field magnet and armature coil, but instead of slip rings we use a *commutator*, a metallic ring split into two insulated parts of equal size. (In more complicated DC generators the ring is split into several commutator segments.) Each end of the

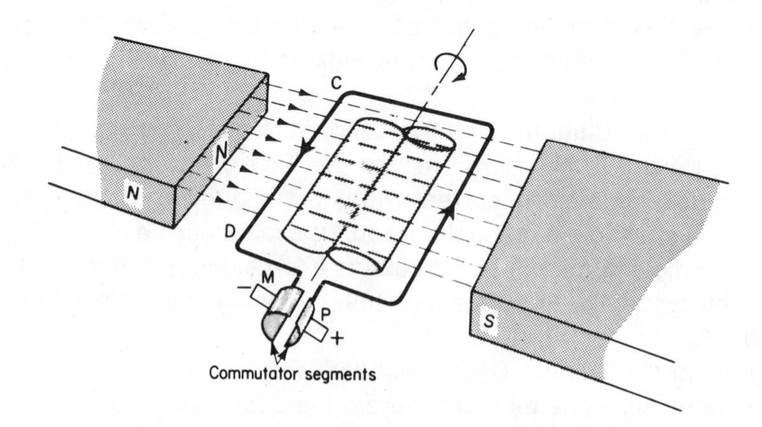


armature coil is connected to a commutator segment. As the armature rotates, the stationary brushes alternately make contact with one segment and then with the other. The brushes are set so that this change in contact occurs just when the direction of the current in the armature coil changes. As a result, the current in the external circuit is always in one direction (DC). With a simple DC generator, however, the current is not steady: it is *pulsating or fluctuating DC.* The fluctuation

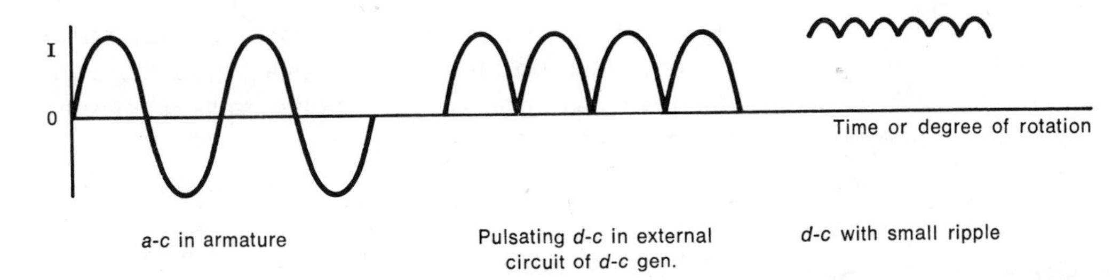


of the DC in the external circuit is reduced in more complicated DC generators by the use of more commutator segments and more armature coils. The residual fluctation is known as *ripple*.

Notice that there is AC in the armature of both the AC and DC generators. The function of the commutator is described as changing AC to DC. Also note that the polarity of the field magnet does not change. This means that if it is an electromagnet, DC must be used. A *magneto* is a dynamo whose field magnet is a permanent magnet.

**8** ** **The Electric Motor** The DC motor is similar to the DC generator except that electrical energy is supplied to the motor which converts it to mechanical energy. The motor is connected to a source of emf, and as a result the armature of the motor turns, gaining kinetic energy.

Look again at the diagram of the DC generator. Imagine a battery connected to the brushes, the negative terminal to brush *M*, the positive terminal to brush *P*. Why will the armature rotate? First note that now the direction of the current is opposite from what it was for the generator, since the direction of electron flow

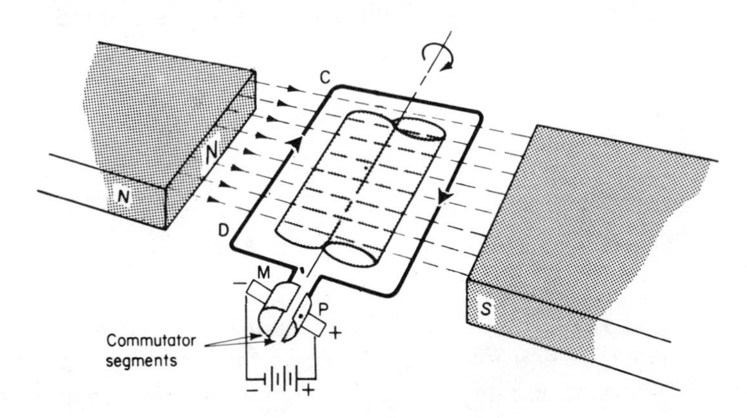


outside the battery is from negative to positive. Using our Left-hand rule on wire *CD* with the thumb pointing in the direction of the current, we notice that the magnetic field produced by the current reinforces the external magnetic field under the wire and weakens it above. This forces wire *CD* up; the armature rotates clockwise. The commutators reverse the direction of the current in the armature at the right instant, to keep the armature rotating in the same direction.

However, because the coil rotates in a magnetic field, an emf is induced in the armature. According to *Lenz's Law*, this emf will be in such a direction as to oppose the rotation. Therefore this induced emf is known as a *counter-emf* or *back emf*. This counter emf tends to reduce the current in the armature. Because the counter emf is greatest when the armature is rotating fastest, the current in the armature is less when the motor is running at full speed than when it is starting. (This is the reason for the momentary dimming of the lamps in a house when the motor of the electric refrigerator starts. The large starting current produces a large voltage drop in the "line," the wires leading up to the refrigerator. The momentary lower voltage available for the lamps produces the dimmer light.)

## 9 Electron Beams

We have given characteristics of the electron which indicate that it is a particle, that it is the elementary negative charge of electricity. It gives us the concept that electricity has a *graininess*. The work of J. J. Thomson in 1897 did a great deal towards giving us this picture of the electron. His experiment on cathode rays is similar to the one described below, but the exact details are not important here.

**Cathode Rays** Under certain conditions there is a flow of electricity through a gas.

**EXAMPLE:** If we have a glass tube filled with a gas at low pressure, with a metal plate at each end, the gas can be made to glow as in a neon tube if a sufficiently high voltage is connected to the two plates. The plate connected to the negative terminal of the voltage source is called the *cathode*, the plate connected to the positive terminal is the *anode*. If the pressure of the gas is low enough, the gas itself does not glow, but the walls of the tube start to fluoresce. This seemed to be due to a radiation emitted by the cathode and was called *cathode rays*. J. J. Thomson showed that the cathode ray is a stream of electrons emitted by the cathode.

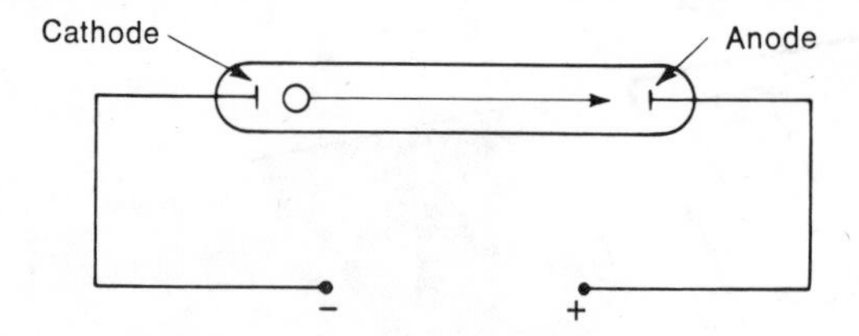


In modern cathode ray tubes, the picture tubes of TV receivers and those that are used in oscilloscopes, electrons are emitted by a cathode which is heated to a high temperature. This emission of electrons by incandescent objects is known as *thermionic emission*.

**10 Control of Electron Beams** In the cathode ray tube the electrons move quickly through what is a very good vacuum. The electrons may be deflected from their straight-line path by either electric fields or magnetic fields. In the diagram below, the cathode ray passes between plates *A* and *B*. Plate *A* has been charged negatively and plate *B* positively. Therefore there is an electric field between *A* and *B*, and the electrons are deflected towards plate *B*. The path in the electric field is curved; outside the electric field, the electrons continue along a straight line because of their inertia, if the gravitational effect is negligible. Notice that in the electric field, the electron beam is deflected by a force which is parallel to the field and directed towards the positive plate. If the screen is coated with a fluorescent material, the point of impact will show up as a bright spot.

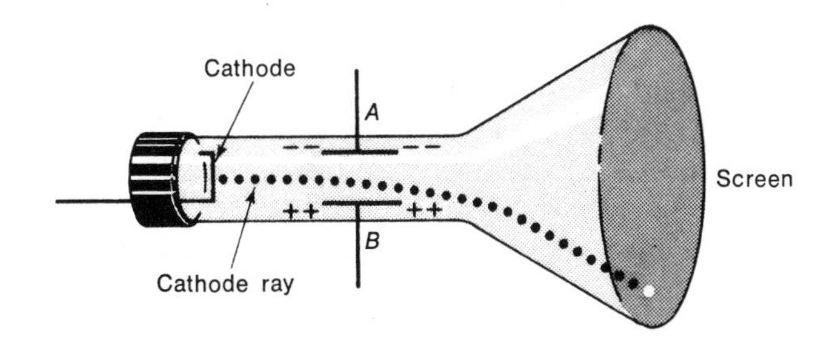


*Deflection by Electric Field*

In the experiment performed by J. J. Thomson, the cathode rays were deflected by a magnetic field. In the diagram below, this magnetic field is represented by the circle marked *M*. The magnetic field is perpendicular to the path of the electron beam; imagine it perpendicular to the paper. The beam leaves the cathode and goes through the hole in the cylindrical anode marked *P*. In the magnetic field the beam is deflected by a force which is perpendicular to both the beam and the field.

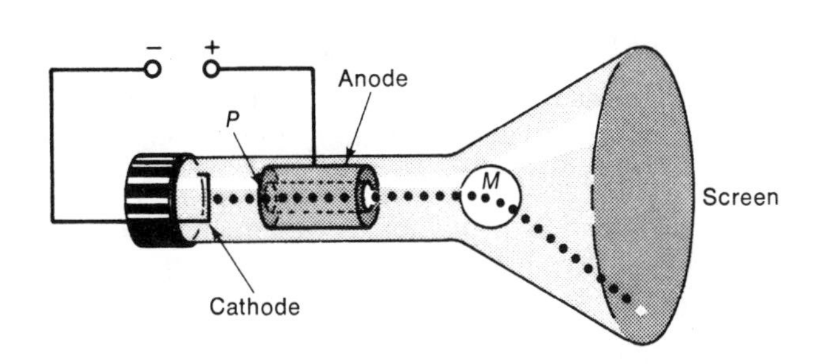


*Deflection by Magnetic Field*

As a result, in the magnetic field the path of the electrons is circular. Outside it continues along a straight line. The point of impact on the fluorescent screen again shows up as a bright spot.

On the basis of his measurements of deflections of the cathode rays, Thomson showed that the rays consist of streams of electrons with a definite ratio of charge to mass. When the charge of these electrons is determined by methods such as *Millikan's Oil Drop Experiment*, the mass of the electron can be calculated. Of course, as we know now, no matter what material is used as a source of the cathode rays, all electrons have the same charge and the same mass.

It is not necessary to memorize the following derivation, but it is instructive, and a good review, to look at the steps.

Let $e$ be the charge of the electron and $m$ its mass. If $V$ is the difference of potential applied between the cathode and the anode, the kinetic energy which the electron gets as a result of being accelerated by the difference in potential is given by:

1. $\frac{1}{2}mv^2 = Ve$, where $v$ is the speed gained by the electron.

When the electron moves in the magnetic field *M*, the magnetic force which acts on it is perpendicular to the magnetic field, and has a value:

2. $F = Bev$, where $B$ is the flux density.

This magnetic force is the centripetal force which makes the electron move in a circular path while it is in the magnetic field:

3. $F = mv^2/r$, where $r$ is the radius of this circular path.

If you combine equations 1, 2, and 3, you can get the expression for the ratio of the charge of the electron to its mass:

$$\frac{e}{m} = \frac{2V}{B^2r^2}$$

## Questions and Problems

1. Name three factors which affect the magnitude of an induced emf.
2. Name two ways in which the direction of an induced emf can be changed.
3. Name three ways in which the magnitude of an induced emf may be doubled.
4. What unit is equivalent to a weber per meter$^2$?
5. A coil has an emf of 0.4 volt induced in it when it moves in a magnetic field whose flux density is 1000 webers per square meter. If the flux density is doubled without changing the speed of motion, what will be the induced emf (when the field has become steady)?
6. Two coils are wound on the same iron core. One coil is connected to a galvanometer, the other to a battery. As the battery is disconnected, the galvanometer reads momentarily. Explain.
7. A wire 20 cm long is moved so as to cut across a magnetic field. If the wire moves with a speed of 3.0 m/sec at right angles to the flux, and an emf of 0.0025 volt is induced, what is the flux density?
8. If the speed of the wire in question 7 were changed to 6.0 m/sec, what would be the induced emf ?
9. An emf of 0.002 volt is induced in a wire when it moves at right angles to a uniform magnetic field with a speed of 4.0 m/sec. Calculate the flux density, if the wire is 60 cm long.
10. In question 9, what would be the induced emf if the length of wire were only 30 cm?
11. In question 9, what would be the effect if part of the wire were outside the magnetic field?
12. Calculate the emf induced in a wire 25 cm long which moves perpendicularly across a magnetic field whose flux density is $4.0 \times 10^{-2}$ weber/m$^2$. Assume the speed of the wire is 30. cm/sec.
13. In a certain cathode ray tube, an electron moves with a speed of $2.0 \times 10^6$ m/sec. Calculate its kinetic energy.
14. An electron in a certain cathode ray tube is accelerated from cathode to plate by means of a difference of potential of 300 volts. Assuming that the electron started from rest, *a.* what kinetic energy does the electron acquire? *b.* what is the speed of the electron when it reaches the plate?
15. To the electron described in question 13, a magnetic field is applied perpendicularly to its velocity. What is the resulting force on the electron. Give both magnitude and direction.
16. An electron is projected perpendicularly into a magnetic field whose flux density is 1.2 weber/m$^2$. As a result it moves in the field in a circular path whose radius is 0.35 meter. Calculate *a.* its speed *b.* its centri-centripetal acceleration *c.* the centripetal force on the electron.

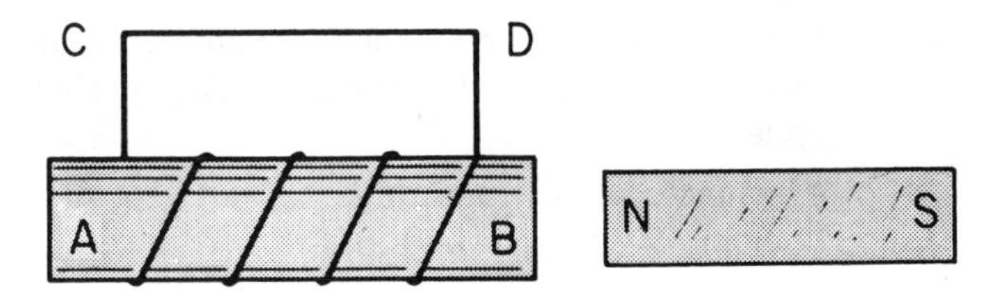


## Test Questions

1. In the above diagram, when the magnet is moved towards the left, a north pole will be produced near point 1. A 2. B 3. C 4. D
2. In question 1, the motion results in electron flow in wire CD from 1. C to D 2. D to C 3. up 4. down.
3. If we wish to ,ncrease the magnitude of the induced voltage in the above coil, we may 1. increase the number of turns on the coil 2. decrease the number of turns 3. use lower resistance wire 4. lower the temperature.
4. An unpolarized beam of electromagnetic waves is one whose vibrations 1. are confined to a single plane 2. occur in all directions 3. occur in all directions perpendicular to the direction of propagation 4. have not passed through a polaroid disk.
5. In a plane polarized wave, the electric and magnetic fields 1. are at right angles to each other 2. are in the same direction 3. are in opposite directions 4. may be in any direction with respect to each other.
6. *Lenz's Law* is a special case of which general law? 1. conservation of momentum 2. conservation of energy 3. *Coulomb's law* 4. *Ohm's law*.

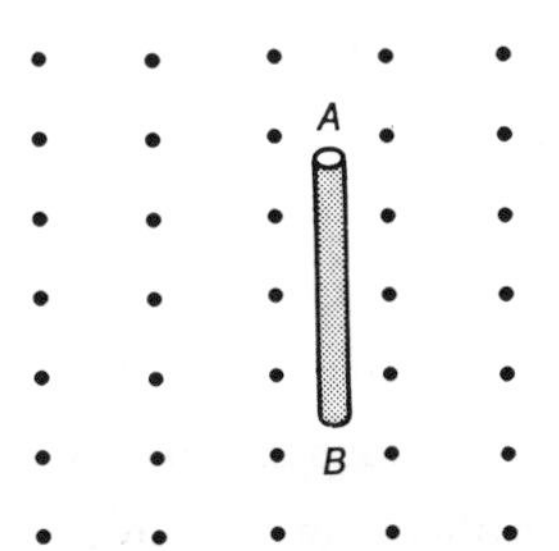


7. Imagine a uniform magnetic field directed perpendicularly out of the plane of the paper. This is represented by the dots in the diagram. Wire *AB* is moved to the right, perpendicularly to the magnetic flux. As a result an emf is induced. To quadruple the induced voltage, we may change the flux density so that it becomes 1. one-fourth as great 2. one-half as great 3. twice as great 4. four times as great.
8. In question 7, the emf may also be quadrupled by changing the speed with which the wire move so that it becomes 1. one-fourth as great 2. one-half as great 3. twice as great 4. four times as great.
9. In question 7, if wire *AB* is made smaller, the induced emf 1. increases 2. decreases 3. remains the same.
10. In question 7, the motion of wire AB produces a force parallel to the wire 1. on its electrons only 2. on its protons only 3. on its elec- electrons and protons.

11. The field magnet of an AC generator has poles which 1. do not change polarity 2. change polarity every cycle 3. change polarity every half cycle.

12. A magneto is best described as a 1. cobalt magnet 2. generator 3. motor 4. cathode ray tube.

13. Cathode rays consist of 1. electrons 2. protons 3. neutrons 4. molecules.

14. Thermionic emission is a phenomenon which is best described by saying that it is the emission of 1. heat 2. temperature 3. electrons 4. molecules.

15. Loop ABCD is shown rotating in a uniform magnetic field provided by a magnet whose north and south poles are shown. The loop rotates counter-clockwise, and at the instant shown, its plane is parallel to the magnetic flux provided by the field magnet. During one rotation of the loop, the emf induced in it will be 1. in one direction with A positive 2. in one direction with A negative 3. alternating giving one cycle 4. alternating giving two cyels.

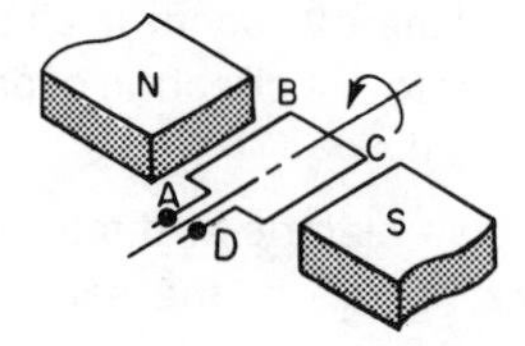


16. In question 15, at the instant shown, the induced emf has reached 1. its maximum value 2. its minimum value 3. an intermediate value.

17. In question 15, at the instant shown, the direction of the current in loop ABCD is 1. from A to B and from C to D 2. from D to C and from B to A 3. from A to B and from D to C.

# Topic Questions

1.1 Under what conditions is an induced emf produced?

1.2 When does an induced emf produce an induced current?

2.1 What are the factors which affect the magnitude of an induced emf?

3.1 Why must work be done in moving a wire across a magnetic field?

3.2 What is *Lenz's Law*?

4.1 If a wire is moved across a magnetic field, why do electrons tend to pile up at one end of the wire?

5.1 *a.* What are the important parts of an AC generator? *b.* What is the function of each of these parts?

5.2 Why is energy required to rotate the armature of a generator?

6.1 What is the relation of the frequency of current in an antenna to the frequency of the electromagnetic waves which radiate from it?

6.2 What is a plane polarized electromagnetic wave?

7.1 How does a commutator differ from a slip ring?

8.1 What are some essential differences between the simple DC motor and the DC generator?

8.2 The current in the armature of a motor is less when it is running at full speed than when it is starting. Why is this so?

9.1 *a.* What are cathode rays? *b.* What is thermionic emission?

10.1 *a.* What are two methods which may be used to deflect moving electrons? *b.* In each of these methods, how can we predict the direction of deflection?

# 16 **Alternating Current Circuits

## 1 Introduction

Power companies supply alternating current (AC) to most homes in the United States. AC can do some things better than DC, and some other things just as well. Until recently, it was possible to generate and transmit AC more efficiently than DC. AC voltages can be changed to higher and lower values more readily than DC.

AC as well as DC can be used conveniently for heating and lighting. The same tungsten light bulbs can be used for both AC and DC. Only DC can be used for charging batteries, electroplating, operation of certain motors, and some applications in electronics. Some efficient motors operate on AC only, but the best variable-speed motors operate on DC only. Rectifiers can be used to change AC to DC. Radio and TV sets which operate on AC contain rectifiers.

**2 Measurement of AC** In the last chapter, we showed that the potential difference at the terminals of an AC generator constantly varies in magnitude and direction. The graph indicated that this voltage varies with time like a sine wave; such an AC voltage is known as a *sinusoidal voltage*. AC voltages can be obtained which vary like a square wave (flat-topped) or like a saw-tooth wave (pointed-top); these are especially useful in radar and television. We shall restrict our description to sinusoidal voltages.

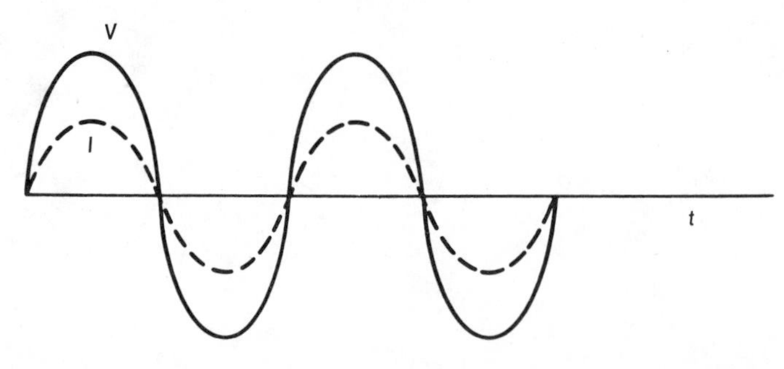


*Current Through Resistor in Phase With Voltage Across it*

When a resistor is connected across an AC generator, the current through the resistor will vary in step with the voltage change: the current in the resistor is always *in phase* with the voltage across it, even when there are other devices in the circuit. This means that when the voltage is zero the current is zero, and when the voltage has a peak value the current has a peak value.

*Ohm's Law* applies directly to a resistor: the current through the resistor at any instant is equal to the voltage across the resistor at that instant divided by the resistance.

In practical work, it is not convenient to talk about instantaneous values. Usually we describe the magnitude of the alternating current and voltage in terms of their *effective values*. An alternating current has an effective value of one ampere if it produces the same heat in one second in a given resistor as direct current of one ampere.

The effective value of a sinusoidal current is approximately 0.707 of the peak current; the effective value of the voltage is approximately 0.707 of the peak voltage. The effective value is sometimes referred to as the *root mean square* (rms) value because of the way in which the 0.707 factor is derived. Usually in describing the magnitude of an alternating current or voltage, the term "effective" is omitted. If we say that we have an AC voltage of 120 volts, it means that the effective voltage is 120 volts. The peak voltage is larger.

$$I = I_{\text{eff}} = I_{\text{rms}} = 0.707\ I_{\text{p}}$$

$$V = V_{\text{eff}} = V_{\text{rms}} = 0.707\ V_{\text{p}}$$

**EXAMPLE:** The AC used in the U.S. is often referred to as 115 *V*, 60 cps. This means that the frequency of the alternating current is 60 cycles per second, and that the effective value of the voltage is 115 volts. The peak value of this voltage, therefore, is 162 volts:

$$V = 0.707\ V_{\text{p}}$$

$$115\ V = 0.707\ V_{\text{p}}$$

$$V_{\text{p}} = (115/0.707) = 162 \text{ Volts}$$

*Ohm's Law* can also be applied directly to resistors using effective values.

**EXAMPLE:** In an AC circuit, 115 volts are applied to a toaster having a resistance of 5 ohms. Calculate the current.

$$I = V/R$$

$$= 115 \text{ volts}/5 \text{ ohms}$$

$$= 23 \text{ amp.}$$

This, of course, is the effective value of the current, which is what we usually want.

*AC meters* are usually calibrated to give effective values. The *hot-wire meter* depends on the fact that alternating current through a wire produces heat in the wire, and that the resulting rise in temperature increases the length of the wire. The increased length allows the needle to move across the scale of the meter.

Some AC meters use the heat produced by the current to generate a DC voltage which can then be measured on the DC meters described in an earlier chapter. This is done with a *thermocouple*: two different wires (e.g. iron and lead) are twisted together at one end; when this end is heated, an emf appears between the other ends.

In some meters, the AC is changed directly to DC by means of rectifiers. This DC is then measured on DC meters.

3 **Capacitors and Capacitance** The use of coils and capacitors in AC circuits produces some strange effects and many useful applications. A *capacitor* consists of two conductors separated by an insulator or *dielectric*. The capacitor was formerly called a condenser. Common dielectrics are paper, mica, air, and ceramics. Capacitors are often named by the dielectric used; e.g. paper capacitors. In some capacitors the relative position of the two conductors can be changed. These

capacitors are known as variable capacitors, the others are fixed capacitors. The *Leyden* jar is an old type of capacitor in which the insulator is a glass jar.

The function of a capacitor is to store an electric charge. This is important in tuning a radio, in reducing sparks at electrical contacts, and in reducing hum in a radio by reducing the ripple or fluctuation of the DC.

How does a capacitor store a charge? Let us first imagine a capacitor in a DC circuit. The diagram shows a capacitor connected to a battery. On the negative terminal of the battery is an excess of electrons. The electrons repel each other, and we have provided a path for these electrons. They start moving along the connecting wire *A* to *B*, the right plate of the capacitor. At the same time, electrons are attracted from the wire *D* and plate *C* of the capacitor to the positive terminal of the battery, where there is a deficiency of electrons. As a result there is a difference of potential across the plates of the capacitor.

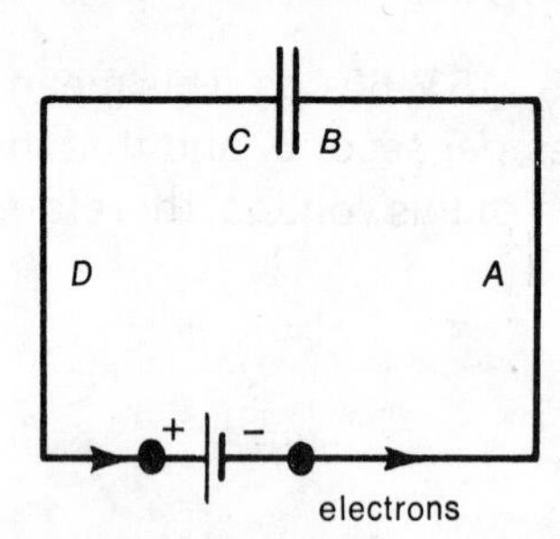


*Battery Charges Capacitor*

The capacitor has been charged: one plate is negative, the other is positive. For every electron that goes to plate *B*, there is another electron that leaves plate *C*. Notice that no electrons move through the dielectric. There is a motion of charge through the wires until the potential difference between the plates of the capacitor is equal to the emf of the battery. It usually takes only a fraction of a second for this to happen. The capacitor is charged almost immediately.

For a given capacitor, the amount of charge which is stored is proportional to the voltage. If we double the emf of the battery, we double the charge on the capacitor, and we double the voltage across the capacitor. If we disconnect the capacitor from the battery, the capacitor can stay charged indefinitely and have a difference of potential between its plates.

$Q = CV$ where $Q$ is the charge on either plate of the capacitor, $C$ is the capacitance of the capacitor, and $V$ is the voltage or potential difference of the capacitor.

If the voltage is expressed in volts and the charge in coulombs, then the capacitance is given in farads. The *farad* is a rather large unit of capacitance. A more practical unit is the *microfarad*. The picofarad is also used.

1 farad $= 10^6$ microfarads
1 farad $= 10^{12}$ picofarads.

The capacitance of a capacitor is proportional to the area ($A$) of its plates. If we double the area of the plates, we provide twice as much space for storing electrons. The capacitance of a capacitor is inversely proportional to the distance ($d$) between the plates. If we double the distance between the plates, we reduce the capacitance to one-half of its original value. The capacitance also depends on the nature of the dielectric.

$$C \propto \frac{A}{d}$$

**4 Capacitors in Series AC Circuits** We saw before, that if a capacitor is connected in a DC circuit, there is a momentary current. The capacitor gets charged quickly. After the initial current, we think of the capacitor as blocking DC. If we connect a capacitor in an AC circuit, the capacitor charges and discharges. The voltage across it goes up and down, and changes direction at the same frequency as the emf of the generator to which it is connected.

However, the current in the circuit is not in phase with the voltage across the capacitor. If we think of one complete cycle as representing 360°, we find that the current "through the capacitor" is 90° ahead of the voltage across it. This means that when the current reaches its peak value, the voltage across the capacitor is just reaching zero. When the current reaches zero, the voltage is just becoming maximum. We speak about the current through the capacitor, but, as in DC, no current actually goes through the dielectric.

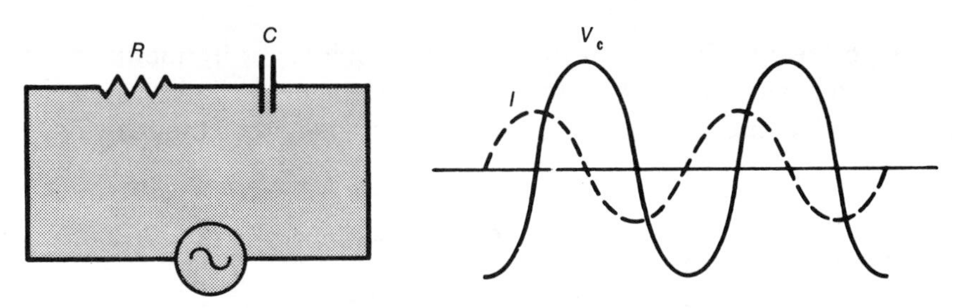


*current through capacitor 90° ahead of voltage*

Notice that, since we have a series circuit, the current through the resistor is the same as through the capacitor. Also, the voltage across the *resistor* is still in phase with the current through it. Therefore, the voltage across the resistor is 90° ahead of the voltage across the capacitor. We can represent this phase relation by means of sine curves. This tends to be a little confusing. We usually use a vector notation to represent the effective values of the current and voltages. We can use different scales for the current and voltages, but, of course, must use the same scale for all the voltages.

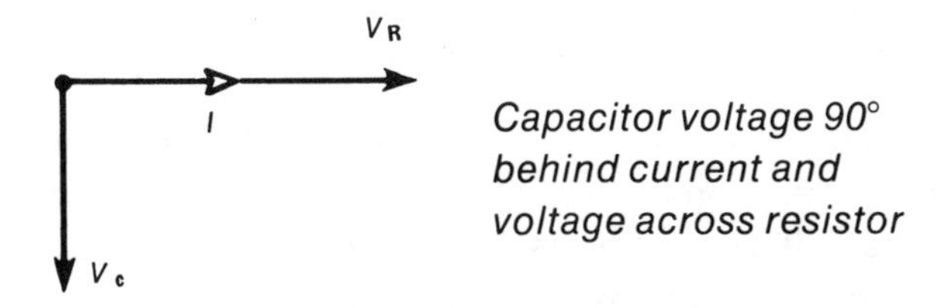


*Capacitor voltage 90° behind current and voltage across resistor*

When a capacitor is charged, we think of it as storing an electric charge. This also means that it stores electrical energy. When the capacitor is discharged, it returns this energy to the circuit.

**5 Coils and Inductance** We have seen that an emf is induced in a coil if there is relative motion between a coil and a magnetic field. If we have alternating current in a coil, the magnetic field which is produced moves away from and back to the coil and also keeps changing direction. As a result an emf is induced in the coil. The direction, or polarity, of this emf is such as to oppose the change in current, in accordance with Lenz's Law. *Self-induction* is the production of an emf in a coil because of the current in the coil. A coil is sometimes called an *inductor*. The emf is sometimes referred to as a self-induced emf.

The induced voltage is proportional to the rate at which the current changes and to a property of the coil called *inductance*. The inductance of a coil depends on its shape, on the number of turns of wire in the coil, and on the nature of the core. If we increase the number of turns, we increase the inductance. If we replace the air core by iron, we also increase the inductance. The unit of inductance is the *henry*. (If a coil has an inductance of 1 henry, a change of current in it of one ampere per second will induce in the coil an emf of 1 volt.) The larger the inductance, the greater the magnetic field produced by the coil with a given current.

If we connect a coil in an AC circuit, the self induced emf goes up and down and changes direction at the same frequency as the generator to which it is connected. However, the current through the coil is not in phase with voltage across it. As in the case of the capacitor, there is a 90° phase difference, if the resistance of the coil is negligible. But this time the voltage ($V_L$) is ahead of the current.

**6 Coils in Series AC Circuit** In AC circuits, a coil ($L$) is frequently connected in series with a resistor to an AC generator. As in the case of the capacitor, the current through the circuit is in phase with the voltage across the resistor.

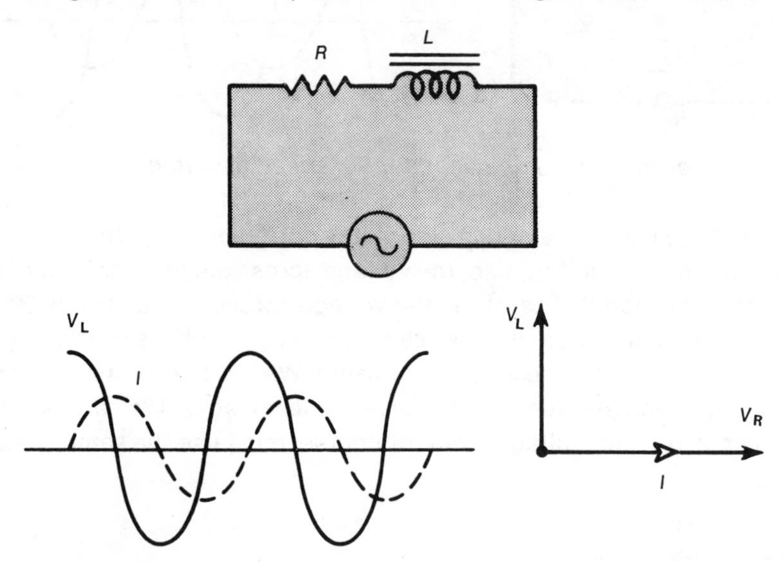


In an AC circuit, the magnitude of the current depends not only on the resistance in the circuit and the applied emf, but also on the inductance of the coil. The larger the inductance, the less is the current. If a capacitor is also in the circuit, things get to be more complicated, as will be indicated below.

**7 Reactance and Impedance** We can try to figure out what the current in an AC circuit is by taking into account the voltages developed across coils and capacitors in the circuit. It is easier, usually, to make calculations by introducing another concept.

**Impedance (Z)** is the total opposition to current produced by resistance, inductance, and capacitance – any of these alone, or any combination of these. The impedance of an ideal capacitor is called *capacitive reactance* ($X_c$). The capacitive reactance varies inversely with capacitance and the frequency of the AC supply:

$$X_c = \frac{1}{2\pi fC}.$$

When the frequency is in cycles per second and the capacitance is in farads, the reactance is in ohms. Most capacitors are close enough to the ideal so that we can ignore any other effect of a capacitor.

The impedance of an ideal coil is called its *inductive reactance ($X_L$)*. This is a coil's opposition to current when its resistance is negligible. It is proportional to the inductance of the coil and the frequency of the AC supply.

$$X_L = 2\pi f L,$$

where $L$ is the inductance of the coil. When the frequency is in cycles per second and the inductance is in henries, and reactance is in ohms.

Actual coils are often far from ideal; that is, coils usually have appreciable resistance in addition to inductance. For the purpose of most calculations, we can think of such a coil as having its resistance in series with its inductance.

The total impedance ($Z$) is not obtained by ordinary addition. Because of the phase differences between voltage and current which we saw earlier, the impedance is obtained by application of the Pythagorean theorem. In a series AC circuit,

$$Z = \sqrt{R^2 + (X_L - X_c)^2}$$
$$X_L = 2\pi f L$$
$$X_c = 1/(2\pi f C)$$
$$I = V/Z.$$

Notice that the last formula is a generalization of *Ohm's Law*. Instead of using resistance, we use impedance to calculate the current in the circuit. This generalization also applies to any part of the circuit. Therefore, the voltage across any part of the circuit is equal to the current times the impedance of that part of the circuit:

$$V_c = IX_c$$
$$V_L = IX_L$$
$$V_R = IR$$

**EXAMPLE:** A series AC circuit consists of a coil, capacitor, and resistor connected in series to an AC generator supplying 120 volts. The reactance of the coil is 90 ohms, that of the capacitor is 60 ohms. The resistor has a resistance of 40 ohms. Calculate the current in the circuit.

$$Z = \sqrt{R^2 + (X_L - X_c)^2}$$
$$= \sqrt{40^2 + (90 - 60)^2}$$
$$= 50 \text{ ohms.}$$
$$I = V/Z$$
$$= 120\ V/50 \text{ ohms}$$
$$= 2.4 \text{ amperes.}$$

If the circuit contains only a coil and resistor, the formula for impedance is simpler: $Z = \sqrt{R^2 + X_L^2}$.

## 8 Resonant Electric Circuits

**Resonant Electric Circuits** We saw in an earlier chapter, that if two tuning forks have the same natural frequency, it is only necessary to set one of them into vibration, and before long the other fork will also start vibrating with rather large amplitude, although the two forks do not touch each other. The two forks are said to be in resonance. Resonant electric circuits are used in tuning radios. What is a resonant electric circuit?

By looking at the expression for impedance,

$$Z = \sqrt{R^2 + (X_L - X_c)^2},$$

we notice that the impedance of a series AC circuit is least if

$$X_L = X_c.$$

When the impedance is least, the current is greatest. When the two reactances are equal,

$$2\pi fL = 1/(2\pi fC).$$

Clearing of fractions and solving for $f$, gives the expression for the *resonant frequency* of the circuit:

$$f_r = \frac{1}{2\pi\sqrt{LC}}.$$

The resonant frequency is also the natural frequency. The circuit is known as a resonant circuit. Notice that it has to have a coil and capacitor, and may have a resistor; it is connected to a generator which supplies an emf at the resonant frequency.

Imagine the circuit connected to an AC generator whose frequency is lower than the circuit's natural frequency. As the frequency of the generator is increased, the current in the circuit increases, and reaches a maximum value when the frequency of the supplied emf is the same as the circuit's resonant frequency. Since it is a series circuit, as the current increases, the voltage across the coil as well as across the capacitor increases. In fact, at resonance each of these voltages may be greater than the voltage supplied by the generator. This is analogous to the way we increase the amplitude of a swing by giving it little pushes at the natural frequency of the swing. (For a more detailed explanation see the author's *How To Prepare for College Board Achievement Tests—Physics*, page 81.)

Usually when we tune a radio, we vary a capacitor so that the natural frequency of the radio circuit is the same as the frequency of the signal sent out by the desired radio station. Then a small signal will produce a relatively large voltage across the capacitor in the radio's resonant circuit.

9 **Induction Coil** We saw before that an emf is induced in a coil if the current in that coil is changing. This is self-induction. We have *mutual induction* if an emf is induced in a coil because of current changes in a second coil. The current may be DC or AC. If it is DC, however, the current must not remain constant. As long as there is a change in the current, there will be a change in the magnetic field surrounding the coil. If this changing magnetic field cuts the second coil (or links with it), an emf is induced in the second coil.

The *primary* coil ($P$) is the coil which is directly connected to the source of the electrical energy, such as a dynamo. The *secondary* coil ($S$) is the coil from which electrical energy may be taken as a result of the emf induced in it. In the *induction coil* the source of energy is DC, usually a battery.

Induction coils are used to supply a high voltage; they get their energy from a source with a low DC voltage. In an automobile, the induction coil is used as the *ignition coil*: it provides the high voltage which, at the spark plug, results in a spark that ignites the fuel mixture.

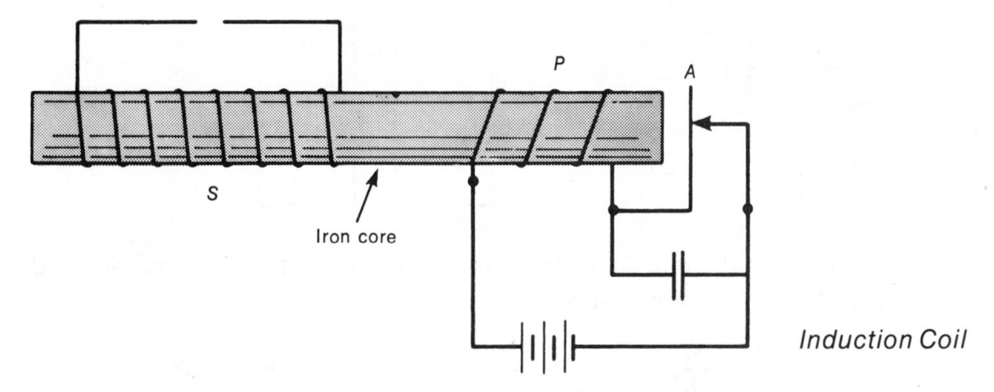

*Induction Coil*

A steady DC would produce no emf in the secondary. A battery supplies a steady DC. An interrupter is used to interrupt the current automatically. This results in a changing magnetic field which then cuts across the many turns of the secondary coil. The greater the number of turns in the secondary, the greater the induced emf.

In the diagram, *A* is a springy, magnetic material. When the iron core is sufficiently magnetized, it will pull the armature *A* to the left, thus breaking, or interrupting, the primary circuit. When, as a result of this, the primary current decreases, the magnetism becomes less and the spring *A* returns to its original position. This completes the primary circuit again, primary current and the magnetic field increase, inducing an emf in the secondary. Usually this emf is greater in one direction than in the other, and the gap in the secondary circuit can be arranged so that there will be current in only one direction in the secondary. A spark at the gap is evidence of current in the secondary coil.

10 **The Transformer** In the transformer, we also depend on mutual induction. Usually AC is applied to one coil, and AC is obtained from a second coil. In the transformers used on low frequency AC (up to a few thousand Hertz), materials with high permeability, such as iron, are used, and the following formulas apply;

$$\frac{V_s}{V_p} = \frac{N_s}{N_p}$$

$$V_s I_s = \text{efficiency} \times V_p I_p$$

Power = $VI$.

where
$V_s$ = voltage induced in the secondary
$V_p$ = voltage applied to the primary
$N_s$ = number of turns in secondary
$N_p$ = number of turns in primary
$V_s I_s$ = power supplied by the secondary
$V_p I_p$ = power supplied to the primary

When the efficiency is 100%, $V_s I_s = V_p I_p$.

The efficiency of practical transformers is high and constant over a wide range of power. When more power is used in the secondary circuit, more power is supplied to the primary circuit automatically. If we want a higher voltage than the generator supplies, we use a *step-up transformer*: more turns on the secondary coil than on the primary. A *step-down transformer* has fewer turns on the secondary than on the primary, and, therefore, produces a lower voltage than the generator supplies.

**EXAMPLE:** A transformer is designed to step up 220 volts to 2200 volts. It has 200 turns on the primary, is 90% efficient, and supplies a current of 1.5 amperes to a resistor connected to the secondary. Calculate *a.* the number of turns on the

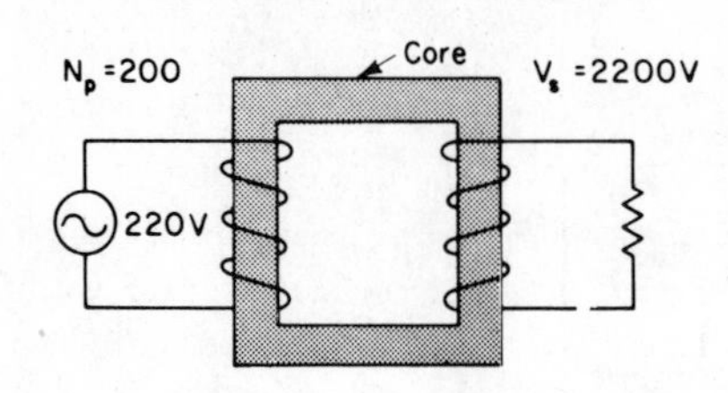


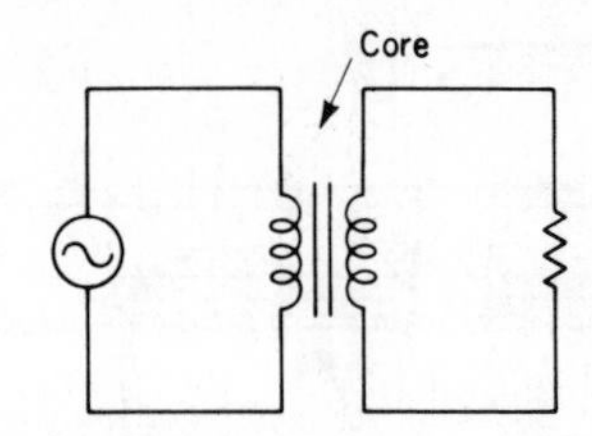


secondary and *b.* the current in the primary. The circuit may be represented in either of the two ways shown above.

*a.* $\frac{V_s}{V_p}=\frac{N_s}{N_p}$; $\frac{2200}{220}=\frac{N_s}{200}$; $N_s = 2000$.

*b.* $V_sI_s = \text{efficiency} \times V_pI_p$

$2200 \times 1.5 = 0.90 \times 220\, I_p$

$I_p = 16.7$ amp.

# Questions and Problems

1. How can capacitance be increased?
2. What is the effect on capacitance, if the voltage applied to a capacitor is decreased?
3. What is the phase relationship between the current and voltage if an AC voltage is connected to *a.* a resistor only *b.* a capacitor only *c.* an inductor (coil) only?
4. A resistor having a resistance of 20 ohms is connected to an AC source supplying 60 volts at 60 cycles per second. Calculate *a.* the peak voltage *b.* the current in the resistor *c.* the power used by the resistor.
5. A coil having an inductance of 5.0 henries is connected to an AC source supplying 62.8 volts at 60 cycles/sec. Calculate *a.* the reactance of the coil *b.* the current in the coil.
6. A coil having a reactance of 30. ohms is connected to an AC source supplying 60. volts at 60 cycles/sec. Calculate *a.* the current in the coil *b.* the inductance of the coil.
7. In question 6, if the frequency of the source were doubled, what would happen to the *a.* reactance of the coil *b.* the current in the coil?
8. A capacitor is connected to a source of AC supplying 80 volts at 60 cy. per sec. If the reactance of the capacitor is 40 ohms, calculate *a.* the current in the capacitor *b.* the capacitance of the capacitor.
9. In question 8, if the frequency of the source were doubled, what would happen to *a.* the reactance *b.* the current?
10. A capacitor having a capacitance of 5.0 microfarads is connected to a source supplying 200. volts at 100. cycles/sec. Calculate *a.* the reactance of the capacitor *b.* the current in the circuit.
11. A series AC circuit contains an inductive reactance of 40 ohms, a resistance of 30 ohms, and a power source supplying 120 volts at 60 cycles/sec. Calculate *a.* the impedance of the circuit *b.* the current in the circuit *c.* the voltage across the resistor.

12. A series AC circuit contains a capacitive reactance of 100 ohms, an inductive reactance of 130 ohms, a resistance of 40 ohms, and a power source supplying 140 volts. Calculate *a.* the impedance of the circuit *b.* the current in the circuit *c.* the voltage across the coil.
13. A coil having an inductance of 80. henries is connected to a capacitor having a capacitance of 5.0 microfarads. What is the resonant frequency of the combination?
14. A resistance of 40 ohms is connected in series with a coil, capacitor, and source of AC supplying 100 volts at 60 cycles/sec. The reactance of the coil and capacitor are 40 ohms each. Calculate *a.* the impedance of the circuit *b.* the current in the circuit *c.* the voltage across the coil.
15. A transformer is designed to supply 2000 volts when operating from a 100-volt, 60 cycle supply. The primary winding has 400 turns. Calculate *a.* the number of turns in the secondary winding *b.* the current in the secondary winding if the transformer is 90% efficient and the current in the primary winding is 3 amperes.
16. A transformer has 50 turns on the primary winding and 250 turns on the secondary. 120 volts AC are supplied to the primary, and 200 watts of power are taken from the secondary winding. *a.* Calculate the voltage supplied by the secondary winding. *b.* Calculate the current in the secondary winding. *c.* If the efficiency is 100%, how much power is supplied to the primary? *d.* How much power is supplied to the primary if the efficiency is only 80%?

## Test Questions

1. In an AC circuit, 100 volts at 60 cycles/second is applied to a 20-ohm resistor. The effective value of the current is ........... amp.
2. In question 1, the peak value of the voltage is ........... volts.
3. In question 1, the phase of the current in the resistor, as compared with the voltage, is 1. 90° behind 2. 90° ahead 3. 0° (in phase).
4. When an iron core is inserted into a coil, its impedance 1. increases 2. decreases 3. remains the same.
5. An incandescent lamp in series with a capacitor is connected first to 110 volts AC and then to 110 volts DC. If the lamp is rated at 110 volts, and the capacitor has a large capacitance, the lamp will light on 1. AC only 2. DC only 3. AC and DC.
6. A unit of inductance is the 1. henry 2. farad 3. ohm.
7. If the plates of a capacitor are moved closer together without letting them touch, the capacitance 1. increases 2. decreases 3. remains the same.
8. If the area of the plates of a capacitor is increased, the capacitance 1. increases 2. decreases 3. remains the same.
9. If a coil and resistor are connected in series to a source of AC, the phase of the current in the coil, as compared with the voltage across it is 1. 90° behind 2. 90° ahead 3. 0° (in phase).
10. A capacitor is connected to an AC voltage. If the frequency of the AC is halved, the reactance of the capacitor 1. remains the same 2. is halved 3. is doubled 4. is quadrupled.

11. If the inductance of a coil is doubled, its reactance 1. remains the same 2. is doubled 3. is quadrupled 4. is halved.
12. In a series AC circuit containing resistance, inductance, and capacitance, if the inductance is increased, the impedance of the circuit 1. increases 2. decreases 3. may increase or decrease.
13. An iron-core transformer has 30 turns on the primary winding and 120 turns on the secondary winding. If 40 volts AC is applied to the primary, the voltage induced in the secondary is, in volts 1. 10 2. 40 3. 120 4. 160.
14. If the above transformer has an efficiency of 100% and 80 watts is supplied to the primary, the power used in the secondary circuit is, in watts, 1. 20 2. 80 3. 320 4. 9600.
15. The current in the primary winding of an induction coil is 1. steady DC 2. fluctuating DC 3. AC.
16. In a circuit containing both inductance and capacitance, if the capacitance increases, the resonant frequency of the circuit 1. increases 2. decreases 3. remains the same.
17. A step-up transformer steps up 1. current 2. voltage 3. power 4. all three.
18. A variable capacitor and an incandescent lamp are connected in series to an AC voltage. As the capacitance decreases, the current in the lamp 1. increases 2. decreases 3. remains the same.
19. AC may not be used for 1. operating toasters 2. operating motors 3. charging storage batteries 4. heating incandescent lamps.
20. DC should not be connected to a(n) 1. transformer 2. induction coil 3. incandescent lamp 4. portable radio.

# Topic Questions

1.1 For what are rectifiers used?

2.1 In an AC circuit containing resistance, the current in the circuit is in phase with the voltage across the resistor. What does that mean?

2.2 In an AC circuit, how is the effective current related to the peak current?

2.3 An alternating current in a circuit usually produces heat. How can you support this statement by observations around the house?

3.1 What is the function of the paper in a paper capacitor?

3.2 In what way does the charge on a capacitor depend on the voltage across it?

3.3 In what way does the capacitance of a capacitor depend on the area of its plates and the distance between them?

4.1 What is the phase relation between the current and voltage of a capacitor?

5.1 What is a self-induced *emf*?

5.2 What is the henry?

6.1 In an AC circuit, if an inductance is introduced into a series circuit containing resistance, what will be the effect on the current?

7.1 On what do the following depend: *a.* inductive reactance *b.* capacitive reactance

7.2 What are some of the important relationships in a series AC circuit?

8.1 In a resonant AC circuit, what is true about the coil and the capacitor?

9.1 *a.* What is the basic function of an induction coil? *b.* For what purpose is it used in an automobile?

# 17 Quantum Theory of Light

**1 Successes of the Wave Theory** What is light? We saw that in the 1600's Newton and Huygens disagreed. Newton said that light is a stream of corpuscles (particles) while Huygens said light is a wave. In physics we try to decide between conflicting theories by considering experimental results. Both theories explained reflection and refraction of light equally well. However, in explaining the refraction of light when it goes from air into water, Newton assumed that light travels faster in water than in air. *Huygens' Explanation* stated that light travels more slowly in water than in air. There was no method known at the time for measuring the speed of light in water, and so it was not possible to test the two predictions. This was first done by Fizeau in 1849. He found that light travels more slowly in water than in air, as predicted by the wave theory.

By the end of the 19th century the wave theory was well established as an excellent explanation for the behavior of light. It had been extremely successful in explaining interference, diffraction and polarization of light, polarization providing the additional clue for describing light as a transverse wave. What is waving? In about 1864 Maxwell developed the electromagnetic wave theory which indicated that light is a wavelike fluctuation of electric and magnetic fields. The fluctuations of these fields are at right angles (transverse) to the direction in which the light is traveling.

**Failure of the Wave Theory** We have described continuous and bright-line spectra. For example, when the white light from an incandescent source, such as a glowing tungsten filament, is dispersed by a prism, we get a continuous display of colors, corresponding to a continuous variation of wavelengths. The light is most intense in the yellowish region. Why is the energy of the light distributed among the different wavelengths the way it is? Some of the most brilliant physicists of the second half of the 19th century tried to answer this question by applying *Maxwell's Electromagnetic Wave Theory*, but failed.

Hertz discovered the photoelectric effect in 1885. It was investigated further by Lenard in 1902. The observations were in contradiction to what is expected on the basis of a wave theory for light. This will be explained below.

**2 Planck's Quantum Theory** In 1900 Planck explained the energy distribution in the continuous spectrum by introducing a startling new concept. In his work he described the radiation from the ideal blackbody (often written as one word), but the theory is more general. The blackbody is one that would absorb all the radiation that falls on it. The body's radiation is a continuous spectrum. Planck's new concept leads to this important conclusion: The body contains atomic oscillators which can emit electromagnetic energy in discrete amounts (or little bundles) only. Each little bundle of energy is called a *quantum*, (plural: quanta).

The quantum is electromagnetic energy; the radiation from the oscillator has a definite frequency. The energy of each quantum is proportional to the frequency of the radiation. We can write this in the form of an equation by introducing a proportionality constant, $h$, called *Planck's Constant*.

$E = hf$ where $E$ is the energy of the quantum,
$h$ is Planck's constant,
$f$ is the frequency of the radiation.

When $E$ is in joules and $f$ in cycles per second, $h$ is $6.63 \times 10^{-34}$ joules-sec; (listed in *Reference Tables*).

**EXAMPLE:** What is the energy of a quantum of light whose wavelength is 5000 Angstrom units?
Let us first calculate the frequency of this light.

$$c = f\lambda.$$
$$3.00 \times 10^{8} \text{ m/sec} = f(5000 \times 10^{-10} \text{ m})$$
$$f = 6.00 \times 10^{14} \text{ cycles/sec.}$$
$$E = hf$$
$$= 6.63 \times 10^{-34} \text{ joules-sec} \times 6.00 \times 10^{14} \text{ cycles/sec}$$
$$= 3.98 \times 10^{-19} \text{ joule.}$$

**3 Photoelectric Effect** The photoelectric effect is the emission of electrons from an object when certain electromagnetic radiation falls on it. The material which emits the electrons is said to be *photoemissive*. The effect is best studied when monochromatic light is used.

If the effect is studied quantitatively, the following facts are observed:

1. If the intensity of the monochromatic light is increased, the number of emitted electrons increases. (You probably expected that.)
2. If the intensity of the monochromatic light is increased, the speed of the emitted electrons does not increase. This is strange.
3. If we decrease the frequency of the light used, we reach a frequency below which no electrons come off. This is known as the threshold frequency. No matter how intense the radiation used, if it is below the threshold frequency, no electrons will be given off by this substance.
4. Different substances have different threshold frequencies.
5. If the frequency which is used is above the threshold frequency, even the weakest radiation will result in emission of electrons.
6. The electrons coming off the photoemissive material do not all have the same speed. The maximum speed, or kinetic energy, of the emitted electrons increases with the frequency of the incident radiation.

Facts 2 and 3 are clearly not in accord with what we might expect on the basis of the wave theory of light, according to which the light has more energy if its intensity is increased. The intense light should be able to release electrons, and increasing the intensity should increase the speed of the released electrons. It will be shown below that the quantum theory explains this well.

*Albert Einstein*
Courtesy of Addison-Wesley

4 **Einstein's Explanation of the Photoelectric Effect** In 1905 Albert Einstein published three brilliant papers. One of these deals with the photoelectric effect. He proposed that electromagnetic radiation be considered as traveling and being absorbed in little bundles of energy, and not as existing that way merely near the oscillators which emit the radiation. In other words, Einstein suggested that we think of electromagnetic radiation as being granular, each grain or bundle having an amount of energy given by Planck's expression for the quantum, $E = hf$. The bundle does not spread out like a water wave as it goes away from the source. The term *photon* is used to refer to this grain of energy. The words photon and quantum are often used interchangeably. Notice that while all electrons are the same, no matter what their source, all photons are not alike. All photons move with the speed of light, but the quantum of energy in each photon depends on the frequency or wavelength.

Energy is required to remove electrons from a material. In a given material some electrons are easier to remove than others; those near the surface are easiest to remove. The minimum energy needed to remove an electron from a given material is known as the material's *work function*, $W$. Different materials have different work functions.

When the light reaches a material, a photon interacts with one electron. If the energy of the photon is greater than the work function of the material, the electron comes off with a kinetic energy ($E_k$) which increases with the frequency of the photon.

$E_k = hf - W$, where $h$ is *Planck's Constant*
$f$ the frequency of the photon (its wave characteristic)
$W$ the work function of the material being used.

Recall that this kinetic energy is the energy of the fastest electron escaping from the material, and that the speed $v$ of the electron is given by the expression

$E_k = \frac{1}{2}mv^2$ where $m$ is the mass of the electron.

Also, since the photon has a wave characteristic, we can calculate its wavelength:

$\lambda = c/f$ where $c$ is the speed of light in vacuum.

The expression for the energy of the photon,

$$E = hf,$$

can also be written as

$$E = hc/\lambda$$

that is, the energy of the photon is inversely proportional to its wavelength.

If we plot a graph of the kinetic energy of the fastest electrons against frequency of the electromagnetic radiation used, we get a straight line; this kinetic energy is a linear function of the frequency of the photon. Compare the photoelectric equation, $E_k = hf - W$, with the general equation for a straight line, $y = mx + b$. You can see that $E_k$ and $f$ take the place of $y$ and $x$, and that $h$ is the slope of the straight line graph. This provides a fairly simple method of getting *Planck's Constant* experimentally. Also note that the work function $W$ is numerically equal to the $y$-intercept.

The graph should also remind you that if the frequency of the electromagnetic radiation falling on the material is below a certain frequency, $f_o$, no electrons come off; $E_k = 0$. If we substitute zero for $E_k$ in the photoelectric equation, we get

$hf_o = W$:

$f_o$ is the *threshold frequency*.

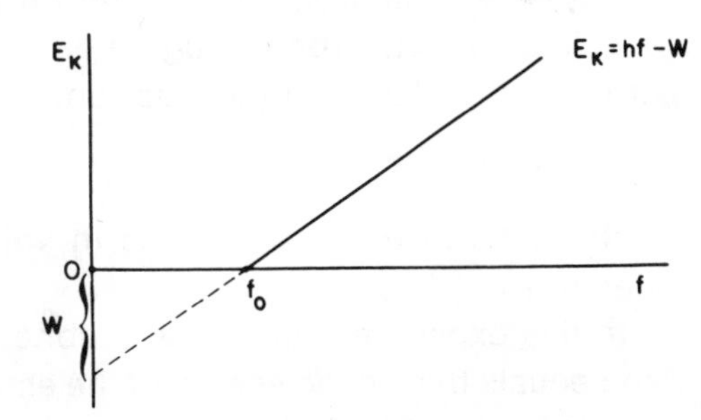


*Photoelectric Effect*

**FOR EXAMPLE:** No electrons are released from zinc as a result of red light shining on it. If we use ultra-violet light (shorter wavelength and therefore higher frequency) instead of red light, electrons are emitted readily.

**EXAMPLE:** Light whose wavelength is 5000 Angstrom units is used to illuminate a metal whose work function is $3.0 \times 10^{-19}$ joules. Calculate the energy of *a.* the photons used *b.* the fastest electrons escaping from the metal.

*a.* $E = hf = hc/\lambda$
$= (6.63 \times 10^{-34}$ joule-sec$)(3 \times 10^8$ m/sec$)/(5000 \times 10^{-10}$ m$)$
$= 4.0 \times 10^{-19}$ joule

b. $E_k = hf - W$
$= 4.0 \times 10^{-19}$ joule $- 3.0 \times 10^{-19}$ joule
$= 1.0 \times 10^{-19}$ joule.

**Experimental Arrangement** The diagram shows the electrodes of a phototube connected in a special way to a source of difference of potential. The phototube is an evacuated tube in which the cathode is made of photosenitive material, material which gives off electrons readily when light shines on it. In normal use the cathode is connected to the negative side of the difference of potential, but in this experiment it is connected to the positive side. Therefore, the electrons emitted by the cathode, when illuminated by the light, will be repelled by the anode. The slow electrons will be stopped by a low voltage. The voltage is increased gradually by means of the variable resistor *R*. When the fastest electrons have been stopped, the galvanometer will read zero.

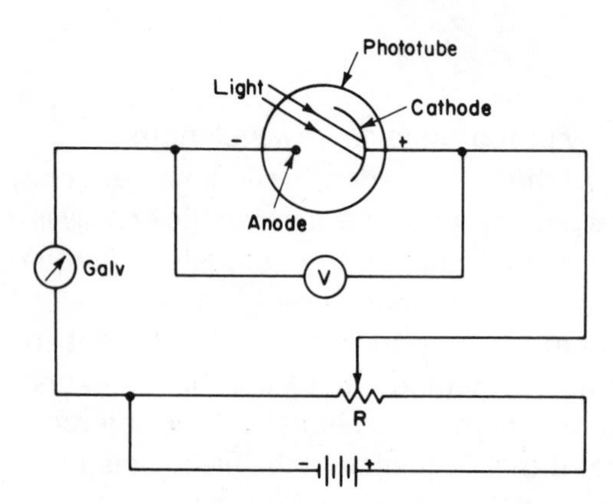


*Finding Cut-off Voltage*

You may recall from the chapter on static electricity, that the work required to accelerate an electron through a difference of potential *V* is equal to the product of *V* and the charge of the electron:

work $= Ve$.

If the difference of potential is in volts, the work done on one electron is an electron volt (*ev*).

In this experiment the work is done to decelerate the electrons, to stop them, and equals the kinetic energy of the emitted electrons: $Ve = 1/2\,mv^2$. The voltage needed to stop the fastest electrons is read from the voltmeter, when the galvanometer reads zero. If we increase the frequency of the monochromatic light used, a higher voltage will be needed to stop the faster electrons. (Remember, $E_k = hf - W$, and *W* remains constant for the tube.) The results agree with the predictions of Einstein's photoelectric equation.

The voltage needed to stop the fastest electrons is called the *stopping voltage* or *cut-off potential*.

**EXAMPLE:** The stopping voltage in a certain experiment was 10 volts. The work function of the cathode is $3.0 \times 10^{-19}$ joule. Calculate *a.* the maximum kinetic energy of the emitted electrons, in joules and ev's *b.* the energy of the photons used for the experiment.

*a.* The work to stop the electrons $= Ve$

$$= (10 \text{ volts})(1 \text{ electron})$$
$$= 10 \; ev.$$

**or,** $\quad$ work $= (10 \text{ volts})(1.6 \times 10^{-19} \text{ coul})$

$$= 1.6 \times 10^{-18} \text{ joule}$$

This work is equal to the kinetic energy of the emitted electron.

*b.*

$$E_k = hf - W$$
$$1.6 \times 10^{-18} \text{ joule} = hf - 3.0 \times 10^{-19} \text{ joule}$$
$$hf = 1.9 \times 10^{-18} \text{ joule.}$$

This is the energy of the photon. Since 1 electon-volt $= 1.6 \times 10^{-19}$ joule, this photon energy is equal to about 12. electron-volts.

# 5 Rutherford Model of the Atom

The quantum theory is also useful in explaining the spectra described in Chapter 10. This requires an understanding of the nature of an atom.

By means of ingenious and patient experimentation, Rutherford had discovered that alpha particles given off by radioactive materials at great speeds are helium nuclei which, we now know, consist of two protons and two neutrons. He then proceeded to use the alpha particles as bullets for many investigations.

Rutherford arranged his equipment to have alpha particles strike thin sheets of goldfoil. With his co-workers he discovered (about 1910) that most of the alpha particles went right through the foil without being deflected. Others were scattered through various angles ranging up to 180°; that is, a few were reflected back along the direction of incidence. The backward deflection was unexpected. As Rutherford said: *It was almost as incredible as if you fired a 15-inch shell at a piece of tissue paper and it came back and hit you.*

The mass of the alpha particle is about 7500 times that of an electron. Rutherford then realized that the backward scattering must be due to a single collision with something very massive. Since most of the alpha particles go right through the foil, with little or no deviation, these massive targets must be widely separated from each other.

Rutherford then decided on the *model of the nuclear atom.* The positive charge (protons) of the atom and most of its mass are concentrated in a small dense core of the atom, called its *nucleus.* Outside this nucleus, a neutral atom has a number of electrons equal to the number of protons in the nucleus. These electrons are quite far from the nucleus. Most of the atom is empty space. (Rutherford calculated that the diameter of the nucleus is about 1/10,000th of the diameter of the atom.) On the basis of this model of the atom, Rutherford made careful calculations which explained the previously obtained data, and made predictions which were verified in later experiments.

*Coulomb's Law* applies in this scattering situation. Notice that the positive alpha particles are repelled by the positive nucleus. In the diagram, the dotted line *C* shows the path an alpha particle would follow for a head-on collision. The closer the particles approach, the greater the force of repulsion. Alpha particles move

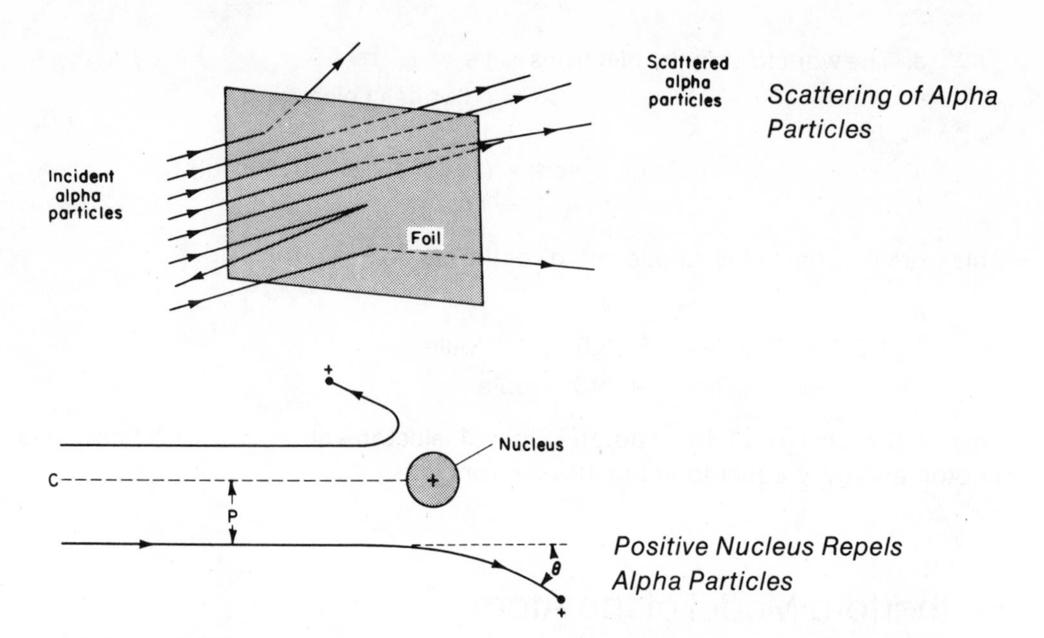


*Scattering of Alpha Particles*

*Positive Nucleus Repels Alpha Particles*

very fast (about $2 \times 10^7$ m/sec) and get quite close to the nucleus. Particles on a path other than that leading to a head-on collision are deflected through various angles. The greater the distance *p* of the actual path from path *C*, the smaller the scattering angle $\theta$. The distance *p* is sometimes referred to as the impact parameter. The alpha particles are deflected into hyperbolic paths.

6 * **Scattering and Atomic Number** We have seen that the scattering of the alpha particles results from the force of repulsion between the positive alpha particles and the positive nucleus, as described by *Coulomb's Law*. Since the nucleus is very small by comparison with the size of the atom, most alpha particles go through the atom at a considerable distance from the nucleus. These are hardly affected by the Coulomb force, which varies inversely as the square of the distance. The closer the alpha particles come to the nucleus in going through the foil, the greater the force of repulsion and the greater the scattering angle.

The scattering angle also depends on the charge of the nucleus. The positive charge of the nucleus is known as the *atomic number* of the element.

**FOR EXAMPLE:** Gold has an atomic number of 79; it has 79 protons in the nucleus of its atoms. Aluminum has an atomic number of 13; it has 13 protons in the nucleus. Alpha particles passing at a certain distance from a gold nucleus will be pushed by a greater Coulomb force than an alpha particle moving at the same distance past the aluminum nucleus. If the alpha particles have the same energy, the one going through the gold foil will be scattered through a greater angle. In general, the number of alpha particles scattered beyond a given angle increases with the atomic number, or charge of the nucleus.

7 **Problems of the Rutherford Model of the Atom** Rutherford's concept of the atom explained the scattering of alpha particles. He himself realized that it raised other questions:

1. What keeps the protons in the nucleus together? Since they are positively charged and close together, the Coulomb force should be very large.

2. What keeps the electrons outside the nucleus from falling into the nucleus? Negative electrons are attracted by the positive nucleus.
3. If, in answering question 2, we assume that the electrons orbit around the nucleus, as planets around the earth, the electrons have a centripetal acceleration. According to *Maxwell's Theory*, accelerated electrons should radiate electromagnetic energy. Why is no such radiation observed?

# 8 The Bohr Model of the Hydrogen Atom

Planck's theory (1900) explained the continuous spectrum. Still unexplained was the bright line spectrum obtained by the dispersion of light given off by glowing gases.

**FOR EXAMPLE:** The light given off by hydrogen through which there is an electric current, is dispersed into a few visible lines corresponding to definite wavelenghts. These wavelengths had been carefully measured, and Balmer had been able to figure out a formula which allowed him to predict the wavelength of some as yet undiscovered lines in the hydrogen spectrum (1886). This was an empirical formula, that is, Balmer had not derived it on the basis of some theory. He devised a formula to fit the experimental results. Could a theory be developed which would lead to a similar formula?

*Niels Bohr*
Courtesy of Addison-Wesley

Niels Bohr (1913) explained the hydrogen spectrum by adopting the Rutherford model of the atom and adding a few radical ideas:

1. The electron in the hydrogen atom is restricted to certain definite stable orbits.
2. The most stable of these orbits is the one closest to the nucleus. In this orbit the atom has the least amount of energy. When the electron is in this orbit, the atom is said to be in its *ground state*. To move into the other orbits, the atom (or the electron) must absorb energy. The atom is then said to be *excited*. In the ground state the atom has the least energy.
3. While it is in these stable orbits or energy states, the electron does not radiate.
4. When an electron jumps from a higher energy state ($E_3$) or outer orbit to a lower energy state ($E_2$) or orbit closer to the nucleus, the difference in energy is radiated in the form of a photon:

   $$E_3 - E_2 = hf.$$

   If we know $f$, the frequency of the radiation, we can calculate its wavelength, since speed of light = frequency × wavelength. An equal quantum of energy must be absorbed by the atom to have the electron jump back to the outer orbit.
5. Only specified orbits are permitted. (Bohr specified these orbits by adding the condition that for all orbits of the electron:

*

$$mvr = \frac{nh}{2\pi}$$ where $m$ is the mass of the electron
$v$ its orbital speed
$r$ the radius of its orbit
$h$ is *Planck's Constant*,
$n = 1, 2, 3....,$ and are known as principal quantum numbers. The product $mvr$ is known as the angular momentum of the electron due to its orbit around the nucleus.)

The Bohr model worked extremely well for hydrogen. It explained the bright line spectrum of hydrogen remarkably, as will be shown later. It worked poorly for more complex atoms. Bohr and others found it necessary to introduce modifications. The idea of circular orbits, especially, has turned out to be too simple. It is often better to speak of energy states and energy levels rather than of orbits. Other modifications will be considered later.

9 **Energy Levels and Atomic Spectrum of Hydrogen** When the atom is in its most stable state, it has the least energy possible. It is then in the ground state. This is the condition it normally tends to achieve. In the diagram this is represented by the line marked $n = 1$. We can take this condition of the atom as our energy reference, and arbitrarily set it equal to zero for the purpose of calculations. This is marked in the right column as 0.00.

The atom can absorb energy only in discrete amounts, in definite quanta. The hydrogen atom in the ground state can absorb nothing less than 10.2 electron volts. If it gets a quantum of 10.2 ev, it jumps from the ground state to the state $n = 2$. The atom is then excited. It can be excited to still higher energy levels.

**FOR EXAMPLE:** If it had absorbed 12.08 electron volts, it could have jumped immediately to the energy level $n = 3$. Only jumps described by whole-number values of

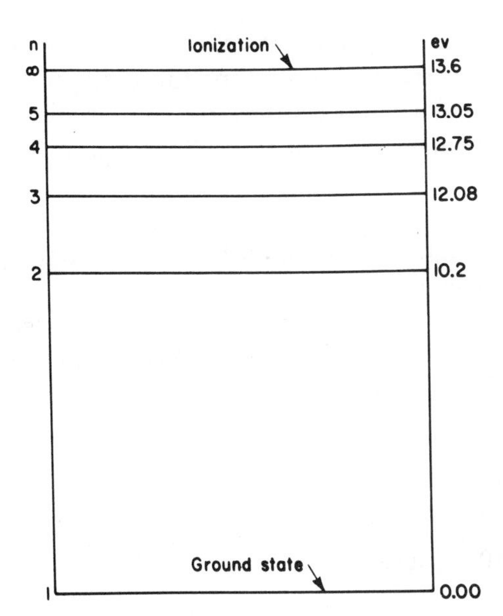


*Energy Levels for Hydrogen*

$n$ are allowed. The values of $n$ are known as the *principal quantum numbers*. We say that the energy jumps are quantized.

When an atom is in an excited state, it tends to return to the ground state. It may do so in one jump, or by stopping momentarily at intermediate energy levels. For each jump, a photon of energy is released with a quantum of energy equal to the difference between the two levels.

**FOR EXAMPLE:** Suppose an atom is in the state $n = 5$. If it returns to the ground state in one jump, it releases a photon with an energy of 13.05 electron volts. It may jump, instead, to $n = 4$, then to $n = 3$, and then to $n = 1$. The corresponding photons released have energies of $(13.05 - 12.75)$ ev, $(12.75 - 12.08)$ ev, and $(12.08 - 0.00)$ ev; or 0.30 ev, 0.67 ev, and 12.08 ev. Notice that the quanta for successive jumps have different values. Instead of speaking of an atom changing energy state, we also speak of the electron gaining or losing energy.

The ionization state ($n = \infty$) or the ground state ($n = 1$) may be taken as the reference level and marked zero. If the ionization state is taken as the reference, the other energy levels are negative. This is shown for hydrogen in the next diagram. Of course, the difference in energy levels is the same no matter which level is selected for reference. (There may be a difference in the last digit because of rounding off of numbers.) Notice that in the *Reference Tables* the ionization state is taken as the reference level.

**10 Balmer Series** We mentioned earlier that long before the Bohr theory, Balmer had been able to write an empirical formula for the wavelengths of some of the lines in the hydrogen spectrum: for all that were known in his day. It turned out that all these lines are due to electrons returning from an excited state to the second energy level, $n = 2$.

**FOR EXAMPLE:** Electrons may jump from $n = 5$ to $n = 2$, from $n = 4$ to $n = 2$, etc. All the lines produced by jumps to the second energy level are said to form the *Balmer Series.* Those known in Balmer's day are visible.

As we mentioned, other jumps are possible. The Bohr Theory led to the prediction of other series, and these were later found.

**FOR EXAMPLE:** Since the electrons end up in the ground state, there should be wavelengths observable corresponding to the release of photons for jumps from the second energy level to the first, $n = 1$. Such a series was found by an American physicist at Harvard University, Professor Lyman. Most of these lines are in the ultraviolet region.

**EXAMPLE:** Calculate the wavelength of the light emitted by a hydrogen atom in which the electron jumps from the energy level $n = 3$ to $n = 2$.

According to the *Energy Level Diagram* in the *Physics Reference Table* (also shown on the next page), the energy released in the jump is (−1.51)−(−3.39) ev. Therefore, the photon which is released has an energy of about 1.9 electron volts. Since, from the Table:

$$1\ ev = 1.60 \times 10^{-19}\ \text{joule},$$
$$\text{energy of photon} = 1.9 \times 1.60 \times 10^{-19}\ \text{joule}$$
$$= 3.04 \times 10^{-19}\ \text{joule}.$$
$$\text{But energy of photon} = hf = hc/\lambda;$$
$$\therefore (hc/\lambda) = 3.04 \times 10^{-19}\ \text{joule}$$
$$(6.63 \times 10^{-34} \times 3.0 \times 10^{8}/\lambda) = 3.04 \times 10^{-19}$$
$$\lambda = 6.6 \times 10^{-7}\ \text{meter}$$
$$\lambda = 6{,}600\ \text{Angstrom units}.$$

Notice that this is a *Balmer Line* and is in the visible range. Also notice that although we get the energy in electron volts from the chart, we must convert it to joules if we are going to use the value of *Planck's Constant* in the table, since that is given in joule-second.

**11 Excitation of Atoms** The process of raising the energy of atoms is called excitation. How is this done? One method is to bombard the atoms of a gas with electrons by passing an electric current through it. This was done in the famous *Franck-Hertz Experiment* (1914). If the bombarding electrons move slowly, they collide with the gas atoms but don't excite them. The collision is elastic. The speed of the bombarding electrons is increased by raising the voltage. If the speed of the electrons is increased sufficiently, they will have the energy needed to raise the atom to the next energy level. The gas atom can accept energy from the bombarding electron only if the latter can supply the quantum needed for the jump to a higher energy level. The gas atoms can accept energy only in these discrete amounts, these quanta. Different gases have different excitation energies.

**FOR EXAMPLE:** The minimum energy needed to excite the atoms in mercury vapor is 4.86 electron volts; we saw that for hydrogen it is 10.2 electron volts. This explains why each element has its own *characteristic spectrum*: the different excitation energies correspond to spectral lines of different wavelengths.

Suppose the electrons used in the Franck-Hertz experiment have an energy greater than 4.86 electron volts when they collide with an atom of the mercury vapor? It still loses only the 4.86 electron volts needed to excite the mercury atom; it retains the rest as kinetic energy.

The experiment shows that atoms of a gas can accept energy only in discrete amounts, only in definite quanta. In other words, atoms have only certain allowable energy levels. Atoms of different elements have different energy levels, and, therefore, emit different spectra. Some of the energy levels of mercury vapor are shown.

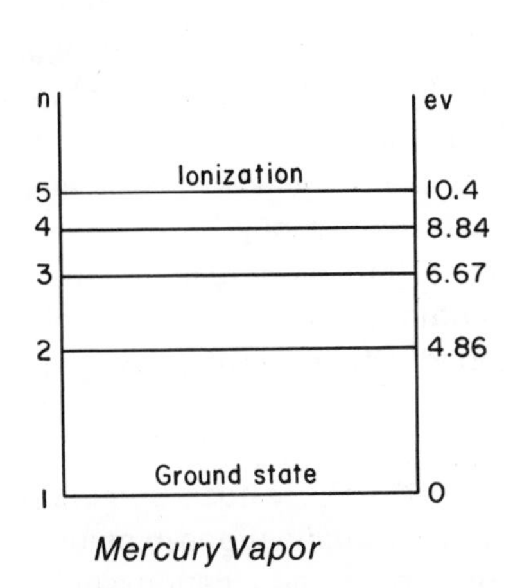


*Mercury Vapor*

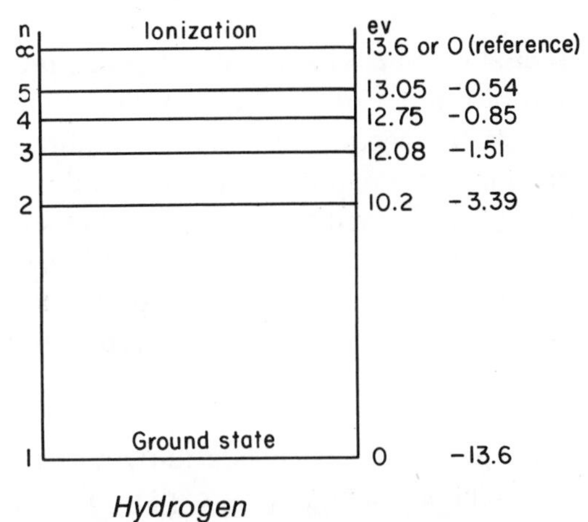


*Hydrogen*

Atoms may also be excited by collision with other atoms and molecules. If a gas is heated, its molecules move faster. Therefore, heating a gas may result in collisions which give an atom the quantum of energy needed to excite it. Another method is to shine light or other electromagnetic radiation of the right frequency on the gas. This will be discussed below under *Absorption Spectrum*.

12 * **Absorption Spectrum** We have seen above, that when an atom returns to the ground state it releases the same amount of energy which it absorbed on getting excited. The energy it absorbs is quantized and so is the energy it releases.

If we study the spectrum of luminous sodium vapor carefully, we see two yellow lines. These lines are characteristic for sodium vapor. They have definite wavelengths corresponding to the quanta of energy which they emit on returning to the ground state. If intense white light is passed through cool sodium vapor, the spectrum of the transmitted light is continuous except for two dark lines occurring in exactly the same yellow region where the two sodium wavelengths would be. The sodium vapor has absorbed photons which have exactly the same energy that it can emit when excited. The rest of the white light consists of photons with the *wrong* quanta of energy and cannot be absorbed.

The sun's spectrum is an absorption spectrum. What appears to be a continuous spectrum on superficial examination, turns out to have many black lines present. It has been possible to show that these lines correspond to wavelengths emitted by elements known on earth. Helium was thus first discovered in the sun's spectrum. Most of the absorption lines are the result of photon absorption by the cooler gases surrounding the hot sun. Some absorption takes place in the earth's atmosphere.

13 **Ionization Potential** We have seen that it is possible to excite a gas by applying a difference of potential to it. If we increase the voltage high enough, the atom can absorb enough energy to remove its electron completely, that is, to ionize the gas. (This corresponds to $n =$ infinity.) The minimum voltage needed to remove an electron from an atom is called its *ionization potential*. For hydrogen, this is 13.6 volts. We can figure this out from the energy level diagram in the *Reference Table*. The ionization energy for hydrogen is given as 13.6 electron volts. This is the work done in removing one electron from the atom. Work = charge × difference of

potential. The unit of charge is the electron. Therefore, the difference of potential is 13.6 volts: 1 electron $\times$ 13.6 volts = 13.6 electron volts.

14 ** **Excitation and Resonance** Do you recall the experiment on sympathetic vibration of tuning forks? If one of two identical tuning forks held near each other is struck, the second fork starts vibrating with a large amplitude. The second fork has the same natural frequency as the first, and is able to build up a large amplitude of vibration if the sound energy reaching it has this frequency. You may see an analogy here with the absorption of photons by a gas. If the light reaching the gas has the same frequency as is required for the quantum to excite the atoms of the gas, then the incident light will be readily absorbed. Otherwise it will be transmitted.

15 ** **The Laser** The laser is a fairly new device (invented 1960) whose tremendous possibilities are still being explored. It has already found applications in medicine, eye surgery, industry, and communications. Some of its applications depend on the fact that the laser provides a narrow beam of intense light. Other applications depend on its being monochromatic coherent light. Ordinary light is not coherent, that is, it does not stay in phase, because the atoms emit the light in short bursts, bursts lasting a very small fraction of a second. Successive bursts occur at random.

Laser light is coherent because excited atoms are stimulated to emit light in regular sequence. The word *laser* is formed from the first letters of the key words in the phrase, light amplification through stimulated emission of radiation.

The first laser was made by using a specially prepared ruby crystal. Many lasers now use gases, such as a mixture of neon and helium. Although the details are too complex to be discussed here, some of the basic principles are similar to those we already described. The gas is sealed in a tube the ends of which are partially silvered. The gas is put in an excited state, for example, by applying a suitable voltage; in the ruby laser a flash-tube surrounds the ruby. Light emitted by the return of some of the atoms to a lower energy state is reflected at one end. As this light moves through the tube, it stimulates other atoms which are in the right energy level to emit photons of the same frequency. Thus the amount of light moving back and forth through the tube builds up in intensity as more and more excited atoms are stimulated to emit photons. Some of the light keeps escaping at the end (or ends) in a fine beam. Where does the energy come from? As mentioned before, many gas lasers use an external voltage source to excite the atoms of the gas.

## 16 Dual Nature of Light

We saw how the wave theory of light gradually replaced the corpuscular theory. Interference, diffraction, and polarization of light can be explained well on the basis of a wave theory. In fact, since light can be polarized, this indicates that light is a transverse wave. This agrees with *Maxwell's Theory*, that light is just part of the electromagnetic spectrum.

On the other hand, the wave theory failed to explain the photoelectric effect. This required that we think of light and other electromagnetic energy as being concentrated in little bundles, the photons, which do not spread out like a wave, but stay together like particles. The same concept was required by Bohr in explaining the bright line spectra. Another phenomenon involving electromagnetic effects which could not be explained on the basis of the wave theory is the Compton effect.

**Compton Effect** When a beam of X-rays is incident on a material such as carbon, some of the scattered X-rays have a frequency lower than the incident X-rays. This effect was explained by A. H. Compton in 1923 by assuming that the X-rays consist of a stream of photons. Each photon has momentum as well as energy. When a photon collides with the electrons in the (carbon) atom, momentum and energy are conserved. In other words, when a photon and a particle collide, the photon behaves like a particle. Since the photon of X-ray loses some of its energy, the rebounding photon has a lower frequency than the incident X-ray photon: the new $hf$ is less than the original $hf$.

17 **Wave Property of Moving Particles** If electromagnetic waves can have properties of both waves and particles, should not ordinary matter also have properties of waves and of particles? In 1924, De Broglie proposed that a moving particle has a wavelength which is inversely proportional to its momentum. It is known as the *De Broglie Wavelength*. He used an expression analogous to the one used to explain the *Compton Effect*. (The wavelength is equal to *Planck's Constant* divided by the momentum of the particle.)

These matter waves can be observed, but only for small particles such as electrons. Davisson and Germer obtained diffraction patterns (1925) of moving electrons. Neutrons have been used since. We see that not only does light have these dual properties of wave and particle, but so does ordinary matter.

** It is easy to remember the expressions for the momentum of a photon, if you recall Einstein's expression for the equivalence of matter and energy: $E = mc^2$, which will be discussed in more detail later. Now, the energy of a photon is $hf$, the product of *Planck's Constant* and the photon's frequency. Therefore,

$$mc^2 = hf.$$

Divide both sides by $c$, the speed of light:

$$mc = hf/c.$$

But $c/f$ is the wavelength, and $mc$ is the product of mass and velocity, analogous to momentum of a particle ($p$). Therefore,

$$p = h/\lambda.$$

We can use the same expression to get the *De Broglie Wavelength* for a particle. Just let $p = mv$, the momentum of the particle. Then the De Broglie wavelength is given by:

$$\lambda = h/p.$$

The wave property of electrons is used in the electron microscope.

**Other Views of the Atom** The quantum theory produced a drastic impact on physics. Einstein and Bohr generalized Planck's novel idea, and it then made possible the explanation of puzzling phenomena not only in light, but also in other areas such as heat.

This quantum theory was expanded still further into the more complex theory of quantum mechanics. Its mathematics is more complex, but it incorporates the duality of matter, the fact that particles such as electrons have a wave characteristic.

We can think of the electron in the atom as a standing wave whose wavelength is given by the De Broglie formula. In the excited atom, the electron is a standing wave with more nodes. The excited states are also known as stationary states.

Another part of quantum mechanics stresses the fact that we can measure the position of the electron with only a limited degree of certainty. The description of the electron around the nucleus as a standing wave is then merely an indication of the probability of finding the electron at a particular position.

** It is possible to show quickly that Bohr's assumption follows from the De Broglie wavelength:

$\lambda = h/p$, where $p$ is the momentum of the electron,
$\therefore \lambda = h/mv$, $m$ its mass,
$v$ its speed in the orbit.

Assume that one wavelength of the standing wave just fits the circumference of the orbit:

$$\lambda = 2\pi r.$$

Therefore $2\pi r = h/mv$;

then $mvr = h/2\pi$.

For the ground state, one wavelength fits the circumference. For the first excited state (second orbit), $n = 2$, two wavelengths fit into the circumference, etc.

**Is The Model Vague?** Do you have trouble visualizing the electron in the atom? Don't let that disturb you. The model of the atom has had to become more cloudy to account for its properties. Remember that no one has ever seen an electron. In trying to visualize it, we have problems similar to those of a person born blind who is trying to visualize a red rose.

Keep in mind the function of the scientist's model or theory. The term *model* used in this connection is not intended to be a scale reproduction of the actual object or system. It is just about hopeless to do this for something we can't see. The model is intended to be a simple usable description which incorporates our concepts and observations. It helps the scientist to systematize and summarize observations, to show connection among many apparently unrelated measurements and phenomena. To be of real value, the model must suggest new experiments and predict the results of these experiments.

The early atomic theory gave good results with a simple picture of the atom: a little sphere which could not be broken down into smaller parts. This model was adequate for the description of the ideal gas. This model is still useful when thinking about many phenomena of actual gases under conditions of moderate pressure and temperature.

More observations indicated that this model of the gas had to be revised: an atom has a nucleus and is surrounded by electrons.

Bohr's picture or model of the atom visualizes the electrons as very tiny spheres which move in definite paths around the nucleus. It was useful in explaining and predicting new lines in the hydrogen spectrum. We saw that this model had to be revised under the pressure of more data. The mathematics of quantum mechanics is clear to the people trained in its use, but the picture it presents is not sharp. However, it has led to many predictions which were later verified. More elementary particles have been discovered, and the nucleus itself has been scrutinized. Particles which are now classified as elementary, may some day turn out to be divisible into still smaller particles. If a cloudy model helps in such discoveries, it serves the scientist's purpose.

Properties of the nucleus will be considered in the next chapter.

## Questions and Problems

1. *a.* Give one property of a photon which shows that it is like a particle. *b.* Give another characteristic of a photon which depends on a wave property.
2. Considering the electromagnetic spectrum, as we go up in frequency, what happens to the energy of the photons?
3. Considering the electromagnetic spectrum, as we increase the wavelength, what happens to the energy of the photon?
4. Calculate the quantum of energy for an electromagnetic wave whose frequency is $10^{15}$ cycles/sec *a.* in joules *b.* in electron volts.
5. Calculate the energy of a photon whose frequency is $4.2 \times 10^{15}$ hertz *a.* in joules *b.* electron volts.
6. Calculate the energy of a photon for light whose wavelength is *a.* $4.0 \times 10^{-7}$ *m* *b.* 5500 Angstrom units.
7. A monochromatic beam of light falling on a certain metal results in the emission of electrons. If a partially silvered piece of glass is placed in the path of the beam, what happens to *a.* the energy of the photoelectrons *b.* the number of photoelectrons?
8. The threshold frequency of a certain metal is $10^{15}$ cycles/sec. Calculate its work function.
9. How large is the work function of a substance whose threshold frequency is $3.5 \times 10^{14}$ cycles/sec?
10. What is the maximum kinetic energy of the photoelectrons in a certain experiment, if the stopping voltage is *a.* 10 volts *b.* 12 volts?
11. For the photoelectrons in question 10, calculate the maximum speed.
12. The metal described in question 8 is illuminated with electromagnetic radiation whose wavelength is 5000 Angstroms. Calculate *a.* the maximum kinetic energy of the emitted electrons *b.* the speed of these electrons.
13. The metal described in question 8 is exposed to a beam of electromagnetic radiation whose wavelength is 2000 Angstroms. Calculate *a.* the maximum kinetic energy of the emitted photoelectrons *b.* the speed of the electrons.
14. Electromagnetic radiation whose wavelength is 2500 Angstroms shines on the metal described in question 9. Calculate *a.* the maximum kinetic energy of the emitted electrons, *b.* the speed of these electrons.
15. Monochromatic radiation whose photon energy is $1.8 \times 10^{-19}$ joule falls on a metal whose work function is $6.0 \times 10^{-19}$ joule. *a.* Calculate the

maximum energy of the emitted photoelectrons. *b.* What would be the energy of photoelectrons if the photon energy were $7.8 \times 10^{-19}$ joule?

16. What observation led Rutherford to the belief that an atom is mostly empty space?

17. In *Rutherford's Scattering Experiment*, why should alpha particles passing close to the nucleus of the metal foil be deflected more than those farther away?

18. If faster moving alpha particles were used in the scattering experiment, what would be the effect on the angle of scattering?

19. *a.* What is meant by an angle of scattering of 180°? *b.* When does this occur? *c.* Why does this happen to only a very small fraction of the alpha particles in the *Rutherford Scattering Experiment*?

20. If a photon having an energy greater than 13.6 *ev* is absorbed by unexcited hydrogen, what happens to the photon energy?

21. If a hydrogen atom is in the excited state $n = 3$, it can return to the ground state in two jumps. Calculate the *a.* energy of the photon which results in a Balmer line, and *b.* the wavelength of this line.

22. If the electron in an excited hydrogen atom jumps from $n = 3$ to $n = 1$, what is *a.* the energy of the released photon *b.* the resulting wavelength *c.* the frequency of the emitted radiation?

23. How much energy is absorbed by an atom of hydrogen which goes from the ground state to energy level $n = 4$?

24. Referring to the energy level diagram in the text for mercury vapor, *a.* what is the least amount of energy required to excite a mercury atom in the ground state? *b.* How much energy is required to ionize a mercury atom in the ground state? *c.* What is the ionization potential of mercury?

25. If an X-ray photon is scattered, with a resulting increase in wavelength, as in the *Compton Experiment*, what happens to *a.* its energy *b.* its momentum. *c.* What gains energy and momentum?

26. What happens to an electron's *De Broglie Wavelength* when the electron's momentum is increased?

27. Assuming that an electron in circular orbit around a nucleus is at a distance of $10^{-8}$ cm from the nucleus, when in the ground state, what is its *De Broglie Wavelength*?

28. Calculate the wavelength of a matter wave of an electron whose speed is $1.5 \times 10^{7}$ m/sec.

29. Using the ionized state as the reference level, one can derive the following expression for the energy of the different levels of the hydrogen atom:

$$E_n = \frac{-13.6\ ev}{n^2},$$

where $n$ is the principal quantum number. Calculate the energy for $n = 1$, 2, and 3, and compare your answers with the values shown in the *Reference Tables*.

# Test Questions

1. The number of neutrons in each alpha particle is 1. 1 2. 2 3. 3 4. 4.
2. As the intensity of monochromatic yellow light increases, the 1. energy of each photon increases 2. energy of each photon decreases 3. number of photons increases 4. number of photons decreases.
3. As compared with blue light, the energy of each quantum of yellow light is 1. greater 2. smaller 3. the same.
4. If the frequency of a photon decreases, its momentum 1. increases 2. decreases 3. remains the same.
5. Photons incident on a certain metal have an energy of 5.0 *ev*. If the work function of the metal is 3.0 *ev*, the maximum kinetic energy of an emitted electron is, in *ev*, 1. 8.0 2. 2.0 3. 3.0 4. 15.
6. If, in question 5, the wavelength of the incident photons is decreased, their energy 1. increases 2. decreases 3. remains the same.
7. If a certain radiation has a frequency of $2.0 \times 10^{15}$ cycles per second, the energy of its photons is, in joules, 1. $3.3 \times 10^{-19}$ 2. $13 \times 10^{-19}$ 3. $8.6 \times 10^{-19}$ 4. $1.6 \times 10^{-19}$.
8. When light shines on certain metals, electron are given off. This phenomenon is known as 1. thermionic emission 2. photoelectric effect 3. piezoelectric effect 4. radioactivity.
9. The speed of a photon, as compared with the speed of an electromagnetic wave, is 1. greater 2. less 3. the same.
10. According to De Broglie, as we increase the mass of a particle, its wavelength 1. increases 2. decreases 3. remains the same.
11. Photoelectrons are emitted only when the light directed at the substance exceeds a certain minimum 1. amplitude 2. frequency 3. wavelength 4. velocity.
12. The energy of each photon of weak red light, as compared with the energy of each photon in the electromagnetic wave of your favorite radio station, is 1. smaller 2. greater 3. the same.

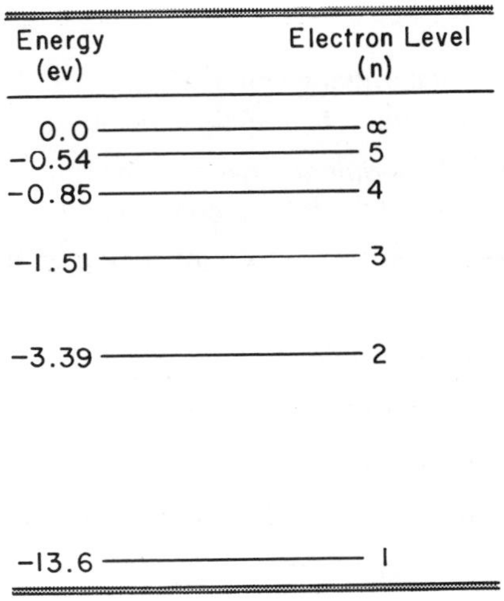

| Energy (ev) | Electron Level (n) |
|---|---|
| 0.0 | $\infty$ |
| −0.54 | 5 |
| −0.85 | 4 |
| −1.51 | 3 |
| −3.39 | 2 |
| −13.6 | 1 |

The diagram above shows some of the energy levels of a hydrogen atom. Refer to it in answering questions 13–18.

13. When an electron jumps from level $n = 5$ to $n = 2$, it will release 1. 13.1 *ev* 2. 10.2 *ev* 3. 2.9 *ev* 4. 0.5 *ev*.

14. A line for the Balmer series will be produced if an electron jumps from level $n = 5$ to $n$ equals 1. 1 2. 2 3. 3 4. 4.

15. If an atom has an electron in level $n = 5$, the minimum energy required to ionize the atom is 1. 0.54 *ev* 2. 13.1 *ev* 3. 13.6 *ev* 4. infinity.

16. An atom in the ground state will not be excited or ionized if it is irradiated with electromagnetic waves whose photon energy is, in *ev*, 1. 10.2 2. 12.5 3. 13.1 4. 14.2.

17. An atom in the ground state will not be ionized or excited if it is bombarded with electrons whose kinetic energy is, in *ev*, 1. 9.8 2. 12.1 3. 13.1 4. 14.1.

18. As compared with the wavelength for the jump from $n = 2$ to $n = 1$, the wavelength for the jump from $n = 3$ to $n = 2$ is 1. greater 2. less 3. the same.

19. When alpha particles are aimed at thin gold foil, the probability of a head-on collision between an alpha particle and a gold nucleus is 1. very high 2. very small 3. 1/2 (an even chance).

20. In the Rutherford scattering experiment, as the atomic number of a nucleus in the foil increases, the angle of scattering of a given alpha particle 1. increases 2. decreases 3. remains the same.

## Topic Questions

1.1 *a.* In what ways was the wave theory of light successful? *b.* In what ways did it fail?

2.1 How is the energy of a quantum related to its frequency?

3.1 What are two features of the photoelectric effect which are not explained well by the wave theory of light?

4.1 What is meant by the work function of a material?

4.2 In the photoelectric effect, what determines the energy of the emitted electrons?

5.1 What experimental evidence made Rutherford decide that *a.* the atom has a core or nucleus? *b.* that this nucleus is very massive?

6.1 In *Rutherford's Experiment*, how did the angle of scattering depend on the atomic number of the foil which he used?

7.1 What questions did the Rutherford model of the atom leave unanswered?

8.1 What is meant by an *empirical formula*?

8.2 What is meant by an *excited atom*?

8.3 What are some of the important ideas in the *Bohr Theory* of the atom?

9.1 In what ways can an excited atom return to the ground state?

10.1 How is the *Balmer Series* produced?

11.1 What does the *Franck-Hertz Experiment* indicate?

11.2 What explains the fact that different elements have different, or characteristic, spectra?

12.1 How do the wavelengths of light emitted by an excited atom compare with those the atom can absorb?

13.1 What is meant by the term *ionization potential*?

15.1 Why is ordinary light not coherent?

16.1 *a.* Is momentum conserved when a photon collides with an electron?
*b.* In the Compton effect, what happens to the frequency of some of the scattered X-rays?

17.1 How can the *De Broglie Wavelength* be calculated?

18.1 What functions are served by a scientist's models?

# 18 The Nucleus of the Atom

## 1 Introduction

**2 The Nuclear Atom** What is the physicist trying to find out about the atom? What tools does he use? Rutherford's scattering experiments with alpha particles provided the data for a picture of an atom with a core or nucleus containing practically all of the mass of the atom. In 1932 Chadwick discovered the neutron. This led to the following picture of the atom. The nucleus of an atom is composed of neutrons and protons. The term *nucleon* refers to either. The *mass number* ($A$) of an atom is the total number of its nucleons, that is, the sum of the number of protons and neutrons. The *atomic number* ($Z$) is the number of protons; in a neutral atom, this is also equal to the number of electrons outside the nucleus. If $N$ is the number of neutrons in the nucleus,

$$A = Z + N.$$

There are over 100 different chemical elements. One way in which they differ from each other is in the charge of the nucleus of the individual atom, the atomic number. All elements whose atomic number is greater than 92 are man-made. They are known as the *trans-uranic* elements; uranium has an atomic number of 92. One of these is lawrencium, which has an atomic number of 103.

All atoms, except those of common hydrogen, contain neutrons in the nucleus in addition to protons. Most chemically pure elements contain atoms that are identical in atomic number (and therefore very similar in chemical properties), but differ in mass number because they differ in the number of neutrons. These are called *isotopes* of the element. For example, there are three known isotopes of hydrogen. The common isotope (sometimes called *protium*) has one proton in the nucleus, the next one (also called *deuterium*) has one proton and one neutron, and the third (called *tritium*) has one proton and two neutrons. These are represented symbolically below in the same order:

$${}_1H^1;\ {}_1H^2;\ {}_1H^3. \text{ or } {}_1^1H;\ {}_1^2H;\ {}_1^3H.$$

Notice that in each case the mass number is written as a superscript and the atomic number as a subscript. Isotopes of the same element have the same atomic number; in this case, 1.

**3 Transmutation of the Elements** The medieval alchemists dreamed of changing cheap metals like lead into the precious gold. As the knowledge of chemistry progressed in the 19th century, this seemed an idle dream; it became "obvious" that elements are made of atoms which can not be broken down further. Of

*Marie Curie*
*Niels Bohr Library*

course, now we know that elements can be transmuted, but it may not be practical to change lead into gold. *Transmutation* is the conversion of one element into another. Becquerel discovered natural radioactivity in 1896, but he didn't fully understand what change was taking place. He did find out that the radiations are given off by the uranium, that this was not affected by external light, and that these radiations can discharge an electroscope. (Remember that Rutherford's nuclear model of the atom came later.)

Naturally radioactive substances emit three types of radiation, as can be shown by the following experiment. A radioactive substance is placed at the end of a long narrow groove in a lead block. This block is placed in an evacuated chamber at one end of which is a photographic plate. A magnetic field is applied perpendicular to the plane of the paper. Three distinct lines will develop on the plate. From the deflection due to the magnetic field it is determined that the radiation deflected to

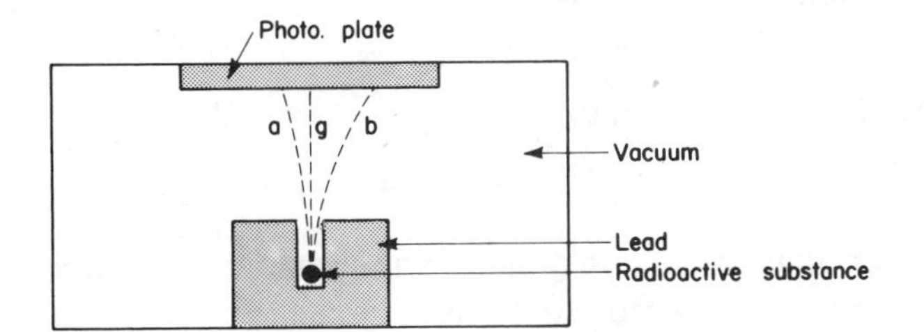


*Radiation from Radioactive Substance*

the left *a* in the diagram, is positively charged; it is called alpha rays. The one deflected to the right *b* is negatively charged and is called beta rays. The one not deflected at all *g* is called gamma rays.

Rutherford showed (1909) that alpha rays consist of positively charged helium. The alpha ray is a stream of helium nuclei. An alpha particle is a helium nucleus. Beta rays were shown to be streams of rapidly moving electrons, also called beta particles. Gamma rays are electromagnetic waves similar to X-rays.

4 **Nuclear Mass and Binding Energy** The mass of the nucleus is less than the sum of the masses of the nucleons which compose it. This difference in mass is sometimes called the *mass defect* of the atom. Why does this mass difference exist? Einstein had predicted (1905) that mass and energy might be converted into one another. (Sometimes we say that they are manifestations of the same thing.)

$E = mc^2$ where $m$ is the mass converted, in kilograms
$c$ the speed of light, in meters per second
$E$ the equivalent amount of energy, in joules.

Of course, other systems of units may be used.

The nucleons are held together in the nucleus by a very strong force, called the *nuclear force*. It is a short range force, meaning that it is significant only at very short distances. By comparison, the gravitational forces and Coulomb force in the nucleus are small, especially the gravitational force.

Imagine trying to separate the nucleus into its nucleons. We would have to do a great deal of work pulling the nucleons away from each other against the attraction of the nuclear force. The *binding energy* of a nucleus is the work that must be done in order to separate it into its nucleons. The energy which we have expended in doing this work is not lost; it is gained by the nucleons in the form of increased mass. In other words, the binding energy is the energy equivalent of the increased mass of the separated nucleons. (The effect of the atom's electrons is negligible.) If these nucleons join again to form a nucleus, this binding energy is released and the mass of the nucleus is less than the total mass of its isolated nucleons. When the binding energy is divided by the number of particles in the nucleus, we get the binding energy per nucleon. The greater this binding energy per nucleon, the more stable the atom. We can thus compare stability of isotopes.

**EXAMPLE:** What is the energy equivalence of 1 gram of mass, in joules and electron-volts?

$E = mc^2$

$= (0.001 \text{ kg})(3 \times 10^8 \text{ m/sec})^2$

$= 9 \times 10^{13}$ joules. (This is a tremendous amount of energy. It turns out to be equal to 25,000,000 kilowatt-hours.)

1 electron volt $= 1.6 \times 10^{-19}$ joule

$9 \times 10^{13}$ joule $= (9 \times 10^{13})/(1.6 \times 10^{-19})$ *ev*

$= 6 \times 10^{32}$ *ev*.

5 * **Atomic Mass Unit** The number of atoms in one gram of any substance is tremendously large. The mass of each atom is therefore extremely small. In talking about nuclear changes it is convenient to introduce a small unit of mass. The *atomic unit of mass (amu)* is defined as one-twelfth of the mass of an atom of *carbon 12*, that is, of the isotope of carbon which has a mass number of 12. Of course, 1 amu is then approximately equal to the mass of one neutron, or $1.7 \times 10^{-27}$ *kg*. The mass

of the proton is slightly less than that of the neutron. You may notice from the Reference Table that 1 *amu* = 931 *Mev*; *Mev* stands for million electron volts. Try to check this by using the method illustrated in the above Example.

6 * **Natural and Artificial Transmutation** Becquerel discovered natural radioactivity. In such cases the nuclei of some of the atoms of high atomic numbers disintegrate spontaneously. Rutherford bombarded lighter elements with alpha particles and discovered that the nuclei of some of the bombarded atoms disintegrated. He thus achieved the first artificial transmutation (1919). This is also called *induced transmutation*.

7 **Conservation of Charge and Mass Number** When a nucleus disintegrates, the total charge before the disintegration must equal the total charge of the products. We add the charges algebraically; for example, the total charge of 2 protons and 1 electron is +1. Also, during such a disintegration, the total mass number of the products before the reaction must equal the total mass number of the products. This will be illustrated below under alpha and beta decay.

8 **Alpha Decay** A radioactive change is also referred to as radioactive decay. *Alpha decay* is the emission of an alpha particle from a nucleus. This occurs when radium (*Ra*) changes to radon (*Rn*). The alpha particle is a helium nucleus. This can be represented by the following equation:

$$_{88}Ra^{226} \rightarrow {}_{86}Rn^{222} + {}_{2}He^{4}.$$

Notice that charge is conserved. The subscript gives the atomic number or charge of the nucleus. We start out with a charge of 88, the atomic number of radium. We end up with a charge of 86, the atomic number of radon, plus a charge of 2, the atomic number of helium.

Notice that mass number, too, is conserved. We started out with 226, the mass number of radium. We get radon and helium, with mass numbers of 222 and 4, respectively. 222 + 4 = 226. Another way to look at it is to say that the emission of an alpha particle decreases the mass number by 4 and the atomic number by 2, since these are appropriate numbers of the alpha particle: it is composed of two protons and two neutrons.

9 **Beta Decay** In some radioactive changes an electron is given off; it is ejected by the nucleus. *Beta decay* is the emission of an electron (beta particle) from a nucleus. You might ask, how can an electron come out of the nucleus if the nucleons are the proton and neutron? Physicists were confronted with this and another problem. The electrons ejected by the nucleus did not come out with a definite speed, as is true of alpha particles in alpha decay. The electrons are ejected with a wide range of speeds; some come out slowly, some fast. With great faith in the important Principle of Conservation of Energy, Pauli suggested in 1931 that there must be another tiny particle given off in beta decay: the *neutrino*. It is electrically neutral, and has practically no mass. The existence of the neutrino was experimentally verified in 1956. (A recent discovery indicates that there are actually two kinds of neutrinos; each should have an anti-particle.)

Where does the electron come from? The present theory is that in the nucleus a neutron can change into a proton, electron, and neutrino: (We shall not distinguish between the neutrino and the anti-neutrino.)

$$\text{neutron (in nucleus)} \rightarrow \text{proton} + \text{electron} + \text{neutrino.}$$

The neutrino hypothesis saved the principles of conservation of energy and conservation of momentum.

Bismuth (*Bi*) undergoes beta decay, producing polonium:

$${}_{83}Bi^{210} \rightarrow {}_{84}Po^{210} + {}_{-1}e^{0} + \text{neutrino.}$$

As in alpha decay, energy is also released. This is not shown. Notice again that charge is conserved. Bismuth has an atomic number of 83. Since an electron with a charge of minus one leaves the nucleus, the charge of the remainder increases by one: as shown above, another proton with a charge of +1 is produced and stays in the nucleus. The resulting atom must have a nucleus with a charge of 84. The element which has an atomic number of 84 is polonium. Since the neutrino is neutral, its zero charge does not affect the calculation.

Notice also that mass number is conserved. The mass number of bismuth is 210. The mass numbers of the electron and neutrino are each zero. Observe that the emission of a beta particle increases the atomic number by one, but does not change the mass number.

**10 Gamma Radiation** In many cases of radioactive decay, gamma radiation is found to accompany the emission of alpha particles and beta particles. Diffraction and interference experiments similar to those with X-rays showed that gamma radiation, too, is part of the electromagnetic spectrum and has short wavelengths similar to those of X-rays. X-rays are produced by changes outside the nucleus.

Study of gamma radiation suggests that the nucleus of an atom may be excited and may have energy levels analogous to those we studied for the hydrogen and mercury atoms. In the case of atoms, the photons which are emitted when the atom returns to a more stable state, have relatively long wavelengths, in the visible range or thereabouts. When the *nucleus* returns to a more stable state, the photons which are emitted have relatively short wavelengths and therefore relatively high energies. Gamma radiation consists of high energy photons originating in the nucleus. The emission of gamma radiation does not change the atomic or mass numbers.

**11 Radioactive Series** Most of the naturally radioactive isotopes can be fitted into one of three radioactive series, uranium, thorium, and actinium. The *Physics Reference Tables* shows the uranium series under the heading of Uranium Disintegration Series. Atomic mass number is plotted against Atomic number and chemical symbol. (Numerically, the atomic mass, if expressed in atomic mass units, is approximately equal to the mass number.) It shows the decay of uranium-238 successively into thorium-234, protactinium (*Pa*)-234, uranium-234, thorium-230, radium-226, etc. In each of these changes, mass number and atomic number are conserved.

**FOR EXAMPLE:** Take the first change:

$${}_{92}U^{238} \rightarrow {}_{90}Th^{234} + ?$$

Another particle must be given off; in order to conserve mass number, the other particle must have a mass number of 4: $234+4=238$. In order to conserve charge, the particle must have a charge of +2: $90+2=92$. The element which has an atomic number of 2 is helium. The reaction can then be written as:

$$_{92}U^{238} \rightarrow {}_{90}Th^{234} + {}_{2}He^{4}$$

In other words, an alpha particle is given off. In the next decay step, an electron is given off. In the series, decay takes place by a succession of alpha and beta particle emissions.

**12 Half Life** Radioactive disintegration does not take place instantaneously. If we have a mass of radioactive substance, we cannot predict which nucleus will disintegrate. It seems to be a completely random event. However, a definite fraction of the remaining nuclei will always disintegrate in the same time. The rate of disintegration is frequently expressed in terms of the *half life* of the isotope: the time required for one-half of the nuclei of a sample of the isotope to disintegrate.

**FOR EXAMPLE:** The half life of radium is 1620 years; the half life of polonium-210 is 1.38 days. If we have 20 grams of this polonium isotope, in 1.38 days only 10 grams will be left. In another 1.38 days only 5 grams will be left. Of course, atoms of other substances will be produced, and mass-energy is conserved.

We can speak about the half life as a length of time. One half life for the above polonium isotope is 1.38 days. Two half lives is 2.76 days ($2 \times 1.38$), etc.

Notice that after one half life, 1/2 of the original mass is left. After two half lives, $1/2^2$ of the original mass is left, etc. In general, after $n$ half lives, $1/2^n$ of the original mass is left:

$$m_f = \frac{1}{2^n} m_i,$$ where $m_f$ is the final mass, or mass of isotope left over, $m_i$ is the initial mass of the isotope.

**EXAMPLE:** If an element has a half life of two days, how much of it will be left after 8 days if you start with 64 grams of it? Eight days is 4 half lives of this isotope. Therefore $n=4$.

$$m_f = \frac{1}{2^4} m_i$$

$$= \frac{1}{16} \times 64 \text{ grams}$$

$$= 4 \text{ grams.}$$

**13 * Discovery of the Neutron** The discovery of alpha particles quickly led to their use by experimental physicists as atomic bullets. Rutherford himself (1919) bombarded nitrogen with alpha particles and produced oxygen-17, an isotope of oxygen. (The common oxygen isotope has a mass number of 16.) This was the first artificial transmutation of an element. It can be represented by the following equation:

$$_{7}N^{14} + {}_{2}He^{4} \rightarrow {}_{8}O^{17} + {}_{1}H^{1}.$$

Notice that a proton is emitted in the reaction.

When beryllium, a metallic element with an atomic number of 4, was bombarded with alpha particles, a strange, penetrating type of radiation was emitted. It was first thought that this was gamma radiation, since it had no charge. But this theory left other things unexplained. In 1932, Chadwick, as a result of many experiments, suggested that this radiation consisted of a stream of neutral particles; they are called neutrons. The beryllium reaction can be represented by the following equation:

$$_4Be^9 + {_2He^4} \rightarrow {_6C^{12}} + {_0n^1}.$$

Notice that carbon is also produced in the reaction.

Neutrons are ideal bombarding particles for many nuclear reactions because they are uncharged and do not experience an electrostatic repulsion or attraction by the nuclei. When they are very close to the nucleus, they are attracted by it because of the strong, short-range nuclear force. When a neutron enters the nucleus, the mass number of the atom increases by 1; the atomic number is not changed. Neutrons moving slowly near the nucleus have a better chance of being pulled into the nucleus than fast neutrons. On the other hand, positively charged particles must move very fast to overcome the electrostatic repulsion by the nucleus.

14 ** **Resonance Again** The study of spectra has given us information about the structure of atoms, molecules, and nuclei. To get data on the structure of atoms and molecules, visible light as well as infrared and ultraviolet radiation have been used. For the study of nuclei, gamma rays have served as a powerful tool. Much still has to be learned.

Gamma rays have a shorter wavelength than ultraviolet. Therefore its photons have a higher energy. Neutrons are also very useful for studying the structure of the nucleus for similar reasons. Remember that a neutron has a *De Broglie Wavelength*, and therefore an equivalent frequency ($c = f\lambda$). The wavelength depends on the neutron's momentum and energy.

If atomic nuclei are bombarded with neutrons, neutrons with a certain energy will be absorbed. Neutrons with somewhat more and somewhat less energy will be transmitted. In other words, the nuclei tend to absorb neutrons with certain frequencies better than with other frequencies. This is similar to, but of course not the same as, resonance in tuning forks, where a tuning fork absorbs best the energy from a passing wave which has the same frequency as the natural frequency of the fork.

It is also similar to what was observed in the *Franck-Hertz Experiment*. There we noticed electrons having certain specific energies give up this energy on collision with atoms. This furnished evidence of energy levels in atoms.

The absorption of neutrons having certain energies is evidence of energy levels in the nucleus. For example, thermal neutrons are readily absorbed by nuclei of uranium 235. If the energy of the neutrons is increased, fewer neutrons are absorbed (and more are transmitted). However, when the energy of the neutrons increases to about 1 electron volt, they are again absorbed strongly. At increased energies, less absorption takes place again, until the energy of the neutrons goes up to about 2 electron volts. Then the absorption increases again. (Thermal neutrons have very little energy; see next paragraph.)

15 ** **Nuclear Fission** The nuclear disintegrations described so far result in the emission of comparatively light particles, such as the proton. In 1939 Hahn and Strassmann discovered *nuclear fission*: the breaking up of a nucleus into two particles of intermediate masses. A study of the binding energy per nucleon of the elements had indicated that in the event of such fission, large amounts of energy should be

released; nuclear fission immediately became the concern of knowledgeable scientists.

Only certain massive nuclei are fissionable. These include uranium-235, thorium-90, and the man-made element, plutonium. Slow neutrons are most effective in producing fission of uranium-235. The term *thermal neutrons* refers to slow neutrons having kinetic energies approximating those of molecules of substances at ordinary temperatures. The kinetic energy of such neutrons is about 0.03 electron volt. Uranium-238 can fission with fast neutrons, but these are not readily captured by the nucleus. Uranium-235 can be fissioned with fast or slow neutrons; slow neutrons are more readily captured and are usually used.

When thermal neutrons are absorbed by uranium-235, the nucleus splits, usually into two nearly equal parts. This is accompanied not only by a release of energy, but also by the release of additional neutrons. This makes a *chain reaction* possible: the neutrons released by one disintegration are used in turn to fission other nuclei.

In a *nuclear reactor* or atomic pile the fissionable material is arranged so that the release of energy can be controlled. *Moderators* are used to slow down neutrons to thermal speeds, speeds at which they have thermal energies. The neutrons slow down by collisions with the atoms of the moderator. Graphite, paraffin, and heavy water have been effectively used as moderator materials. Heavy water uses the deuterium isotope instead of the common hydrogen. A nuclear reactor also incorporates *control rods*. These are used to absorb neutrons when the reaction threatens to go too fast. Boron and cadmium are effective absorbers of neutrons and are used in control rods. These may be moved at various distances into the reactor. Nuclear reactors are used throughout the world for many peaceful purposes. They are being used to produce electric power as well as radioactive isotopes for use in medicine and industry. Much research is still being done with the aid of the reactor.

An atomic bomb or A-bomb depends on the fission reaction, but there are two additional requirements. There must be at least a minimum or *critical mass*, and with the fission of each atom, on the average, there must be more than one neutron made available for fission of other atoms. Then the chain reaction proceeds rapidly. During such an explosion, the temperature of the material reaches several million degrees, the pressure rises to a few million atmospheres, and a large amount of radioactive material and gamma rays are produced.

16 * **Fusion** We saw that in fission, atoms with very large mass numbers split. In *fusion*, the nuclei of some light elements like lithium and hydrogen combine. Again there is a loss of some mass with the consequent release of energy. The fusion reaction requires a temperature of millions of degrees. This is therefore called a *thermonuclear reaction*, one that requires high temperature. The sun is believed to derive its energy primarily from fusion reactions.

In the H-bomb, fusion has been achieved in an uncontrollable way. An A-bomb is first used to produce the required high temperature. Scientists have been working hard towards achieving a controllable fusion reaction. One of the difficulties is that all known materials vaporize long before the required temperature is reached, and some other method has to be devised to contain the reactants. If it is achieved, tremendous amounts of energy will become available for peaceful purposes everywhere by the use of hydrogen present in the water molecules.

An example of the fusion reaction is the combining of common hydrogen (mass number 1) with its isotope tritium (mass number 3) to produce the common isotope of helium (mass number 4). Of course, large quantities of energy are released ($Q$).

$$_1H^1 + _1H^3 \rightarrow _2He^4 + Q.$$

**17 Particle Accelerators** We have mentioned some sub-atomic particles such as the proton, neutron, and electron. These are sometimes referred to as fundamental or elementary particles. Many others have been discovered, some of which you probably know: neutrino, positron, mesons, and hyperons. At least 30 different particles have been discovered. Then there is the intriguing discovery of antimatter. (The positron is the antiparticle of the electron. When two anti-particles meet, they annihilate each other with the resultant release of the equivalent energy.)

What is the significance of all these particles? Are they all fundamental? Are any of them? Are there more? Are there any undiscovered particles with a mass smaller than that of the electron, at present our lightest particle? We don't know. (A recent theory predicts a particle with a charge smaller than that of the electron; it is called the *quark*.) A scientist often has to admit ignorance. One purpose of experimental research is to remove this ignorance, to get facts and to develop a generalization, a *law*, and a new theory.

Particle accelerators have been built to help get new facts, and to check on theories. Accelerators are used to give charged particles sufficient kinetic energy to overcome the electrostatic forces of the nucleus. It may be helpful to think of the *De Broglie Wavelength* of the accelerated particle. The greater the kinetic energy or momentum of the particle, the smaller its *De Broglie Wavelength*: $\lambda = h/p$. The smaller the object we want to examine, the smaller must be the wavelength of the radiation we use.

**FOR EXAMPLE:** An ordinary microscope using light may be used to examine cells; an electron microscope using electrons, must be used to examine a virus. The wavelength of the radiation used must be no larger than the diameter of the object we are examining. The greater the energy of the particles used to bombard or *smash* the nucleus, the more detail the physicist expects to find about the structure of the nucleus and its particles. This has led to the discovery of some of the particles mentioned above.

Electrons, protons, or deuterons are commonly used as the particles or bullets of these accelerators. A *deuteron* is the nucleus of a deuterium atom. These particles can be accelerated by being placed in an electric field. In the *van de Graaff*

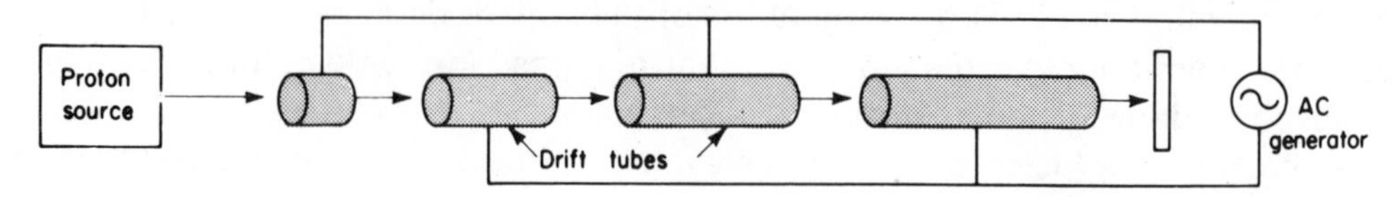


*Linear Accelerator*

machine and in the *linear accelerator* the particle gains speed along a straight path. The electric field is produced by a large difference of potential. In the linear accelerator, a series of cylindrical electrodes is used, and the particles acquire more and more energy as they pass between successive pairs of electrodes or drift tubes. (You don't need to bother with details.) In the van de Graaff generator, the accelerating voltage is produced by building up a large charge on a hollow sphere. The charge is brought to the sphere, which is at the end of a hollow column, by a moving belt within the column. To achieve high energies, long tubes and columns are needed.

In order to avoid the need for such long pipes, circular particle accelerators have been constructed. The great length of path is achieved by having the particles go around the circular path many times. This is done in the cyclotron, synchrotron, bevatron, and cosmotron. The charged particles are forced to go in a circular path by a magnetic field directed at right angles to the path of the particles.

The first of these was the *cyclotron*, invented by E. O. Lawrence in 1932. In the cyclotron, a flat, evacuated cylindrical box is placed between the poles of a strong electromagnet. Inside the box are two hollow, D-shaped electrodes called *dees*. The protons or deuterons to be accelerated are fed into the space between the electrodes. An AC voltage of high frequency (about $10^7$ cycles per second) is applied to the dees, and the particles are accelerated across the gap. The magnetic field bends the moving particles into a circular path. The voltage reverses in time

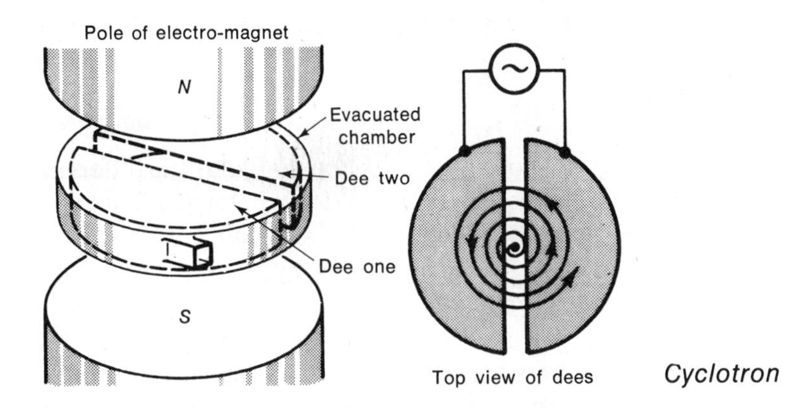


*Cyclotron*

to accelerate the particles emerging from the dee. After 100 trips around the cyclotron the particles may acquire an energy of several million electron volts (*Mev*). Successive semicircular paths in the dees are of larger radius, but each of these is covered in equal lengths of time. This is so because the greater speed just compensates for the greater length of path.

When a proton has acquired a speed corresponding to an energy of about 10 *Mev*, a relativity effect sets in: its mass starts to increase noticeably with further increases in energy. The same electric field produces a smaller acceleration, and the particle takes more time to cover each successive semicircle. The frequency of the AC voltage applied to the dees must therefore change to compensate for this increase in mass. In some accelerators the magnetic field as well as the frequency of the electric field is changed.

**18 Tools to Observe Nuclear Changes** How is a change in an atomic nucleus observed? In addition to knowledge of physics and mathematics, the experimental physicist needs equipment, equipment to produce a change and equipment to observe a change. Becquerel discovered radioactivity by observing a change in a photographic plate produced by uranium salts. Photographic plates are still used, and so is the simple electroscope. For some investigations more complex equipment is needed. This includes geiger counters, scintillation counters, cloud chambers, and bubble chambers. (You need not know details of construction.)

You are already familiar with the electroscope. By careful design and shielding, an electroscope can be made very sensitive. If a scale is added, it becomes an *electrometer* and can be used for more precise measuring. A charged electroscope will gradually discharge if a radioactive material is brought near it. Gamma rays and other radiation from radioactive substances collide with gas molecules in the air and ionize them. An ionized gas becomes a conductor.

*Geiger counters* detect charged particles by using their ionizing effect on a gas in a special tube. The tube is shown schematically, but there are different types. The tube contains a gas at relatively low pressure (about 10 cm of mercury). A wire parallel to the axis of the tube and insulated from it, extends into the tube. A difference of potential is maintained between the outside of the tube and the wire. The window at the end of the tube is made of special material which will allow the particles to enter. Assume it is a positive particle, such as the alpha particle.

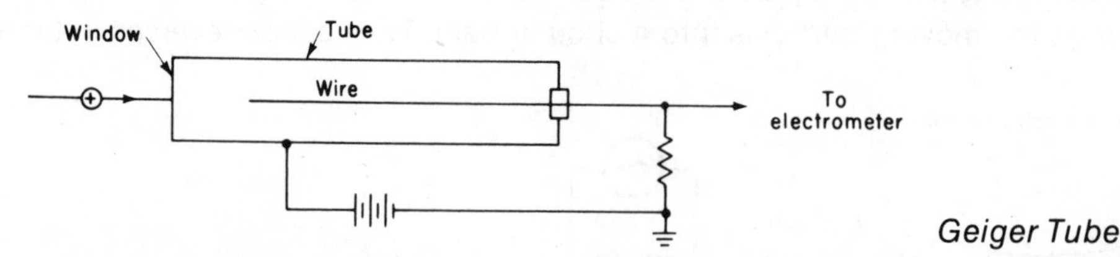


*Geiger Tube*

The entering particle ionizes the gas. These ions are accelerated by the difference of potential, collide with other gas molecules and ionize them, these are accelerated, etc. An avalanche of ions is rapidly produced, and a large current is obtained. The tube and circuit are designed to give only a momentary current, so that new entering particles will produce a separate momentary current, and produce a separate indication.

Cloud chambers and bubble chambers are similar. In both, tracks are produced by fast-moving charged particles. In the *cloud chamber* a super-cooled vapor is used. The charged particle produces ions in the air by colliding with the gas molecules. The super-cooled vapor condenses around these ions and leaves a visible trail. The *bubble chamber* uses a super-heated liquid, such as liquid hydrogen.

In the *scintillation counter* use is made of the fact that some radiation can produce fluorescence in certain materials.

**FOR EXAMPLE:** Rutherford made use of the fact that when alpha particles strike zinc sulfide, visible radiation is given off. Under the microscope this looks like a brilliant flash or scintillation. The flash is used to activate photoelectric tubes to magnify the effect.

The tracks obtained in cloud chambers and bubble chambers are photographed and studied. What do they reveal? Often, nothing new. Below is sketched a set of tracks as they might appear on the photograph of a bubble chamber operating in a strong magnetic field. At first glance the spiralling tracks may seem of the greatest interest. Actually they are of no special value. They are tracks of electrons ejected from atoms during collisions. They spiral inward because they are bent by the external magnetic field, and they are slowed on moving through the liquid. Of interest may be a track which changes path abruptly, as at point *A* or in tracks which suddenly start, as at *B*. Careful investigation shows that at point *A* two particles are produced as a result of the decay of the particle on track 2: a

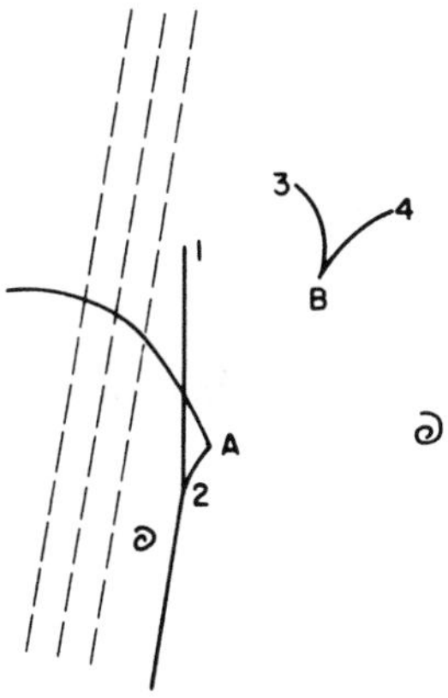

*Tracks in Bubble Chamber*

charged one which goes off to the left, and a neutral particle which leaves no track. This neutral particle continues up and to the right; at *B* it decays into two charged particles, which is shown by tracks 3 and 4.

In the study of cosmic rays, photographic plates are often arranged in fairly thick stacks. Particles going through these stacks may leave exposed areas on the photographic plates. When these plates are analyzed, tracks similar to those shown above are pieced together.

19 **What is Ahead?** By the late 1800's it seemed that practically all important questions in physics had been answered. In fact, one outstanding physicist said, in effect, that there was nothing left to be discovered in physics. One could merely refine the techniques of measurement. He could not have been more wrong. The tremendous discoveries and developments of new fields in physics have shown how rash and incorrect predictions can be, even if made by an expert scientist.

However, it does seem that there are many more discoveries ahead. Is there some small number of yet undiscovered particles that are really basic? What is time.... Is it quantized? Are there some as yet undiscovered forms of energy and energy conversions? Does the astronomical red shift really indicate that the universe is expanding? Perhaps some of you young students will participate in the new discoveries.

# Questions and Problems

1. If two neutral atoms have the same atomic number, what is true about their number of *a.* protons *b.* neutrons *c.* electrons?
2. How do isotopes of the same element compare with respect to *a.* atomic number? *b.* mass number?
3. Explain why mass numbers are always whole numbers but atomic mass usually is not. For a given atom, why is the mass, expressed in atomic mass units, nearly the same number as the mass number?
4. If an atom undergoes alpha decay, what happens to its *a.* mass number *b.* atomic number *c.* atomic mass?
5. If an atom undergoes beta decay, what happens to its *a.* mass number *b.* atomic number *c.* atomic mass?
6. Why was the neutrino hypothesis needed for the explanation of beta decay, but not for alpha decay?

7. What effect does the emission of gamma radiation have on an atom with respect to its *a.* mass number *b.* atomic number *c.* atomic mass?

8. It has been said that the physicist can predict exactly what will happen in nature. In the case of radioactive disintegration, give one example to show that this is *a.* true *b.* false.

9. What evidence is there that there is a force which *a.* in the nucleus is greater than the *Coulomb Force* *b.* outside the nucleus is weaker than the *Coulomb Force*?

10. If a positron is emitted from the nucleus, what happens to the atomic number of the nucleus?

11. What is the principle by which the speed of protons is increased in *a.* the linear accelerator *b.* the cyclotron?

12. *a.* What is the function of the magnetic field in the cyclotron? *b.* How is it arranged with respect to the dees? *c.* Why doesn't it result in an increase in the particle's speed? *d.* Does it result in the particle's acceleration?

13. Why does a fusion bomb also require the fission reaction?

14. What advantage might there be in using the term *unbinding energy* of the nucleus instead of *binding energy*?

15. Give the mass number of each of the following atoms:

    *a.* $_{9}F^{20}$ *b.* $_{22}Ti^{48}$ *c.* $_{42}Mo^{98}$ *d.* $_{54}Xe^{134}$

16. Give the atomic number of each of the atoms listed in question 15.

17. Give the number of neutrons in each of the atoms listed in question 15.

18. Name two conservation laws which were saved by the discovery of the neutrino.

19. Using data in the *Reference Tables*, calculate the energy equivalence of a proton in *a.* joules *b.* electron volts *c. Mev.*

20. A table gives the mass of a neutral atom of $_{42}Mo^{98}$ as 97.93610 *amu*. If the mass of a proton is 1.00760 *amu*, and the mass of a neutron is 1.00899 *amu*, and the mass of an electron is 0.00055 *amu*, calculate *a.* the binding energy per atom *b.* the binding energy per nucleon.

21. In each of the following reactions, give the mass number and atomic number of the product identified as *X*.

    *a.* $_{11}Na^{24} \rightarrow X + _{-1}e^{0}$

    *b.* $_{7}N^{14} + _{0}n^{1} \rightarrow _{6}C^{14} + X$

    *c.* $_{3}Li^{7} + _{1}H^{1} \rightarrow _{2}He^{4} + X.$

22. The half life of a certain isotope of radon is 3.82 days. How much radon will be left in 4 half lives if we start with 32 milligrams?

## Test Questions

1. Thermal neutrons are 1. hot to the touch 2. insulated 3. slow 4. fast.

2. If a neutron is emitted from the unstable nucleus of an atom, the atomic number 1. decreases 2. increases 3. remains the same.

3. Uranium-235 has an atomic number of 92. This means that the number of protons in its nucleus is 1. 92 2. 143 3. 235 4. 327.

4. The number of neutrons in uranium-235 is 1. 92 2. 143 3. 235 4. 327.

5. Isotopes of the same element have the same number of protons but different numbers of 1. electrons 2. neutrons 3. ions 4. eons.

6. When a radioactive substance is heated, its half life 1. increases 2. decreases 3. remains the same.

7. The particle whose charge is equal and opposite to that of the electron is the 1. proton 2. alpha particle 3. neutron 4. hyperon.

8. A uranium-235 nucleus differs from a uranium-238 nucleus in that the latter contains 1. 3 more protons 2. 3 more neutrons 3. 3 more electrons 4. 3 more alpha particles.

9. If graphite is used in the nuclear reactor for the fission of uranium, its function is 1. to slow down neutrons 2. to absorb neutrons 3. to absorb electrons 4. to slow down electrons.

10. Gamma rays are 1. electrons 2. protons 3. neutrons 4. radiation similar to X-rays.

11. If the speed of a neutron increases, its de Broglie wavelength 1. increases 2. decreases 3. remains the same.

12. Ejection of an alpha particle from a nucleus results in its atomic number being 1. increased by 2 2. decreased by 2 3. increased by 4 4. decreased by 4.

13. An increase in the atomic number of a nucleus by 1, results from the emission from the nucleus of a (*n*) 1. proton 2. neutron 3. alpha particle 4. electron.

14. Neutrons penetrate matter readily chiefly because they 1. occupy a much smaller volume than protons 2. occupy a much smaller volume than electrons 3. are electrically neutral 4. are pointed.

15. The radiation which is not significantly deflected by a magnetic field is 1. gamma rays 2. beta rays 3. alpha rays.

16. The particle which can not be accelerated by an accelerator of the cyclotron type is the 1. proton 2. electron 3. neutron 4. deuteron.

17. The accelerator which requires a moving belt is the 1. van de Graaff generator 2. cyclotron 3. synchrotron 4. Cockroft-Walton linear accelerator.

18. According to the *Physics Reference Tables*, uranium-238 changes to $Th^{230}$ in how many steps? 1. 1 2. 2 3. 3 4. 4.

19. The change of $U^{238}$ to $Th^{230}$ is accompanied by the emission of 1. one alpha particle only 2. two electrons only 3. two electrons plus one alpha particle only 4. two electrons plus two alpha particles.

20. As the speed of an electron beam perpendicular to a magnetic field decreases, the radius of the electron's path 1. increases 2. decreases 3. remains the same.

21. A certain radioactive isotope has a half life of 11.2 days. If we start with a mass $m$ of this isotope, the quantity which will not have disintegrated after 33.6 days is 1. $m/2$ 2. $m/4$ 3. $m/6$ 4. $m/8$.

## Topic Questions

1.1 What is the definition of the following terms? *a.* nucleon *b.* mass number *c.* atomic number *d.* isotopes.

2.1 What are *a.* alpha rays? *b.* beta rays? *c.* gamma rays?

2.2 How can a mixture of alpha, beta, and gamma rays be separated?

3.1 What is meant by binding energy?

3.2 How is binding energy of a nucleus related to its stability?

4.1 To what is one atomic mass unit (*amu*) equivalent?

5.1 What is another name for induced transmutation?

6.1 What is meant by conservation of charge?

7.1 When an alpha particle is emitted from a nucleus, how are mass number and atomic number affected?

8.1 What are two nuclear changes produced by beta decay?

8.2 What are the mass numbers of *a.* the neutrino? *b.* the electron?

9.1 *a.* What is gamma radiation? *b.* How does its wavelengths compare with those of light?

10.1 In the *Uranium Disintegration Series*, what particles are given off in the successive steps of the disintegration?

11.1 What is meant by the half life of an isotope?

12.1 What makes neutrons such good particles for bombarding nuclei?

13.1 What deduction is made from the fact that neutrons of only certain energies are absorbed by nuclei?

14.1 What are some of the important parts of an atomic pile? What is their function?

15.1 What are two differences between a thermonuclear reaction and a fission reaction?

16.1 What are two purposes of experimental research?

16.2 What is an advantage of circular accelerators over linear accelerators? How are the particles made to go in a circular path?

17.1 How are cloud chambers and bubble chambers similar? How are they different?

# Answers
## to Test Questions

### CHAPTER 1

*Page 7*

1. volume
2. 1000
3. 1000
4. 25.4
5. 16.4
6. 3.25
7. 4.25
8. 6.45
9. 1.52
10. 5
11. 4
12. 3
13. $3.27 \times 10^4$
14. 4500
15. 0.0045
16. *a.* 3
    *b.* 3
17. 14.25
18. 306,000.7
19. *a.* 9.18
    *b.* $4.24 \times 10^1$

### CHAPTER 2

*Page 20*

1. 2
2. 98
3. 5
4. opposite
5. 150
6. 0
7. 3
8. 3
9. decreases
10. 4
11. 2
12. 1
13. 3
14. 71
15. 1
16. 17 *Nt*, west
17. 5
18. 25 *Nt* at 127°
19. increases
20. decreases
21. increases
22. equilibrant
23. direction
24. 1
25. 1
26. 1

### CHAPTER 3

*Page 27*

1. 50 (*cm*)
2. 0
3. 50+ lb
4. 49 lb
5. 20-*cm* mark (30 *cm* from fulcrum)
6. c
7. 40
8. 4
9. 1

### CHAPTER 4

*Page 40*

1. 1
2. 3
3. 3
4. 3
5. 4
6. 2
7. 3
8. 3
9. 4
10. 1
11. 3
12. 2
13. 1
14. 3
15. 4
16. 2
17. 1
18. 2
19. 1
20. 2

### CHAPTER 5

*Page 56*

1. 2
2. 2
3. 1
4. 4
5. 40 *kg*·m/sec
6. 196
7. 800 lb
8. 1
9. 2
10. 1
11. 4
12. the impulse
13. 6 lb
14. the distance
15. increases
16. rem. the same
17. decreases
18. gravity
19. increases
20. 1/2

### CHAPTER 6

*Page 70*

1. 2
2. 1
3. 3
4. 1
5. 2
6. 2
7. 1
8. 1
9. 3
10. 0
11. 4
12. 1
13. 1
14. 0
15. rem. the same
16. decreases

### Chapter 7

*Page 84*

1. 65
2. 554°
3. 20
4. −4
5. 553
6. 4°C
7. a
8. a
9. c
10. b
11. 400
12. c
13. 5
14. 2700
15. b
16. absolute (or Kelvin)
17. c
18. a
19. 120
20. c

### CHAPTER 8

*Page 103*

1. 2
2. 3
3. 1
4. 4
5. 3
6. 3
7. 1
8. 1
9. 2
10. 3
11. 3
12. 2.5 vib/sec
13. 2
14. 1
15. 4
16. 2
17. 1
18. 0.002 sec
19. 0.35 m
20. 32°C
21. 525 m
22. F
23. B
24. D
25. 4
26. 1

### CHAPTER 9

*Page 119*

1. 2
2. 3
3. 3
4. 3
5. $4.7 \times 10^6$
6. $20 \times 10^{-8}$
7. $1.8 \times 10^8$
8. 2
9. 0
10. $2.2 \times 10^8$
11. 3
12. 3
13. umbra
14. 3
15. 3
16. diamond
17. 0.621
18. 2
19. 4
20. 40

## CHAPTER 10

*Page 134*

1. 3
2. 2
3. 2
4. 1
5. 2
6. 2
7. 3
8. 1
9. 4
10. 4
11. 2
12. 17
13. 2.7
14. 3
15. 4
16. 1
17. color
18. 2
19. reflects
20. diaphragm opening

## CHAPTER 11

*Page 147*

1. 3
2. 2
3. 2
4. 1
5. 1
6. 1
7. 2
8. 4
9. 3
10. 1
11. 4
12. 2
13. 4
14. 2
15. 2
16. 2
17. 1
18. 1
19. 2
20. 2

## CHAPTER 12

*Page 163*

1. 3
2. 3
3. 1
4. 2
5. 4
6. 3
7. 4
8. 1
9. 1
10. 2
11. 2
12. 3
13. 1
14. 2
15. 3
16. 4 (zero)
17. 4
18. 4
19. 3
20. 1

## CHAPTER 13

*Page 178*

1. 2
2. 3
3. 1
4. 2
5. 3
6. 2
7. 1
8. 0.5 amp
9. 70 V
10. 8
11. 2
12. 4
13. 1
14. 1
15. 2
16. 4
17. 1
18. 3
19. 5/6
20. 4

## CHAPTER 14

*Page 190*

1. 1
2. 2
3. 2
4. 1
5. N (outside the magnet)
6. 2
7. 2
8. 1
9. 1
10. 1
11. 2
12. 2
13. 3
14. N
15. 3
16. 2
17. 2
18. 1
19. 2
20. series
21. 1
22. 1
23. 2
24. 2
25. 1
26. 3

## CHAPTER 15

*Page 203*

1. 2
2. 1
3. 1
4. 3
5. 1
6. 2
7. 4
8. 4
9. 2
10. 3
11. 1
12. 2
13. 1
14. 3
15. 3
16. 1
17. 1

## CHAPTER 16

*Page 215*

1. 5
2. 141
3. 3
4. 1
5. 1
6. 1
7. 1
8. 1
9. 1
10. 3
11. 2
12. 3
13. 4
14. 2
15. 2
16. 2
17. 2
18. 2
19. 3
20. 1

## CHAPTER 17

*Page 235*

1. 2
2. 3
3. 2
4. 2
5. 2
6. 1
7. 2
8. 2
9. 3
10. 2
11. 2
12. 2
13. 3
14. 2
15. 1
16. 2
17. 1
18. 1
19. 2
20. 1

## CHAPTER 18

*Page 250*

1. 3
2. 3
3. 1
4. 2
5. 2
6. 3
7. 1
8. 2
9. 1
10. 4
11. 2
12. 2
13. 4
14. 3
15. 1
16. 3
17. 1
18. 4
19. 4
20. 2
21. 4

# Examination June, 1971 Physics

**PART ONE** *A number of questions on this examination require the use of the Physics Reference Tables. Make sure that you have a copy for your use.*

DIRECTIONS (**1-60**): *In the space provided, write the* number *preceding the word or expression that, of those given, best completes the statement or answers the question.* [70]

**1.** Which is a vector quantity? (1) distance (2) speed (3) displacement (4) time 1.........

**2.** Forces *A* and *B* have a resultant *R*. Force *A* and resultant *R* are shown in the diagram below.

Which vector below best represents force *B*?

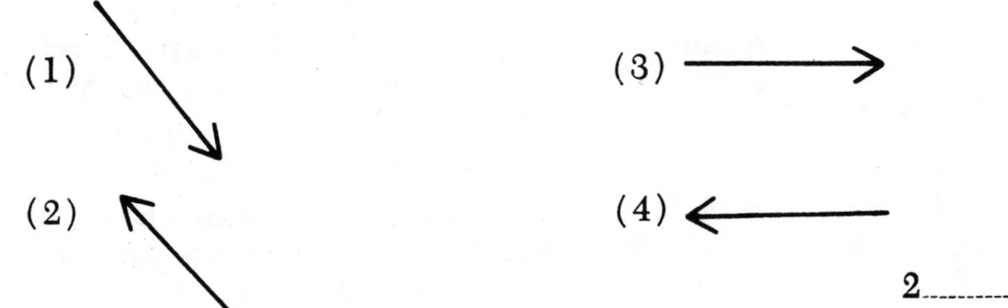


2.........

**3.** Which could be expressed in units of mass only? (1) force/acceleration (2) power × time (3) momentum/time (4) energy/force 3.........

**4.** A 40-newton object is released from rest at a height of 10 meters above the earth's surface. Just before it hits the ground, its kinetic energy will be closest to (1) 0 joules (2) 400 joules (3) 800 joules (4) 1,200 joules 4........

**5.** An object is allowed to fall freely near the surface of a planet. The object has an acceleration due to gravity of 24m./sec.$^2$. How far will the object fall during the first second? (1) 24 meters (2) 12 meters (3) 9.8 meters (4) 4.9 meters 5........

**6.** Which graph best represents an object in equilibrium?

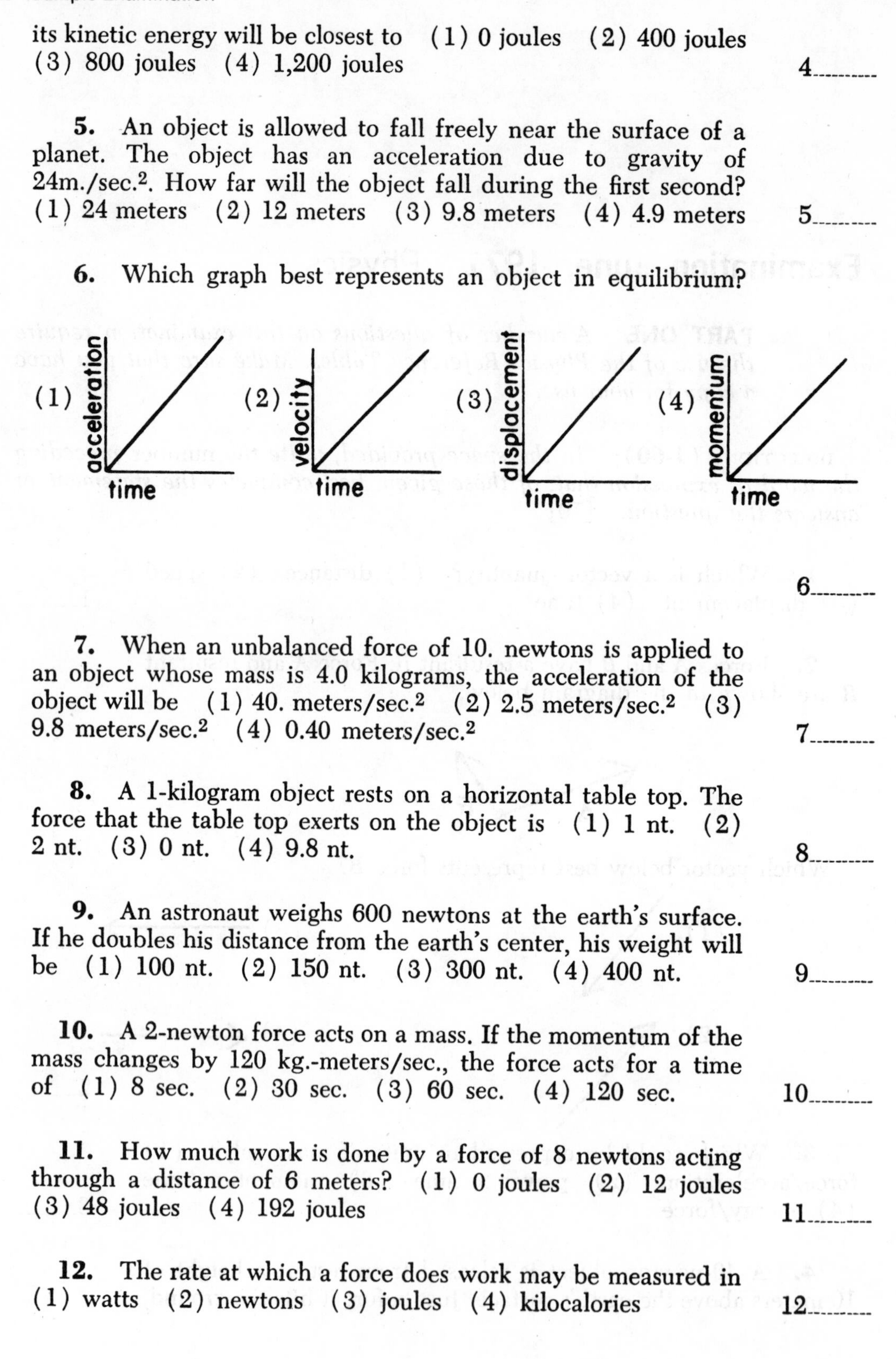


6........

**7.** When an unbalanced force of 10. newtons is applied to an object whose mass is 4.0 kilograms, the acceleration of the object will be (1) 40. meters/sec.$^2$ (2) 2.5 meters/sec.$^2$ (3) 9.8 meters/sec.$^2$ (4) 0.40 meters/sec.$^2$ 7........

**8.** A 1-kilogram object rests on a horizontal table top. The force that the table top exerts on the object is (1) 1 nt. (2) 2 nt. (3) 0 nt. (4) 9.8 nt. 8........

**9.** An astronaut weighs 600 newtons at the earth's surface. If he doubles his distance from the earth's center, his weight will be (1) 100 nt. (2) 150 nt. (3) 300 nt. (4) 400 nt. 9........

**10.** A 2-newton force acts on a mass. If the momentum of the mass changes by 120 kg.-meters/sec., the force acts for a time of (1) 8 sec. (2) 30 sec. (3) 60 sec. (4) 120 sec. 10........

**11.** How much work is done by a force of 8 newtons acting through a distance of 6 meters? (1) 0 joules (2) 12 joules (3) 48 joules (4) 192 joules 11........

**12.** The rate at which a force does work may be measured in (1) watts (2) newtons (3) joules (4) kilocalories 12........

**13.** If the kinetic energy of a 10-kilogram object is 2,000 joules, its velocity is (1) 10 meters/sec. (2) 20 meters/sec. (3) 100 meters/sec. (4) 400 meters/sec. 13.........

**14.** Which point in the wave shown in the diagram is in phase with point *A*?

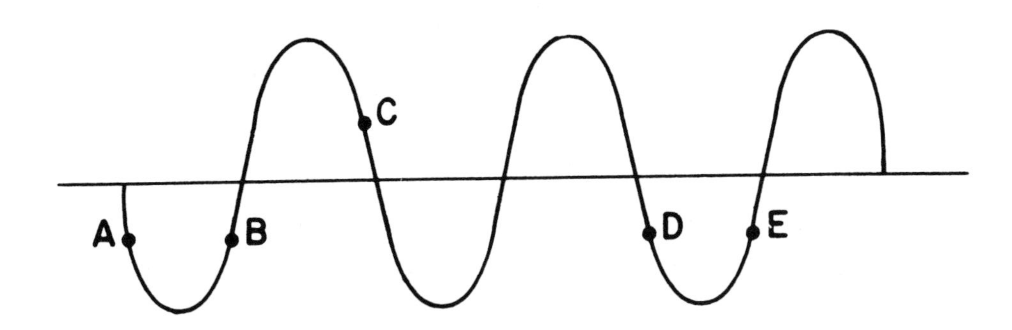


(1) *E* (2) *B* (3) *C* (4) *D* 14.........

**15.** Which color of light has the greatest period? (1) violet (2) green (3) orange (4) red 15.........

**16.** Which wave phenomenon is *not* common to both light and sound waves? (1) reflection (2) refraction (3) polarization (4) diffraction 16.........

**17.** All electromagnetic waves in a vacuum have the same (1) frequency (2) wavelength (3) speed (4) energy 17.........

**18.** The speed of light in water is closest to the speed of light in (1) a vacuum (2) benzene (3) carbon tetrachloride (4) alcohol 18.........

**19.** The wave nature of light is best shown by the phenomenon of (1) diffraction (2) reflection (3) refraction (4) dispersion 19.........

**20.** As a wave enters a different medium with no change in velocity, the wave will be (1) reflected but not refracted (2) refracted but not reflected (3) both reflected and refracted (4) neither reflected nor refracted 20.........

**21.** The diagram at right shows light ray *R* entering air from water. Through which point is the ray most likely to pass?

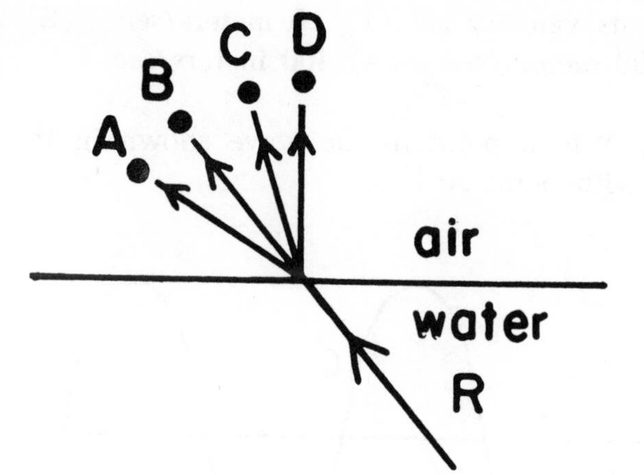


(1) *A*

(2) *B*

(3) *C*

(4) *D*

**22.** If the speed of light in a medium is approximately $2 \times 10^8$ meters per second, the medium could be (1) water (2) benzene (3) alcohol (4) diamond 22.........

**23.** Maximum constructive interference will occur at points where the phase difference between two waves is (1) 0° (2) 90° (3) 180° (4) 270° 23.........

**24.** Compared to the object, real images formed by a lens are always (1) larger (2) smaller (3) inverted (4) erect 24.........

**25.** When ray *I* emerges from the lens shown in the diagram, it will travel along a path toward point

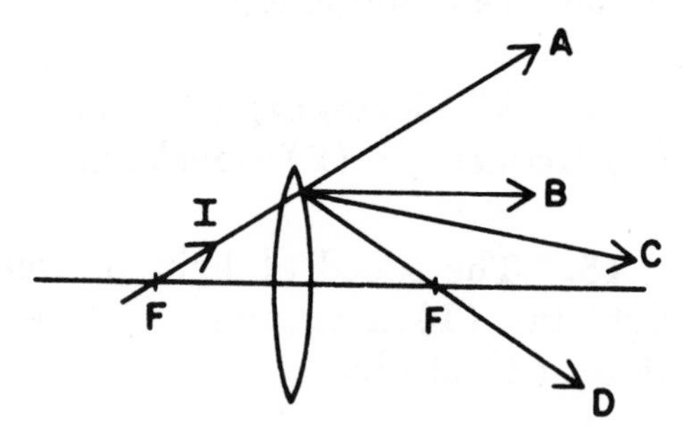


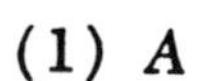

(1) *A*

(2) *B*

(3) *C*

(4) *D* 25.........

**26.** A positively charged body must have (1) an excess of neutrons (2) an excess of electrons (3) a deficiency of protons (4) a deficiency of electrons 26.........

**27.** Negatively charged rod *A* is used to charge rod *B* by induction. Object *C* is then charged by direct contact with rod *B*. The charge on object *C* is (1) neutral (2) positive (3) negative (4) not able to be determined

27.........

**28.** Two point charges attract each other with a force of $8.0 \times 10^{-5}$ newton. If the distance between the charges is doubled, the force will become (1) $16 \times 10^{-5}$ newton (2) $2.0 \times 10^{-5}$ newton (3) $64. \times 10^{-5}$ newton (4) $4.0 \times 10^{-5}$ newton

28.........

**29.** A volt is defined as a (1) joule/coulomb (2) joule/second (3) coulomb/second (4) joule-second/coulomb

29.........

**30.** The two large metal plates shown in the diagram are charged to a potential difference of 100 volts. How much work is needed to move 1 coulomb of negative charge from point *A* to point *B*?

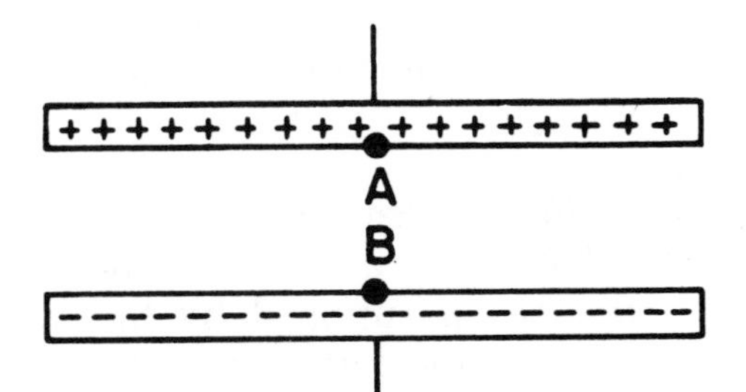


(1) 1 joule
(2) 100 joules
(3) 1 electron-volt
(4) 100 electron-volts

30.........

**31.** Which diagram best represents the relationship between the length of a metal conductor and its resistance?

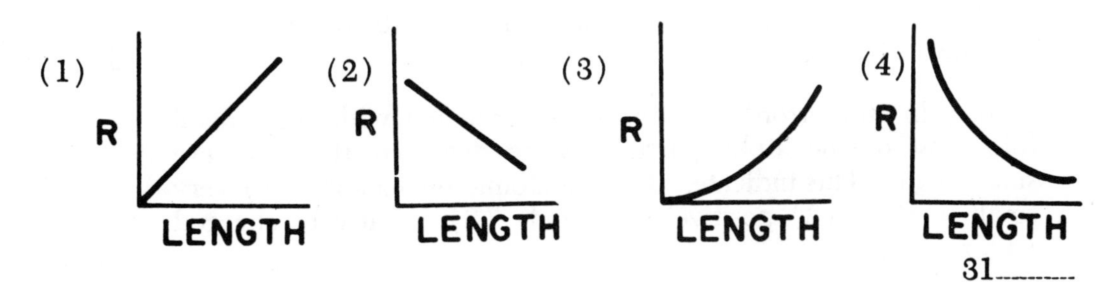


31.........

**32.** If the potential difference across a 30.-ohm resistor is 10. volts, what is the current through the resistor? (1) 0.25 amp. (2) 0.33 amp. (3) 3.0 amp. (4) 0.50 amp.

32.........

**33.** The diagram represents an electric circuit.

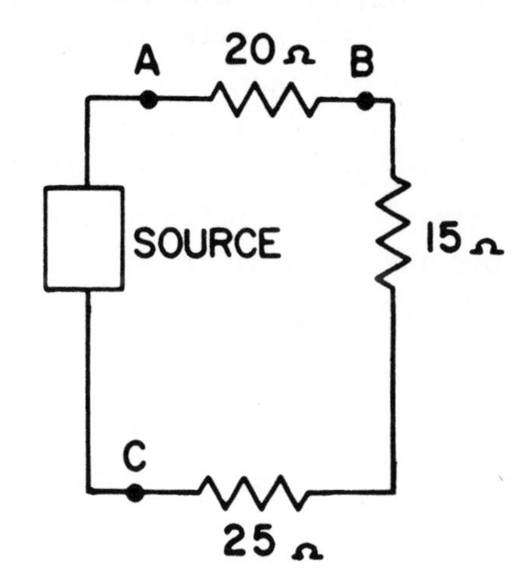


If the voltage between *A* and *B* is 10 volts, the voltage between *B* and *C* is (1) 5 volts (2) 10 volts (3) 15 volts (4) 20 volts 33........

**34.** Three equal resistances are connected in parallel across a battery. Compared to the power dissipated in one resistance, the power supplied by the battery is (1) one-ninth as great (2) one-third as great (3) three times as great (4) nine times as great 34........

**35.** The algebraic sum of all the potential drops and applied voltages around a complete circuit is equal to zero. This is an application of the law of conservation of (1) mass (2) energy (3) charge (4) momentum 35........

**36.** A 120-volt toaster is rated at 600 watts. Under normal conditions, the current in the toaster is (1) 0.20 amp. (2) 5.0 amp. (3) 10. amp. (4) 25 amp. 36........

**37.** In Rutherfords' scattering experiments with thin metal foil, most of the alpha particles were deflected through very small angles. This indicated that the atomic nucleus is (1) very small in size (2) positively charged (3) negatively charged (4) neutral 37........

**38.** Which phenomenon is evidence of the quantum nature of light? (1) interference (2) diffraction (3) polarization (4) photoelectric effect 38........

**39.** Which color light has photons of the greatest energy? (1) red (2) yellow (3) green (4) blue 39........

**40.** A lithium nucleus contains 3 protons and 4 neutrons. What is the atomic number of the nucleus? (1) 1 (2) 7 (3) 3 (4) 4 40........

**41.** The energy needed to ionize a hydrogen atom in the ground state is (1) 2.9 e.v. (2) 3.2 e.v. (3) 13.06 e.v. (4) 13.6 e.v. 41........

**Note that questions 42 through 60 have only three choices.**

**42.** Two forces are in equilibrium at an angle of 180 degrees. If one force increases in magnitude while the other force decreases, their resultant (1) decreases (2) increases (3) remains the same

**43.** A constant force is exerted on a box as shown in the diagram.

As angle $\theta$ decreases to 0°, the magnitude of the horizontal component of the force

(1) decreases
(2) increases
(3) remains the same

Force
θ
horizontal

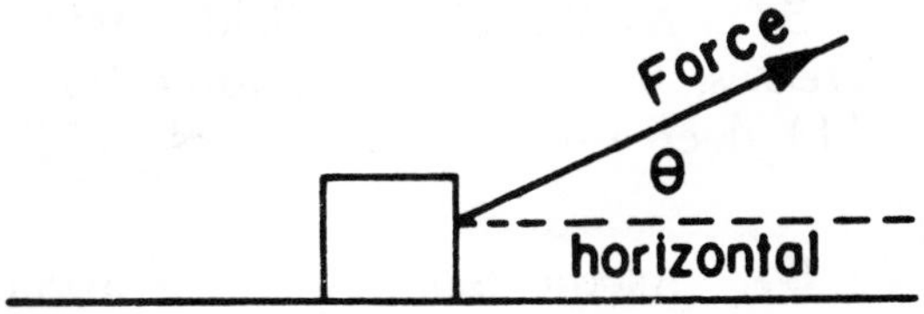

43........

**44.** An object rests on an incline. As the angle between the incline and the horizontal increases, the force needed to prevent the object from sliding down the incline (1) decreases (2) increases (3) remains the same 44........

**45.** As a space ship from earth goes toward the moon, the force it exerts on the earth (1) decreases (2) increases (3) remains the same 45........

**46.** When water at 10° Celsius is heated to 20° Celsius, its internal energy (1) decreases (2) increases (3) remains the same 46........

**47.** When a rising baseball encounters air resistance, its total mechanical energy (1) decreases (2) increases (3) remains the same 47........

**48.** As a wave enters a medium of higher refractive index, its wavelength (1) decreases (2) increases (3) remains the same 48........

**49.** As the frequency of a wave increases, its period (1) decreases (2) increases (3) remains the same 49.........

**50.** A pulse traveling along a stretched spring is reflected from the fixed end. Compared to the pulse's speed before reflection, its speed after reflection is (1) less (2) greater (3) the same 50.........

**51.** As the frequency of an electromagnetic wave increases, its speed in a vacuum (1) decreases (2) increases (3) remains the same 51.........

**52.** A rod is rubbed with wool. Immediately after the rod and wool have been separated, the net charge of the rod-wood system (1) decreases (2) increases (3) remains the same 52.........

**53.** As the electric field intensity at a point in space decreases, the electrostatic force on a unit charge at this point (1) decreases (2) increases (3) remains the same 53.........

**54.** As the temperature of a coil of copper wire increases, its electrical resistance (1) decreases (2) increases (3) remains the same 54.........

**55.** A resistor is connected to a source of constant voltage. If a lamp is connected in a parallel with the resistor, the potential difference across the resistor will (1) decrease (2) increase (3) remain the same 55.........

**56.** A straight conductor in a magnetic field is perpendicular to the lines of force. If the current in the conductor increases, the magnetic force exerted on it (1) decreases (2) increases (3) remains the same 56.........

**57.** As the binding energy of a nucleus increases, the energy required to separate it into nucleons (1) decreases (2) increases (3) remains the same 57.........

**58.** As the mass number of an isotope increases, its atomic number (1) decreases (2) increases (3) remains the same 58.........

**59.** As the speed of an electron increases, its wavelength (1) decreases (2) increases (3) remains the same 59.........

**60.** As the intensity of monochromatic light on a photoemissive surface increases, the maximum kinetic energy of the photoelectrons emitted (1) decreases (2) increases (3) remains the same 60.........

**PART TWO** *This part consists of four groups, each group testing a major area in the course. Choose two of these four groups. Write your answers in the spaces provided.*

## GROUP 1 — Mechanics

*Answer all fifteen questions, 61 through 75, in this group. Each question counts 1 credit. For each of the fifteen questions, write the number of the word or expression that best completes that statement or answers that question.*

Base your answers to questions 61 through 63 on the graph and information below.

Cars *A* and *B* both start from rest at the same location at the same time.

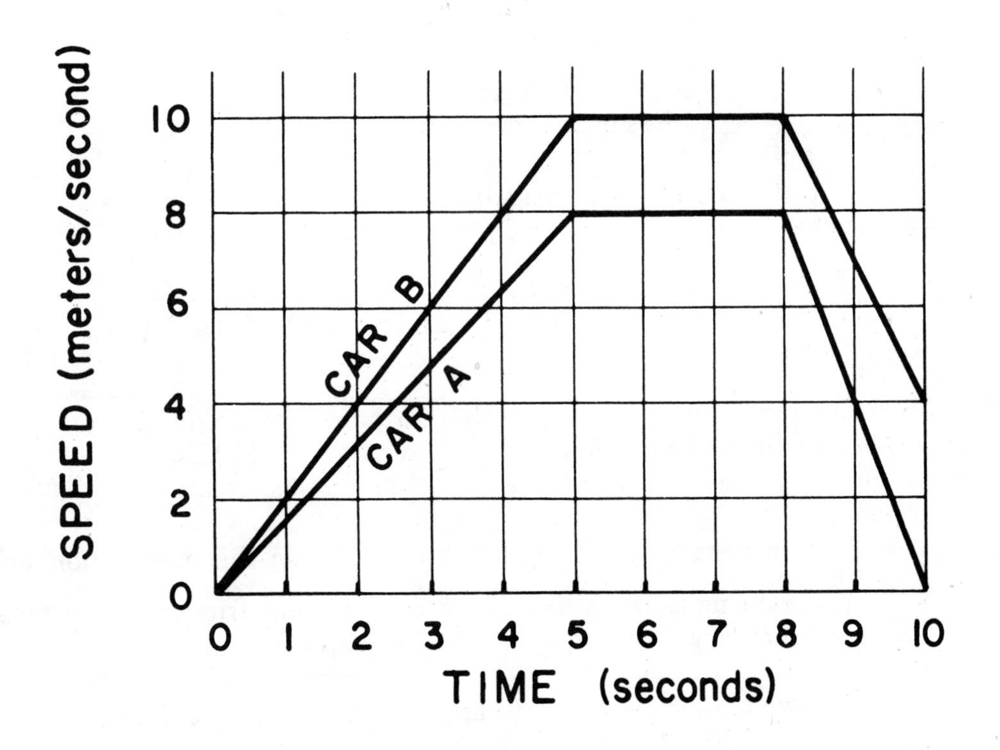


**61.** What is the magnitude of the acceleration of car *A* during the period between $t = 8$ seconds and $t = 10$ seconds? (1) 20 m./sec.$^2$ (2) 16 m./sec.$^2$ (3) 8 m./sec.$^2$ (4) 4 m./sec.$^2$ 61.........

**Note that questions 62 and 63 have only three choices.**

**62.** Compared to the speed of car *B* at 6 seconds, the speed of car *A* at 6 seconds is (1) less (2) greater (3) the same 62________

**63.** Compared to the total distance traveled by car *B* during the 10 seconds, the total distance traveled by car *A* is (1) less (2) greater (3) the same 63________

Base your answers to questions 64 through 66 on the information below and the diagram which represents the moon in circular orbit around the earth.

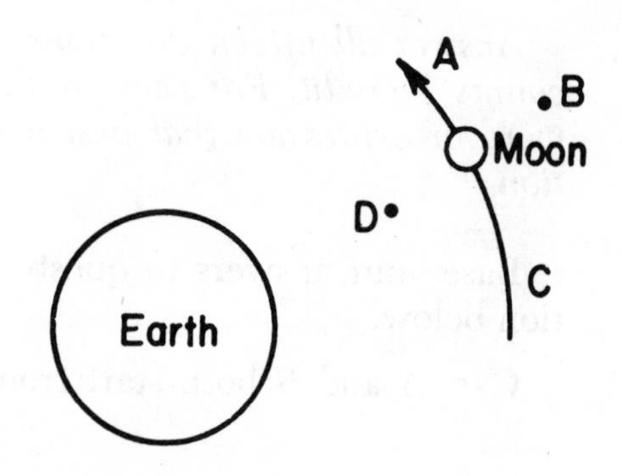


Mass of the earth $= 6.0 \times 10^{24}$ kg.
Mass of the moon $= 7.3 \times 10^{23}$ kg.

**64.** The direction of the centripetal force on the moon is toward point (1) *A* (2) *B* (3) *C* (4) *D* 64________

**65.** If an astronaut has a mass of 91 kilograms on the earth, his mass on the moon would be (1) 11 kg. (2) 45 kg. (3) 70 kg. (4) 91 kg. 65________

**66.** Compared to the force of the earth on the moon, the magnitude of the force of the moon on the earth is (1) $10^{-2}$ as great (2) the same (3) $10^{1}$ as great (4) $10^{2}$ as great 66________

Base your answers to questions 67 through 69 on the information below.

A 2.0-kilogram mass is pushed along a horizontal frictionless surface by a 3.0-newton force which is parallel to the surface.

**67.** The weight of the mass is approximately (1) 9.8 nt. (2) 2.0 nt. (3) 20. nt. (4) 4.0 nt. 67________

**68.** How much work is done in moving the mass 1.5 meters horizontally? (1) 4.5 joules (2) 2.0 joules (3) 3.0 joules (4) 30. joules 68________

**69.** How much gravitational potential energy would be gained by the mass if it is moved 2 meters horizontally? (1) 0 joules (2) 6 joules (3) 40 joules (4) 4 joules

69........

**70.** Which substance has the highest specific heat? (1) copper (sol.) (2) ammonia (liq.) (3) water (liq.) (4) steam (gas)

70........

**71.** Equal amounts of heat energy are added to 1 kilogram of lead and 1 kilogram of metal *X*. If the increase in temperature for metal *X* is one-half the temperature increase for the lead, then metal *X* could be (1) iron (2) tungsten (3) zinc (4) silver

71........

**72.** How many kilocalories of heat are needed to raise the temperature of one kilogram of platinum from 500° C. to 600° C.? (1) 3,000 (2) 100 (3) 3 (4) 30

72........

**73.** If equal masses of the metals listed were at an initial temperature of 100° C., which metal would require the most heat to reach a temperature of 200° C.? (1) silver (2) lead (3) aluminum (4) iron

73........

**74.** Which graph best represents the change of phase of a substance?

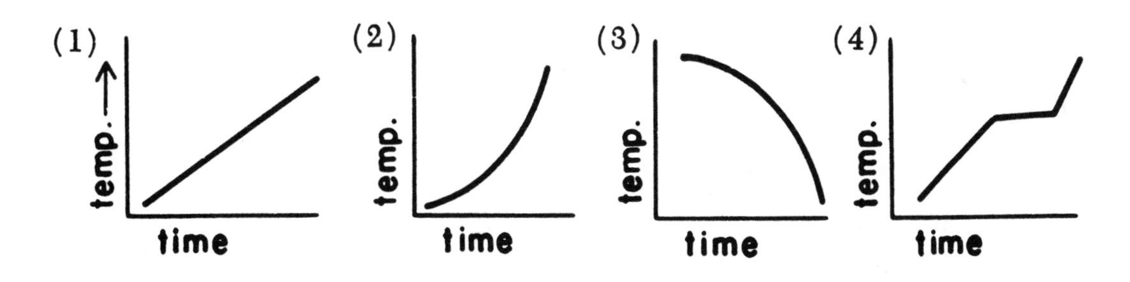


74........

**75.** If the pressure on a gas is doubled and its absolute temperature is halved, the volume of the gas will be (1) quartered (2) doubled (3) unchanged (4) halved

75........

## GROUP 2 — Wave Phenomena

*Answer all fifteen questions, 76 through 90, in this group. Each question counts 1 credit. For each of the fifteen questions, write the number of the word or expression that best completes that statement or answers that question.*

Base your answers to questions 76 through 78 on the diagram below, which represents periodic water waves in a ripple tank. The speed of a wave decreases as it moves from the deep to the shallow portion of the tank.

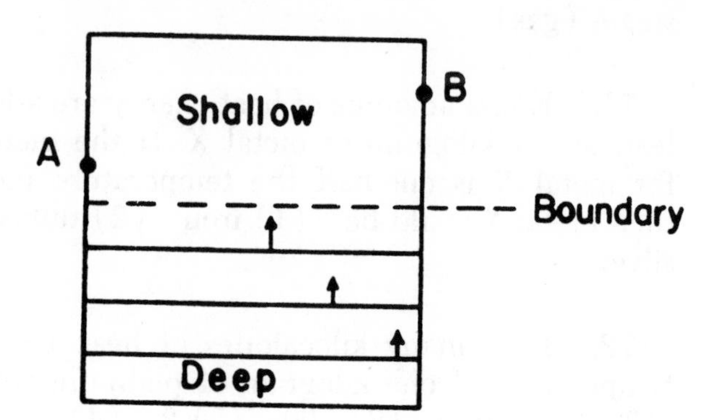


**76.** Which drawing best represents the waves after they enter the shallow section?

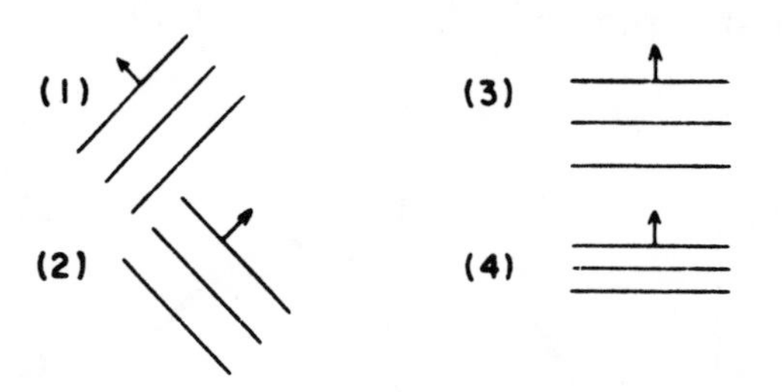


76.________

**77.** Which diagram best represents the pattern produced when the waves are reflected from the boundary between the deep and shallow sections of the ripple tank?

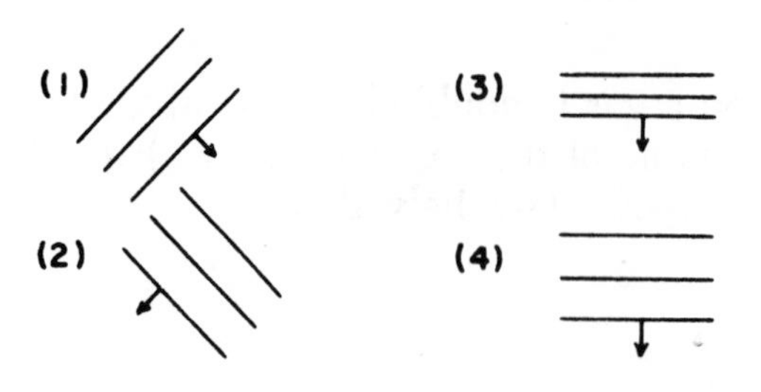


77.________

**78.** If a barrier is placed in the ripple tank connecting points *A* and *B*, which drawing best represents the waves after reflection from the barrier?

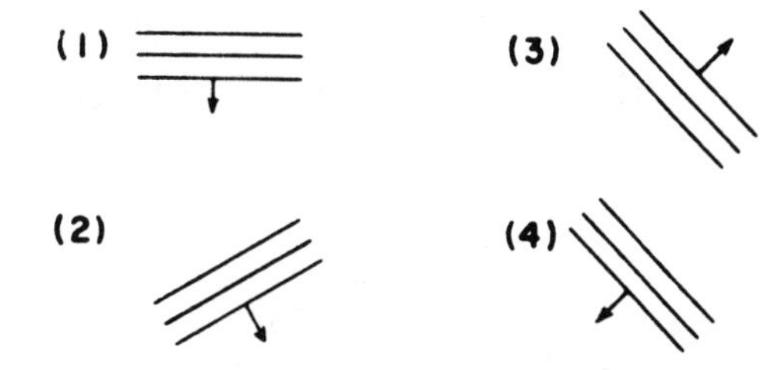


78.........

Base your answers to questions 79 through 81 on the diagram below, which shows a vibrating source moving at a constant speed producing the wave pattern shown.

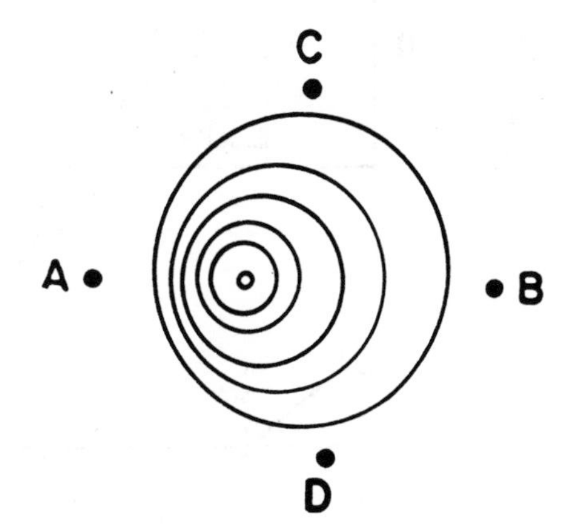


**79.** The above diagram illustrates (1) interference (2) diffraction (3) the Doppler effect (4) polarization 79.........

**80.** The source is moving toward point (1) *A* (2) *B* (3) *C* (4) *D* 80.........

**81.** If the source were to accelerate, the wavelength immediately in front of the source would (1) increase, only (2) decrease, only (3) increase and then decrease (4) remain the same 81.........

Base your answers to questions 82 through 87 on the information below.

A ray of monochromatic orange light passes into a glass tank containing a lucite block that is submerged in a liquid. The wavelength of the light in air is $6.0 \times 10^{-7}$ meter.

**82.** Which diagram shows the path that the light could follow if the liquid were benzene?

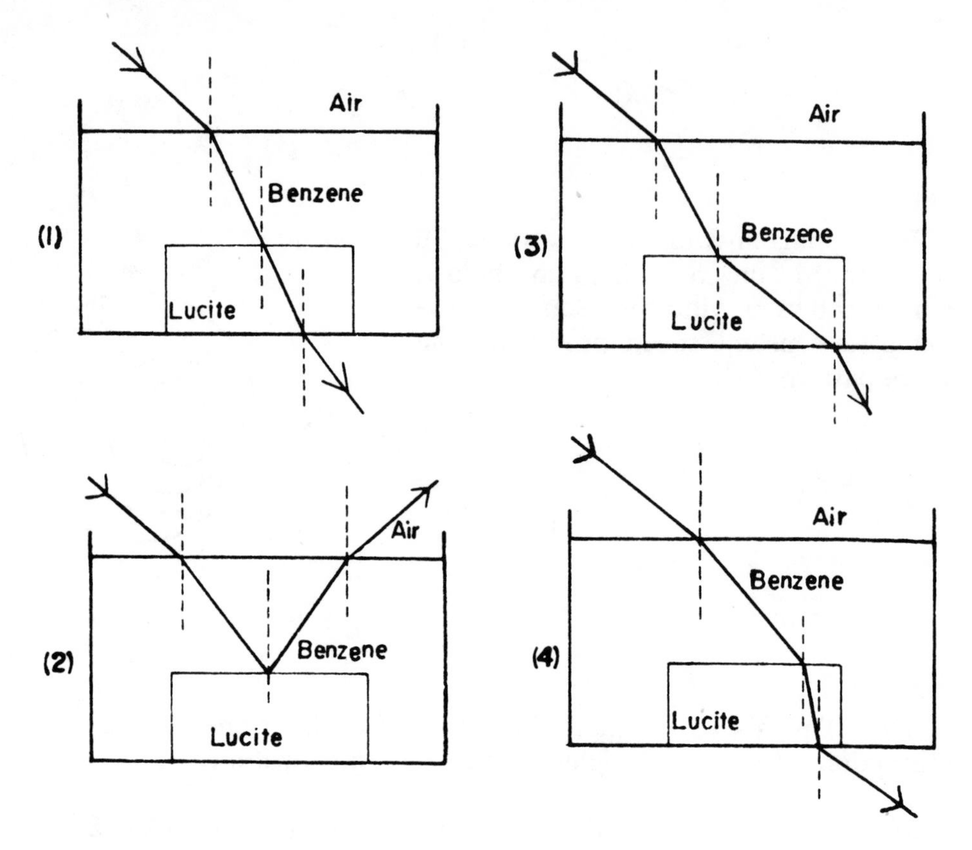


82________

**83.** Which diagram shows the path that the light could follow if the liquid were Canada balsam?

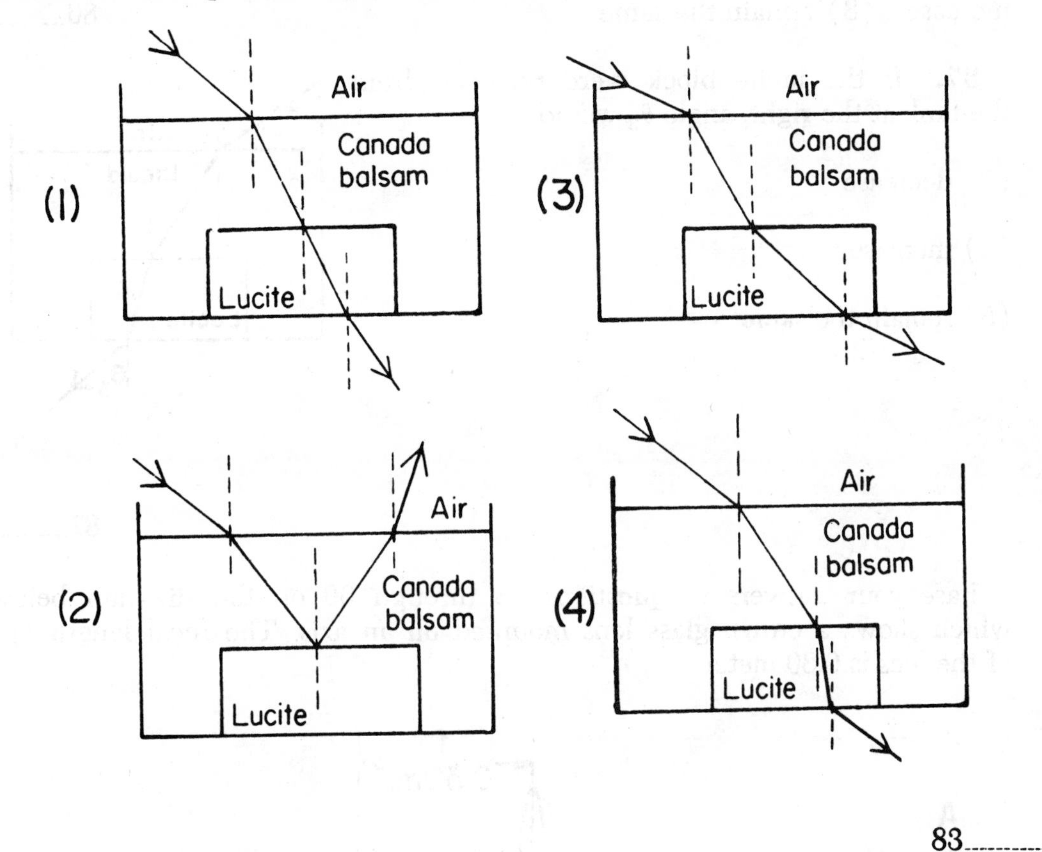


83________

**84.** What is the frequency of the light? (1) $1.5 \times 10^1$ c.p.s. (2) $9.0 \times 10^{14}$ c.p.s. (3) $2.0 \times 10^{16}$ c.p.s. (4) $5.0 \times 10^{14}$ c.p.s.

84________

**Note that questions 85 through 87 have only three choices.**

**85.** If angle $\theta_1$ in the diagram at the right were increased, angle $\theta_2$ would

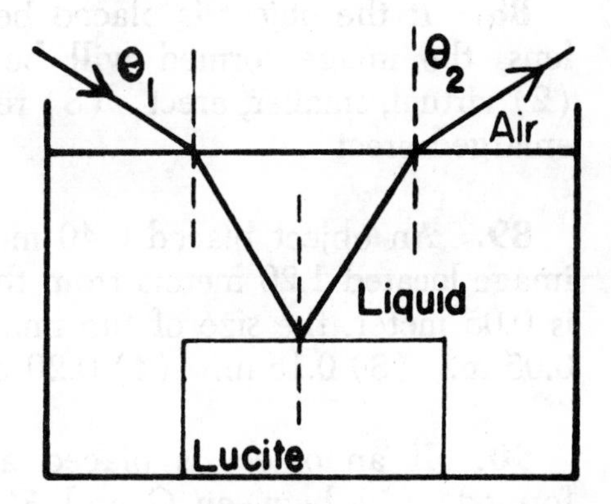


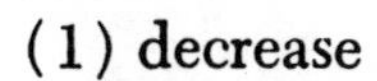

(1) decrease

(2) increase

(3) remain the same

85________

**86.** As the frequency of the incident light is increased in the diagram for question 85, angle $\theta_2$ would (1) decrease (2) increase (3) remain the same

86________

**87.** If the lucite block were removed from the tank at the right, angle $\theta_3$ would

(1) decrease

(2) increase

(3) remain the same

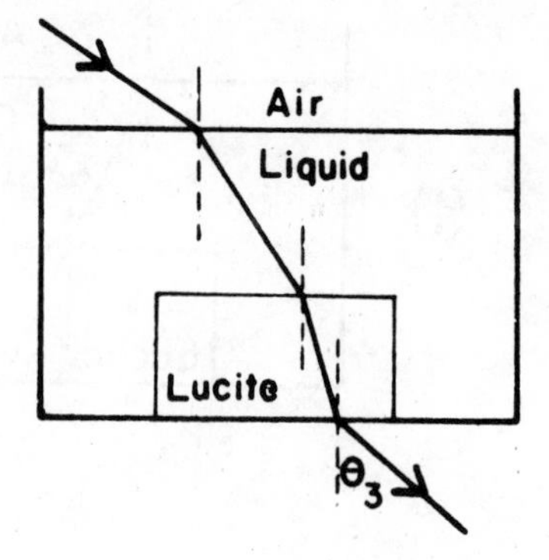


87________

Base your answers to questions 88 through 90 on the diagram below which shows a crown glass lens mounted on an axis. The focal length ($f$) of the lens is 0.30 meter.

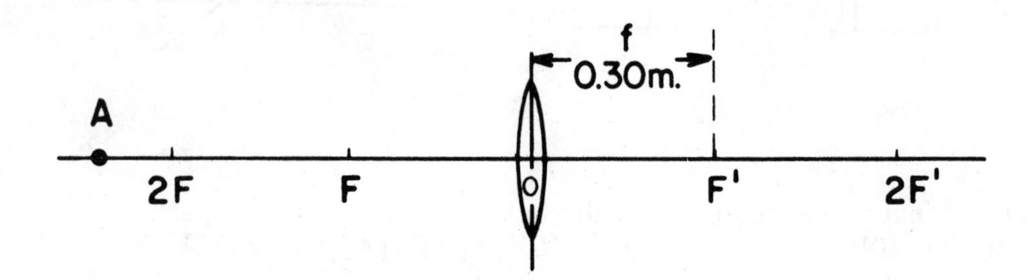


**88.** If the object is placed between the focal point and the lens, the image formed will be (1) virtual, enlarged, erect (2) virtual, smaller, erect (3) real, smaller, inverted (4) real, enlarged erect

88________

**89.** An object placed 0.40 meter from the lens produces an image located 1.20 meters from the lens. If the size of the object is 0.05 meter, the size of the image will be (1) 0.03 m. (2) 0.05 m. (3) 0.15 m. (4) 0.20 m.

89________

**90.** If an object is placed at point *A*, an image will be formed (1) between *O* and *F'* (2) *at F'* (3) between *F'* and 2*F'* (4) beyond 2*F'*

90________

## GROUP 3 – Electricity

*Answer all fifteen questions, 91 through 105, in this group. Each question counts 1 credit. For each of the fifteen questions, write the number of the word or expression that best completes that statement or answers that question.*

Base your answers to questions 91 through 95 on the diagram below which shows three small metal spheres with different charges.

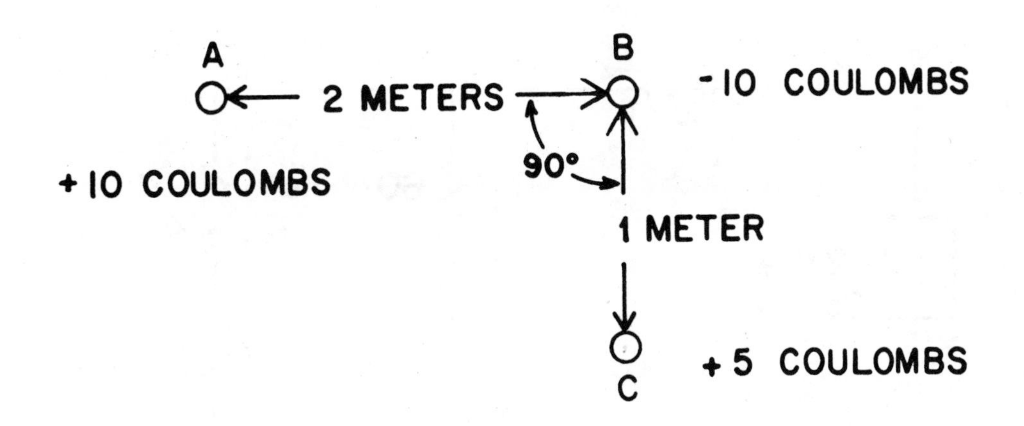


**91.** Which vector best represents the net force on sphere *B*?

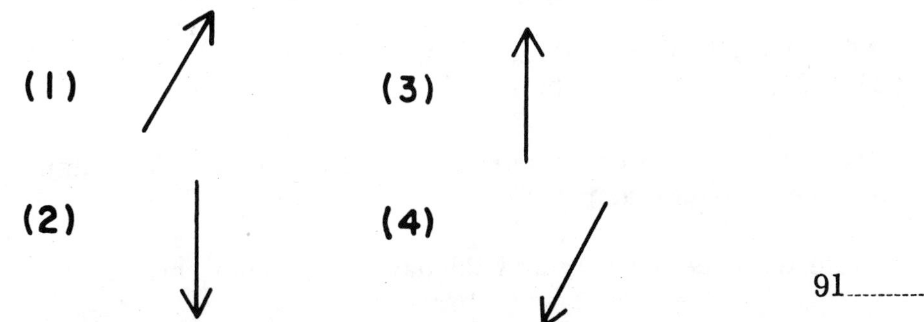


91.........

**92.** Compared to the force between spheres *A* and *B*, the force between spheres *B* and *C* is (1) one-quarter as great (2) twice as great (3) one-half as great (4) four times as great

92.........

**Note that questions 93 through 95 have only three choices.**

**93.** If sphere *A* is moved further to the left, the magnitude of the net force on sphere *B* would (1) decrease (2) increase (3) remain the same

93.........

**94.** If the charge on sphere *B* were decreased, the magnitude of the net force on sphere *B* would (1) decrease (2) increase (3) remain the same 94........

**95.** If sphere *B* were removed, the force of sphere *C* on sphere *A* would (1) decrease (2) increase (3) remain the same 95........

Base your answers to questions 96 through 100 on the circuit diagram below.

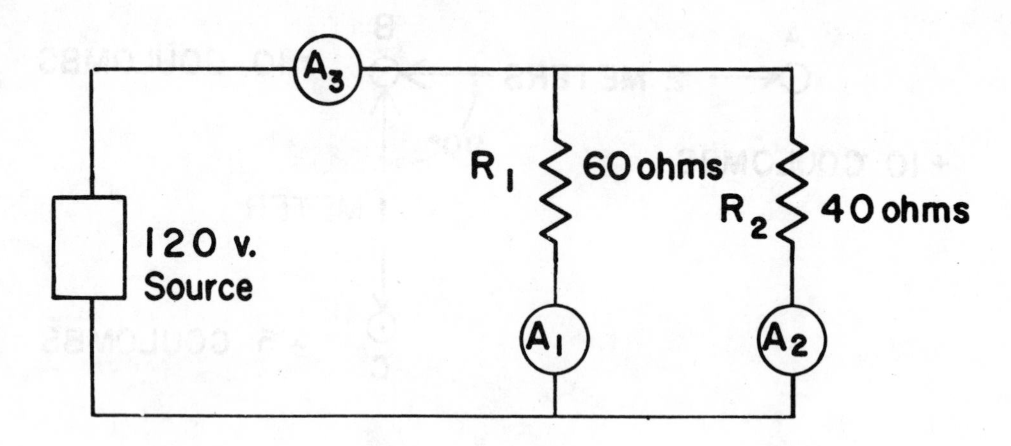


**96.** The effective resistance of the circuit is (1) 600 ohms (2) 100 ohms (3) 24 ohms (4) 20 ohms 96........

**97.** The reading of ammeter $A_1$ is (1) 5 amp. (2) 2 amp. (3) 3 amp. (4) 8 amp. 97........

Note that questions 98 and 99 have only three choices.

**98.** Compared to the current in ammeter $A_3$, the sum of the currents in $A_1$ and $A_2$ is (1) greater (2) less (3) the same 98........

**99.** If a third resistor is connected in parallel to the circuit, the total resistance will (1) decrease (2) increase (3) remain the same 99........

**100.** If a third resistor were connected in parallel to this circuit, the potential difference across the third resistor would be (1) 20 volts (2) 40 volts (3) 60 volts (4) 120 volts 100........

Base your answers to questions 101 through 105 on the diagram below, which shows a circuit containing a solenoid with an iron core.

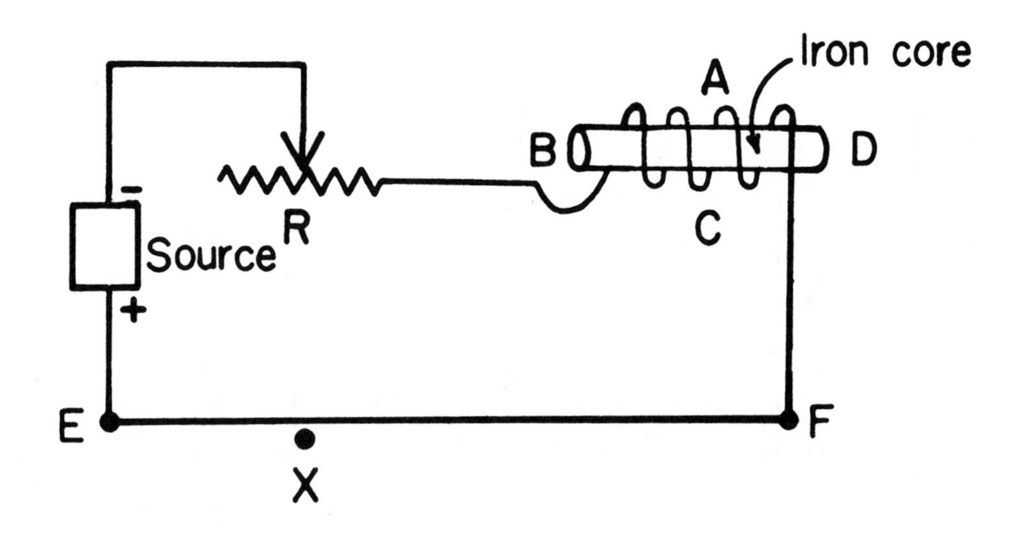


**101.** The north pole of the solenoid is nearest to point (1) *A* (2) *B* (3) *C* (4) *D* 101........

**102.** The direction of the magnetic field caused by conductor *EF* at point *X* is (1) into the page (2) out of the page (3) to the left (4) to the right 102........

**Note that questions 103 through 105 have only three choices.**

**103.** If the resistance of the variable resistor *R* is decreased, the magnetic field strength of the solenoid will (1) decrease (2) increase (3) remain the same 103........

**104.** If the iron core is removed, the magnetic field strength in the solenoid will (1) decrease (2) increase (3) remain the same 104........

**105.** If the number of turns in the coil is increased and the current kept constant, the magnetic field strength of the solenoid will (1) decrease (2) increase (3) remain the same 105........

## GROUP 4 — Atomic and Nuclear Physics

*Answer all fifteen questions, 106 through 120, in this group. Each question counts 1 credit. For each of the fifteen questions, write the number of the word or expression that best completes that statement or answers that question.*

Base your answers to questions 106 through 108 on the graph below which shows the maximum kinetic energy of the photoelectrons ejected when photons of different frequencies strike a metal surface.

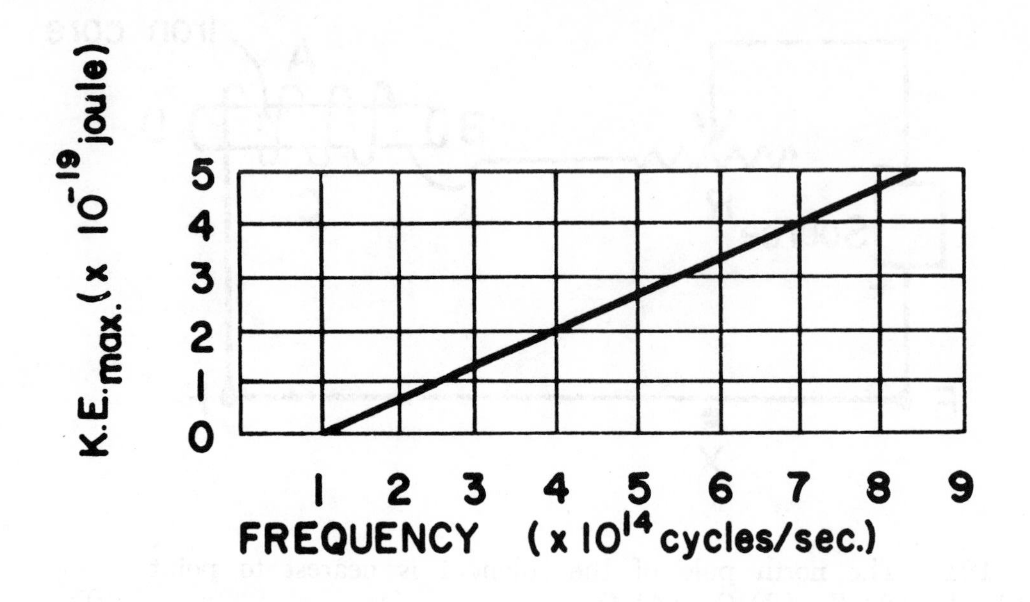


**106.** Which is the lowest frequency photon that will produce photoelectrons? (1) 0 c.p.s. (2) 1.0 c.p.s (3) $1.0 \times 10^{14}$ c.p.s. (4) $1.5 \times 10^{14}$ c.p.s. 106.........

**107.** Photons with a frequency of $4 \times 10^{14}$ c.p.s. will produce photoelectrons with a maximum kinetic energy of (1) $4.0 \times 10^{14}$ joules (2) 1.3 joules (3) $1.3 \times 10^{-19}$ joule (4) $2.0 \times 10^{-19}$ joule 107.........

**Note that question 108 has only three choices.**

**108.** Compared to the energy of the bombarding photon, the energy of the emitted photoelectron is (1) less (2) greater (3) the same 108.........

**109.** The energy change is greatest for a hydrogen atom which changes state from (1) $n = 2$ to $n = 1$ (2) $n = 3$ to $n = 2$ (3) $n = 4$ to $n = 3$ (4) $n = 5$ to $n = 4$ 109.........

**110.** What is the energy of the emitted photon when a hydrogen atom changes from an energy state of $n = 5$ to $n = 4$? (1) 13.06 electron-volts (2) 1.39 electron-volts (3) 0.54 electron-volt (4) 0.31 electron-volt 110.........

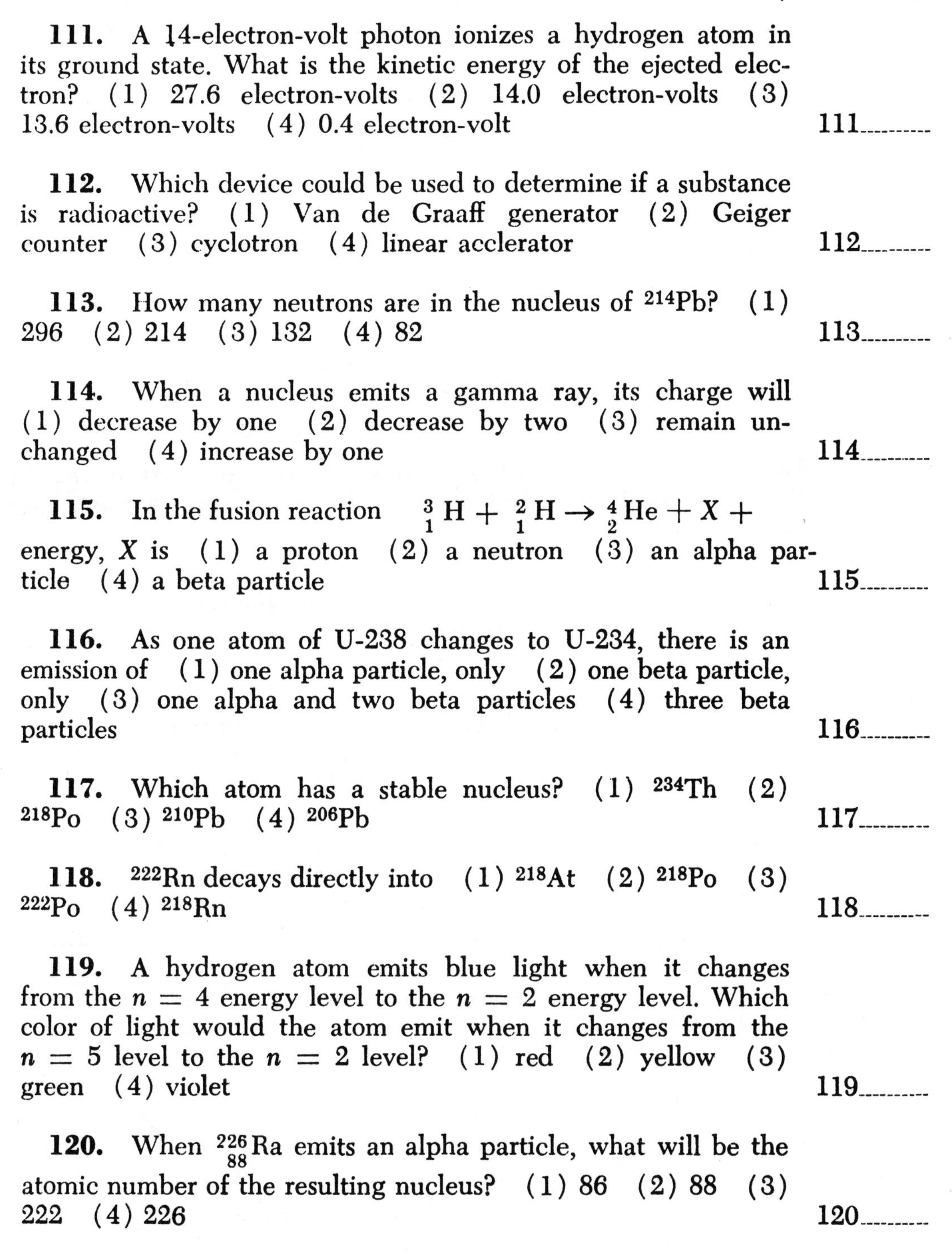

**111.** A 14-electron-volt photon ionizes a hydrogen atom in its ground state. What is the kinetic energy of the ejected electron? (1) 27.6 electron-volts (2) 14.0 electron-volts (3) 13.6 electron-volts (4) 0.4 electron-volt 111........

**112.** Which device could be used to determine if a substance is radioactive? (1) Van de Graaff generator (2) Geiger counter (3) cyclotron (4) linear acclerator 112........

**113.** How many neutrons are in the nucleus of $^{214}Pb$? (1) 296 (2) 214 (3) 132 (4) 82 113........

**114.** When a nucleus emits a gamma ray, its charge will (1) decrease by one (2) decrease by two (3) remain unchanged (4) increase by one 114........

**115.** In the fusion reaction $^{3}_{1}H + ^{2}_{1}H \rightarrow ^{4}_{2}He + X +$ energy, $X$ is (1) a proton (2) a neutron (3) an alpha particle (4) a beta particle 115........

**116.** As one atom of U-238 changes to U-234, there is an emission of (1) one alpha particle, only (2) one beta particle, only (3) one alpha and two beta particles (4) three beta particles 116........

**117.** Which atom has a stable nucleus? (1) $^{234}Th$ (2) $^{218}Po$ (3) $^{210}Pb$ (4) $^{206}Pb$ 117........

**118.** $^{222}Rn$ decays directly into (1) $^{218}At$ (2) $^{218}Po$ (3) $^{222}Po$ (4) $^{218}Rn$ 118........

**119.** A hydrogen atom emits blue light when it changes from the $n = 4$ energy level to the $n = 2$ energy level. Which color of light would the atom emit when it changes from the $n = 5$ level to the $n = 2$ level? (1) red (2) yellow (3) green (4) violet 119........

**120.** When $^{226}_{88}Ra$ emits an alpha particle, what will be the atomic number of the resulting nucleus? (1) 86 (2) 88 (3) 222 (4) 226 120........

# Examination June, 1972 Physics

**PART ONE** *A number of questions on this examination require the use of the Physics Reference Tables. Make sure that you have a copy for your use.*

DIRECTIONS (1–60): *In the space provided, write the* number *preceding the word or expression that, of those given, best completes the statement or answers the question.* [70]

**1.** A car travels a straight-line distance of 160 meters in 8 seconds. What is the magnitude of the average velocity of the car? (1) .05 m./sec. (2) 10 m./sec. (3) 20 m./sec. (4) 1,280 m./sec. 1........

**2.** The speed of an object increases uniformly from 10. meters/second to 20. meters/second in 5.0 seconds. The magnitude of the object's acceleration during this interval is (1) 50. m./sec.$^2$ (2) 2.0 m./sec.$^2$ (3) 100 m./sec.$^2$ (4) 4.0 m./sec.$^2$ 2........

**3.** Assume that an astronaut drops a stone while standing on the surface of the moon. Which graph best represents the motion of the stone? 3........

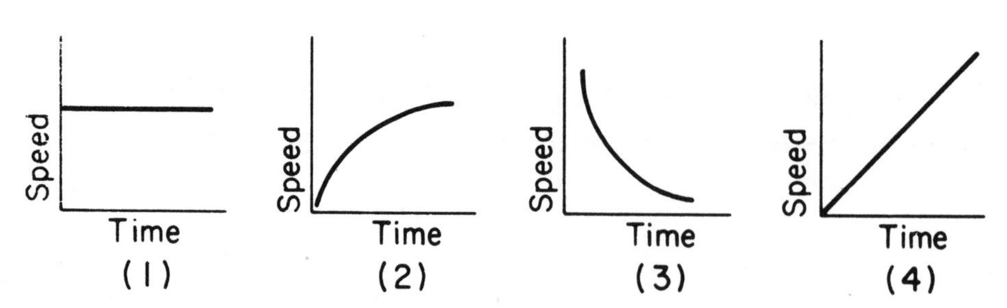


**4.** A rocket rises vertically with a constant acceleration of 20 meters/second$^2$. How much time will be required for the rocket to reach an altitude of 4,000 meters? (1) 10 sec. (2) 20 sec. (3) 200 sec. (4) 400 sec. 4........

**5.** The smallest possible resultant of a 4-nt. force and a 7-nt. force acting at a point is (1) 11 nt. (2) 5 nt. (3) 3 nt. (4) 0 nt. 5........

**6.** Two vectors act on point $P$ as shown at the right. Which vector will most likely produce equilibrium?

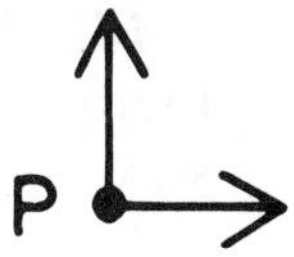

P

(1) (2) (3) (4) 6........

**7.** A force applied to a 5-kilogram mass produces an acceleration of 10 meters per second$^2$. The magnitude of the force is (1) 0.5 newton (2) 2.0 newtons (3) 50 newtons (4) 100 newtons 7........

**8.** An object weighs 1 newton at a point where the acceleration due to gravity is 9.8 m./sec.$^2$. The mass of the object is approximately (1) 1.0 kg. (2) 0.01 kg. (3) 0.1 kg. (4) 10 kg. 8........

**9.** A student exerts a 20-newton force on a desk. The force that the desk exerts on the student is (1) 0 nt. (2) between 0 and 20 nt. (3) 20 nt. (4) more than 20 nt. 9........

**10.** The force of gravity on an elevator is $10^4$ nt. If the upward force acting on the elevator is $10^4$ nt., the elevator can *not* be

(1) at rest
(2) moving upward with constant acceleration
(3) moving upward with constant speed
(4) moving downward with constant speed 10........

**11.** A 2-nt. force acts on an object. If the momentum of the object changes by 120 kg.-m./sec., the force must have acted for (1) 8 sec. (2) 30 sec. (3) 60 sec. (4) 120 sec. 11........

**12.** A 4-kg. cart moving east at 3 m./sec. collides with a 6-kg. cart moving west at 2 m./sec. The combined momentum of the two carts at the moment of collision is (1) 0 kg. m./sec. (2) 6 kg. m./sec. (3) 12 kg. m./sec. (4) 24 kg. m./sec. 12........

**13.** Two bodies of mass $m_1$ and $m_2$ 100. meters apart attract each other with a gravitational force of 5.00 newtons. What will be the force of attraction if the distance between the two masses is tripled? (1) 0.56 nt. (2) 1.10 nt. (3) 1.25 nt. (4) 2.50 nt. 13........

**14.** A 20.-newton block falls from a shelf 3.0 meters above the ground. Its kinetic energy at 1.5 meters above the ground is (1) 20. joules (2) 30. joules (3) 60. joules (4) 120 joules 14........

**15.** If the kinetic energy of a 10-kg. mass is 80 joules, its velocity is (1) 1 m./sec. (2) .125 m./sec. (3) 8 m./sec. (4) 4 m./sec. 15........

**16.** A box is dragged up an incline a distance of 8 meters with a force of 50 newtons. If the increase in potential energy of the box is 300 joules, then the work done against friction must be (1) 0 joules (2) 100 joules (3) 300 joules (4) 400 joules 16........

**17.** Which graph best represents the relationship between the absolute temperature (Kelvin) of a gas and the average kinetic energy of the gas molecules? 17........

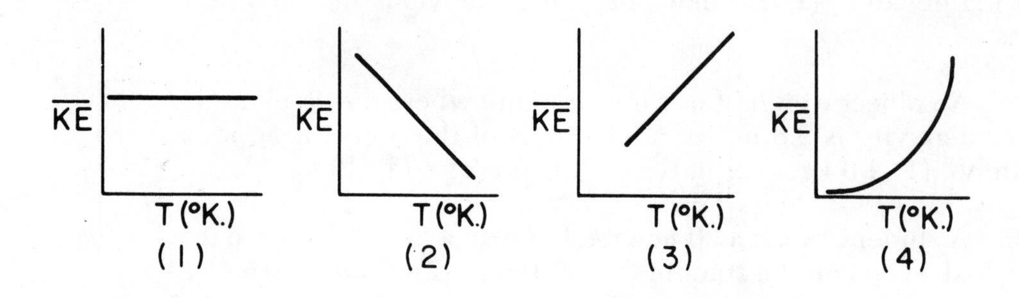


**18.** The Celsius temperature of an object is increased 90°. Its increase on the Kelvin scale will be (1) 50° (2) 90° (3) 162° (4) 273° 18........

**19.** Waves in which the vibrations occur parallel to the direction of motion of the wave are called (1) electromagnetic waves (2) transverse waves (3) longitudinal waves (4) nonperiodic waves 19........

**20.** As a wave passes into a new medium, which characteristic of the wave *never* changes? (1) velocity (2) period (3) intensity (4) direction 20........

**21.** A sound wave has a wavelength of 3 meters and a frequency of 100 cycles/second. What is its velocity? (1) 0.3 m./sec. (2) 33.3 m./sec. (3) 150 m./sec. (4) 300 m./sec. 21........

**22.** Maximum destructive interference occurs where the phase difference between two waves is (1) 0° (2) 90° (3) 180° (4) 270° 22........

**23.** Two points on a wave are in phase. The distance between these two points is (1) 1 wavelength (2) $\frac{1}{4}$ wavelength (3) $\frac{1}{2}$ wavelength (4) $\frac{3}{4}$ wavelength 23........

**24.** A logitudinal wave can *not* be (1) polarized (2) reflected (3) refracted (4) diffracted 24........

**25.** The spreading of a wave into the region behind an obstruction is called (1) interference (2) diffusion (3) dispersion (4) diffraction 25........

**26.** In a dispersive medium, the speed of a light wave depends upon
(1) its frequency, only
(2) its amplitude, only
(3) both its frequency and its amplitude
(4) neither its frequency nor its amplitude 26........

**27.** In the diagram below, which best represents the emergent ray for incident ray I? 27........

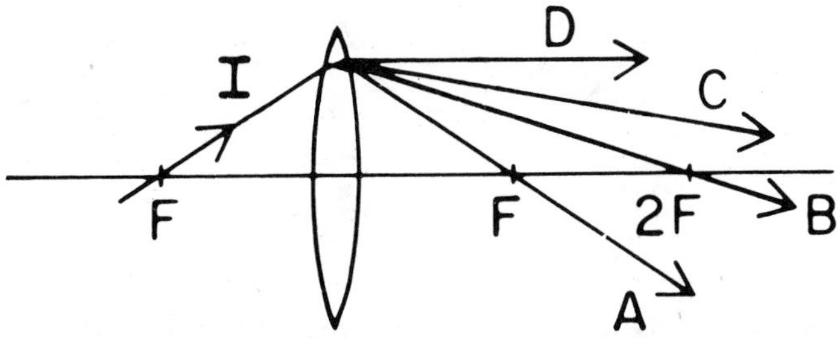

(1) *A*
(2) *B*
(3) *C*
(4) *D*

**28.** Total internal reflection occurs when the angle of incidence is (1) less than the critical angle (2) greater than the critical angle (3) equal to the angle of refraction (4) ninety degrees 28........

**29.** All electromagnetic radiations traveling in a vacuum have the same (1) frequency (2) wavelength (3) speed (4) amplitude 29........

**30.** Which radiation has the shortest wavelength? (1) red light (2) violet light (3) radio waves (4) X-rays 30........

**31.** Doubling the amount of charge on one of two positively charged objects causes the electrostatic force between the two objects to be (1) halved (2) doubled (3) quartered (4) quadrupled 31........

**32.** When neutral materials *A* and *B* are rubbed together, there is a transfer of electrons. As a result, the net charges on *A* and *B* are

(1) like and unequal (2) like and equal (3) unlike and unequal (4) unlike and equal 32........

33. Which is a vector quantity? (1) electric charge (2) electric field intensity (3) electric power (4) electrical energy 33........

34. A proton is accelerated by a potential difference of 50,000 volts. The energy acquired by the proton is (1) 25,000 e.v. (2) 50,000 e.v. (3) 100,000 e.v. (4) 200,000 e.v. 34........

35. Which graph best represents the relationship between the length ($L$) of a conductor of uniform cross section and composition and its resistance ($R$)? 35........

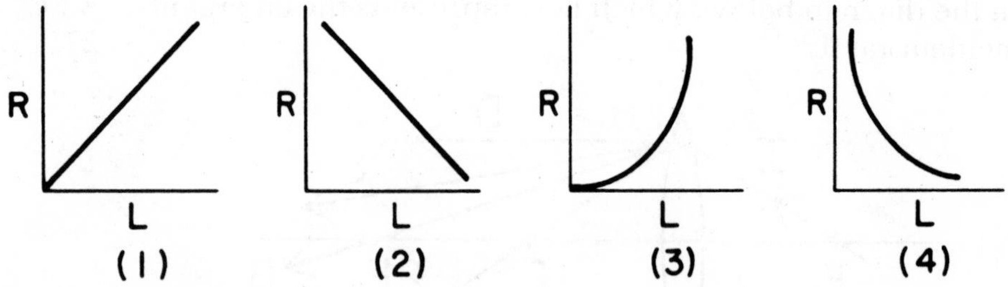

R L (1) R L (2) R L (3) R L (4)

36. A 5-ohm resistor has a current of 4 amperes. The potential difference across the resistor must be (1) 0.8 volt (2) 1.3 volts (3) 10 volts (4) 20 volts 36........

37. Five 50-ohm lamps connected in parallel would have a combined resistance of (1) 10 ohms (2) 50 ohms (3) 55 ohms (4) 250 ohms 37........

38. Compared to the potential drop across the 10-ohm resistor shown at the right, the potential drop across the 20-ohm resistor is
(1) half as great
(2) the same
(3) twice as great
(4) four times as great

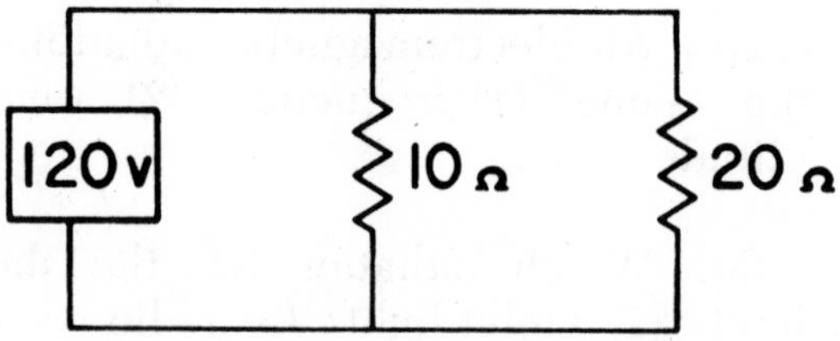

120 v 10 Ω 20 Ω

38........

39. In an ionized gas, a current may be carried by (1) electrons, only (2) positive ions, only (3) negative ions, only (4) electrons, positive ions, and negative ions 39........

**40.** A 120-volt toaster is rated at 600 watts. Under normal conditions, the current in the toaster would be (1) 0.2 amp. (2) 5 amp. (3) 10 amp. (4) 25 amp. 40........

**41.** The rate of flow of charge past a given point is measured in (1) amperes (2) ohms (3) volts (4) watts 41........

**42.** A cross section of wire is shown with the electron flow directed out of the page.

A

The direction of the magnetic field at point *A* is (1) to the right (2) to the left (3) toward the top of the page (4) toward the bottom of the page 42........

**43.** The diagram at the right represents a wire in a magnetic field carrying an electron current perpendicularly into the page. The direction of the magnetic force acting on the wire will be toward letter

(1) *A*

(2) *B*

(3) *C*

(4) *D*

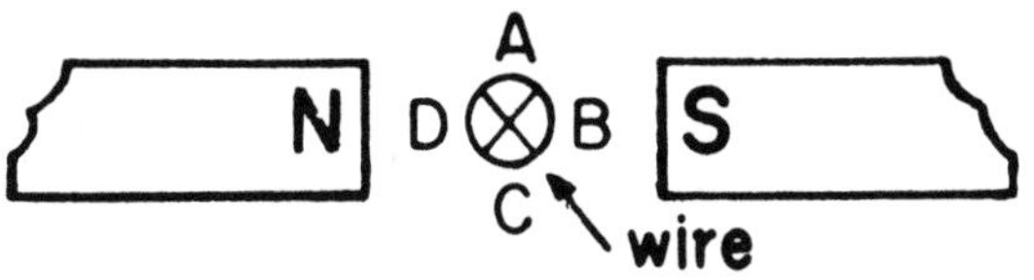

A
N D B S
C wire

43........

**44.** The metal ring in the diagram at the right is located on the axis of a bar magnet as shown. If the ring is suddenly moved away from the magnet along the axis, electrons in the ring will move from

(1) *A* to *B* and from *C* to *D*

(2) *A* to *B* and from *D* to *C*

(3) *B* to *A* and from *C* to *D*

(4) *B* to *A* and from *D* to *C*

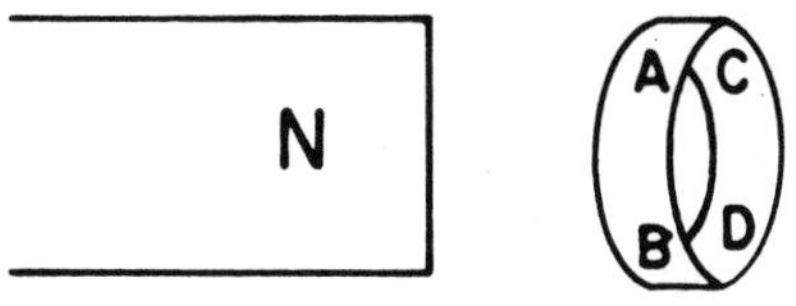

N
A C
B D

44........

**45.** Which is an isotope of $^{12}_{6}C$? (1) $^{10}_{5}C$ (2) $^{12}_{7}C$ (3) $^{13}_{6}C$ (4) $^{12}_{5}C$ 45........

**46.** The ratio of energy to frequency of a photon is (1) greater for photons of higher energy (2) greater for photons of higher frequency (3) greater for photons of higher speed (4) the same for all photons 46........

**47.** Which Bohr postulate regarding the atom violated classical electromagnetic theory? (1) An electron could revolve around the nucleus without radiating energy. (2) Coulomb's inverse square law held for atomic dimensions. (3) The Conservation of Momentum principle held for the atomic structure. (4) The Conservation of Energy principle held for the atomic structure. 47........

**48.** A lithium nucleus contains 3 protons and 4 neutrons. What is its atomic number? (1) 1 (2) 7 (3) 3 (4) 4 48........

**49.** The largest amount of energy is emitted by a hydrogen atom when the atom changes from energy level

(1) $n = 2$ to $n = 1$ (3) $n = 4$ to $n = 3$
(2) $n = 3$ to $n = 2$ (4) $n = 5$ to $n = 4$ 49........

**50.** According to the Bohr theory, when an electron with 6.0 e.v of energy is incident upon a hydrogen atom in the ground state, the atom will be (1) ionized (2) raised partially to the $n = 2$ energy level (3) raised to an energy level from which it can emit a 3.4-e.v. photon (4) unchanged 50........

**51.** When an alpha particle acquires an orbital electron, its nuclear formula is (1) $^{4}_{2}He$ (2) $^{4}_{1}He$ (3) $^{4}_{1}H$ (4) $^{5}_{2}He$ 51........

**52.** When alpha particles are directed at a thin metallic foil, it can be observed that most of the particles (1) pass through with virtually no deflection (2) bounce backward (3) are absorbed (4) are widely scattered 52........

**53.** Radioactivity can *not* be detected by a (1) cloud chamber (2) cyclotron (3) Geiger counter (4) photographic plate 53........

**Note that questions 54 through 60 have only three choices.**

**54.** As a spacecraft moves from the earth toward the moon, the force that the spacecraft exerts on the earth (1) decreases (2) increases (3) remains the same 54........

**55.** As the power of a hoist is increased, the time required to lift an object a fixed distance (1) decreases (2) increases (3) remains the same 55........

**56.** Two pulses on a spring are traveling in the same direction as shown in the diagram at the right. As they proceed toward the wall, the distance between them

(1) decreases

(2) increases

(3) remains the same

56........

**57.** As the frequency of a periodic wave increases, the period of the wave (1) decreases (2) increases (3) remains the same 57........

**58.** As a light wave passes from a medium of 1.3 index of refraction into a medium of 1.5 index of refraction, its wavelength will (1) decrease (2) increase (3) remain the same 58........

**59.** Nucleus *A* has a higher binding energy per nucleon than nucleus *B*. Compared to the energy necessary to separate nucleus *B* into component nucleons, the energy needed to separate nucleus *A* into nucleons is (1) less (2) greater (3) the same 59........

**60.** As an electron of an atom changes from an excited state to the ground state, emitting a photon, the total energy of the atom (1) decreases (2) increases (3) remains the same 60........

**PART TWO** *This part consists of four groups, each group testing a major area in the course. Choose two of these four groups. Write your answers in the spaces provided.*

## GROUP 1—Mechanics

*Answer all fifteen questions, 61 through 75, in this group. Each question counts 1 credit. For each of the fifteen questions, write the* number *of the word or expression that best completes that statement or answers that question.*

Base your answers to questions 61 through 65 on the graph below that represents the motion of a car moving in a straight line.

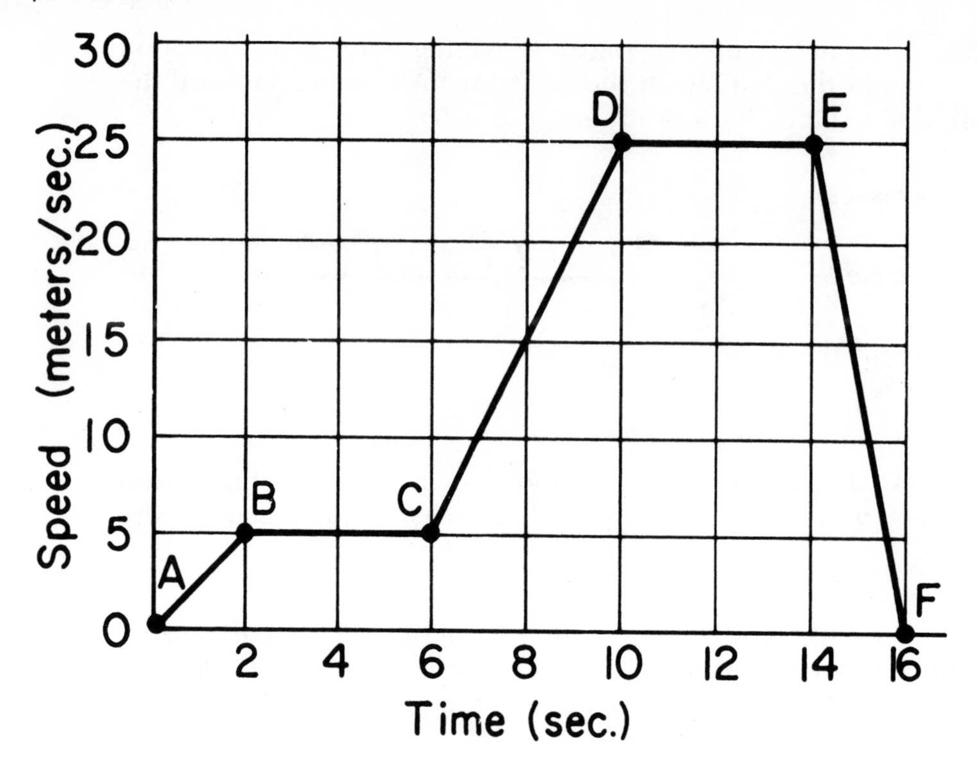


**61.** The average speed of the car during interval *CD* was (1) 5.0 m./sec. (2) 8 m./sec. (3) 15 m./sec. (4) 25 m./sec. 61........

**62.** During which part of the trip was the car moving with a constant speed? (1) *AB* (2) *CD* (3) *DE* (4) *EF* 62........

**63.** During which part of the trip was the greatest distance covered? (1) *BC* (2) *CD* (3) *DE* (4) *EF* 63........

**64.** During which part of the trip was zero net force acting on the car? (1) *AB* (2) *BC* (3) *CD* (4) *EF* 64........

**65.** What was the acceleration during interval *CD*? (1) 5 m./sec.$^2$ (2) 10 m./sec.$^2$ (3) 15 m./sec.$^2$ (4) 25 m./sec.$^2$ 65........

Base your answers to questions 66 through 70 on the diagram and information below which shows a force that acts horizontally on a mass for an interval of 2.0 seconds along *AB*. The mass starts from rest and slides along the horizontal frictionless surface *AB* and up the incline *BC* to point *C*. [Neglect friction.]

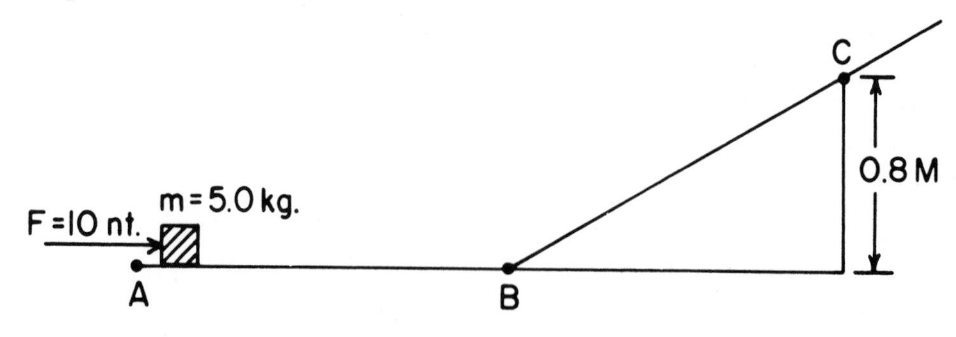

**66.** What is the momentum of the body along section *AB* after 2.0 seconds have elapsed? (1) 10 kg. m./sec. (2) 20 kg. m./sec. (3) 50 kg. m./sec. (4) 100 kg. m./sec. 66........

**67.** What acceleration along *AB* is produced by the force? (1) 4.9 m./sec.$^2$ (2) 2.0 m./sec.$^2$ (3) 10 m./sec.$^2$ (4) 20 m./sec.$^2$ 67........

**68.** The total amount of potential energy that the mass gains between points *B* and *C* is closest to (1) 49 joules (2) 40 joules (3) 8 joules (4) 4 joules 68........

**Note that questions 69 and 70 have only three choices.**

**69.** If a greater force acted on the mass for the same length of time, the maximum vertical height attained by the mass would (1) decrease (2) increase (3) remain the same 69........

**70.** If a greater mass were acted on by the same force for the same length of time, the maximum vertical height attained by the mass would (1) decrease (2) increase (3) remain the same 70........

Base your answers to questions 71 through 75 on the graph below, which represents the temperature change of a substance as it absorbs heat at a constant rate.

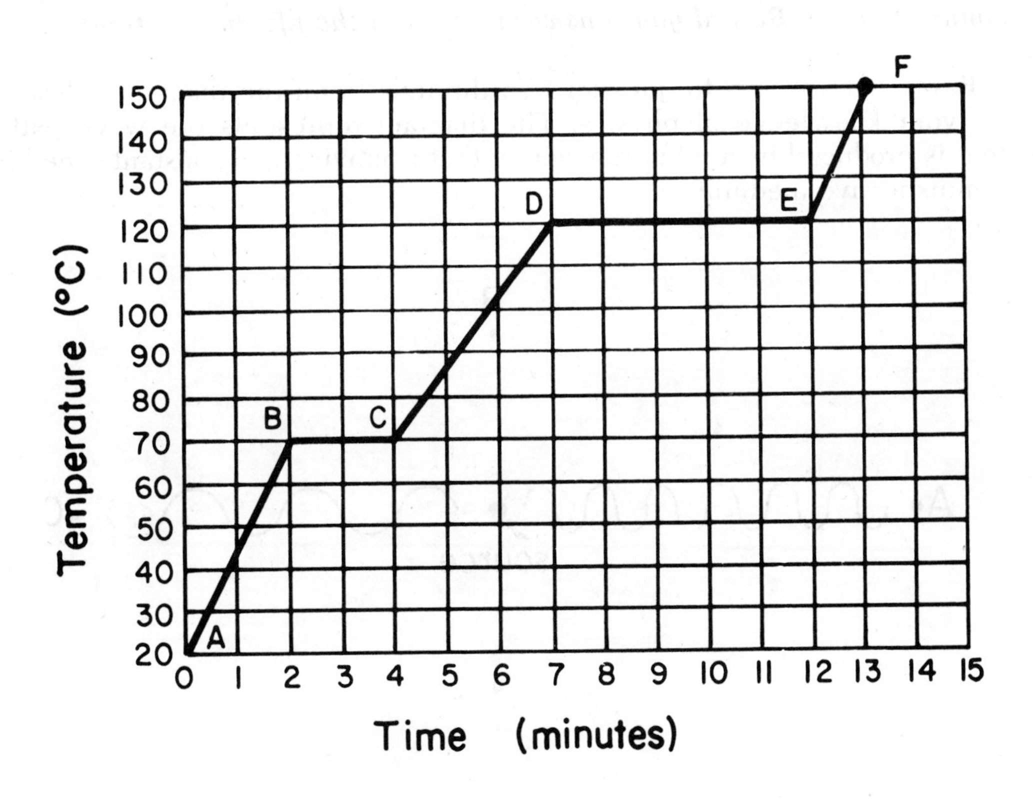

**71.** The heat of vaporization of the substance can be determined by measuring the total number of calories added during the interval (1) *BC* (2) *CD* (3) *DE* (4) *EF* 71........

**72.** The temperature at which the substance can be in both the solid and liquid phase is (1) 20° C. (2) 70° C. (3) 120° C. (4) 150° C. 72........

**73.** The internal potential energy of the substance is increasing during interval (1) *AB* (2) *BC* (3) *CD* (4) *EF* 73........

**Note that questions 74 and 75 have only three choices.**

**74.** Compared to the internal energy of this substance at point *D*, its internal energy at point *E* is (1) less (2) greater (3) the same 74........

**75.** As the temperature of an ideal gas increases, the product of its pressure and volume (1) decreases (2) increases (3) remains the same 75........

GROUP 2—**Wave Phenomena**

*Answer all fifteen questions, 76 through 90, in this group. Each question counts 1 credit. Record your answer to each of the fifteen questions.*

Base your answers to questions 76 through 78 on the diagram below and on your knowledge of physics. The diagram represents the wave pattern that is produced by a vibrating source that is moving at a constant speed in a nondispersive medium.

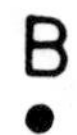

D

**76.** The diagram illustrates (1) interference (2) diffraction (3) the Doppler effect (4) reflection 76........

**77.** The source is moving toward point (1) *A* (2) *B* (3) *C* (4) *D* 77........

**Note that question 78 has only three choices.**

**78.** Compared to the frequency of the waves observed at *C*, the frequency of the waves observed at *A* is (1) less (2) greater (3) the same 78........

Base your answers to questions 79 through 82 on the diagram below.

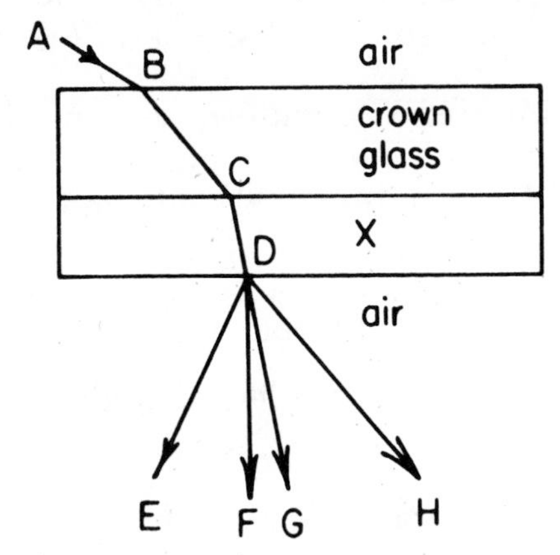


**79.** If the angle of incidence in air is 30°, the sine of the angle of refraction in crown glass is (1) 0.152 (2) 0.161 (3) 0.329 (4) 0.805 79........

**80.** Medium *X* most likely is (1) air (2) lucite (3) crown glass (4) flint glass 80........

**81.** After leaving medium *X*, the most probable path of the ray is (1) *DE* (2) *DF* (3) *DG* (4) *DH* 81........

**82.** Ray *ABC* would be a straight line if the crown glass were replaced with (1) lucite (2) air (3) water (4) alcohol 82........

Base your answers to questions 83 and 84 on the diagram below which represents a convex lens with a focal length of .2 meter. A movable object is on side *A* of the lens.

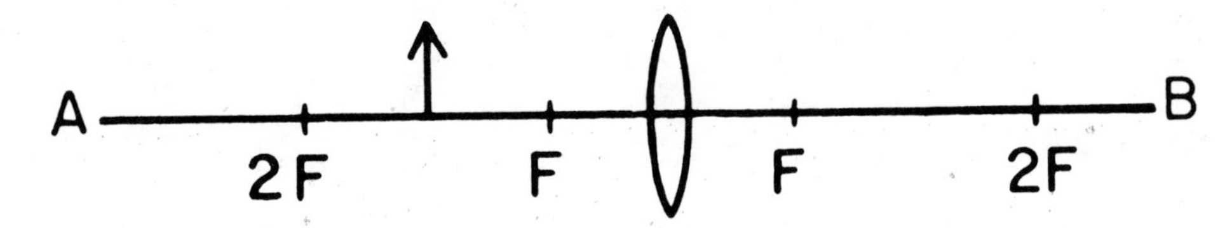

**83.** As the object is moved from $2F$ toward $F$ on side $A$, the image formed on side $B$ will move from (1) $2F$ toward $F$ (2) $2F$ toward infinity (3) $F$ toward $2F$ (4) $F$ toward the lens 83........

**84.** If the object is .3 meter from the lens, how far from the lens will the image be located? (1) .30 m. (2) .40 m. (3) .50 m. (4) .60 m. 84........

**85.** The distance between the bright lines of an interference pattern will increase when there is a decrease in the (1) wavelength of the light (2) energy of the light (3) distance between the slits and the screen (4) distance between the slits 85........

**86.** An interference pattern is produced by 2 slits because light is (1) reflected (2) refracted (3) dispersed (4) diffracted 86........

**87.** If a stationary interference pattern were produced by two monochromatic light sources, then the light from the two sources would have to be (1) coherent (2) reflected (3) dispersed (4) polarized 87........

**Note that question 88 has only three choices.**

**88.** As the length of a wave in a uniform medium increases, its frequency will (1) decrease (2) increase (3) remain the same 88........

**89.** A ray of monochromatic yellow light has the lowest apparent speed in (1) Canada balsam (2) diamond (3) lucite (4) alcohol 89........

**90.** Which is the correct mathematical expression for the relationship between the velocity ($v$), frequency ($f$), and wavelength ($\lambda$) for waves? (1) $f = v\lambda$ (2) $\lambda = v/f$ (3) $v = f/\lambda$ (4) $f = \lambda/v$ 90........

## GROUP 3 – Electricity

*Answer all fifteen questions, 91 through 105, in this group. Each question counts 1 credit. Record your answer to each of the fifteen questions on the* separate answer sheet in accordance with the directions on the front page of this booklet.

Base your answers to questions 91 through 95 on the diagram and information below and on your knowledge of physics.

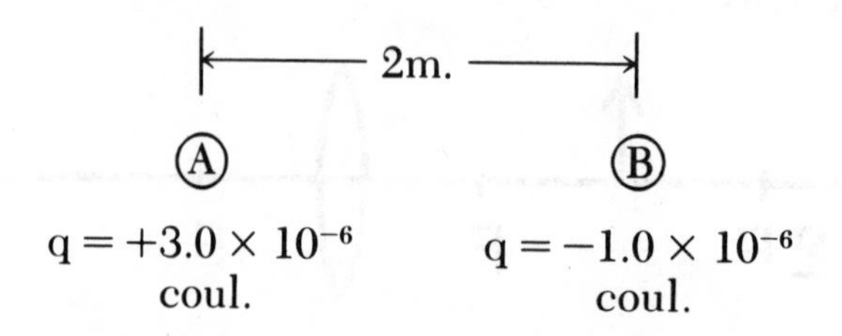

**91.** Which vector diagram best represents the force of sphere *A* on sphere *B*?

(1) Ⓐ Ⓑ↑

(2) Ⓐ Ⓑ→

(3) Ⓐ Ⓑ↓

(4) Ⓐ ←Ⓑ

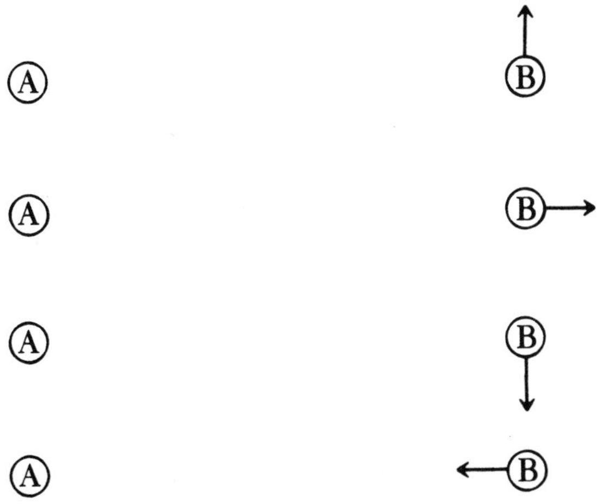

91........

**92.** If the distance between the spheres were doubled, the force between them would be (1) quartered (2) doubled (3) halved (4) quadrupled

92........

Base your answers to questions 93 through 95 on the additional information below.

The two conducting spheres are brought into contact so that charges can move freely between them. They are then separated again by 2 meters.

**93.** Which diagram best represents the charge on each sphere?

(1) A ⊕ B ⊖

(2) A ⊖ B ⊕

(3) A ⊖ B ⊖

(4) A ⊕ B ⊕

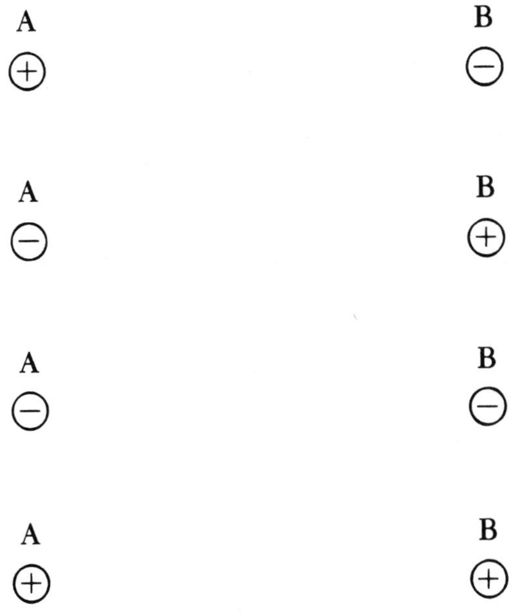

93........

**94.** The net charge on the two spheres after contact is (1) $+1 \times 10^{-6}$ coul. (2) $+2 \times 10^{-6}$ coul. (3) $-3 \times 10^{-6}$ coul. (4) $+4 \times 10^{-6}$ coul. 94........

**Note that question 95 has only three choices.**

**95.** Compared to the force between the spheres before contact, the force between them after contact is (1) less (2) greater (3) the same 95........

Base your answers to questions 96 through 100 on the diagram below.

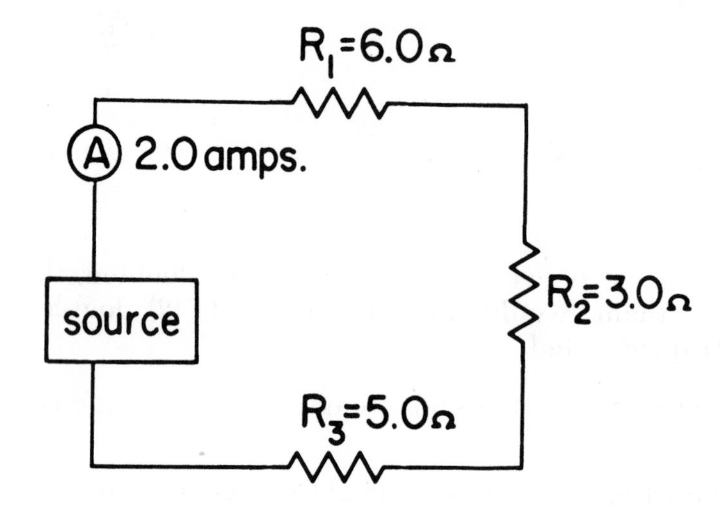


**96.** The potential difference across the ends of resistor $R_2$ is (1) 0.67 volt (2) 1.5 volts (3) 6.0 volts (4) 28 volts 96........

**97.** What amount of work is done as each coulomb passes through resistor $R_1$? (1) 18 joules (2) 12 joules (3) 3.0 joules (4) 0.33 joule 97........

**98.** The total resistance of the circuit is (1) 8 ohms (2) 9 ohms (3) 11 ohms (4) 14 ohms 98........

**99.** Compared to the potential difference across $R_2$, the potential difference across $R_1$ is (1) one-half as great (2) twice as great (3) the same (4) four times as great 99........

**Note that question 100 has only three choices.**

**100.** Compared to the rate of heat production in $R_2$, the rate of heat production in $R_1$ is (1) less (2) greater (3) the same 100........

Base your answers to questions 101 through 105 on the diagram below which shows electrons being emitted from cathode $C$ and accelerated to anode $A$. Some electrons pass through a hole in the anode and continue forward to enter a uniform magnetic field which is directed as shown.

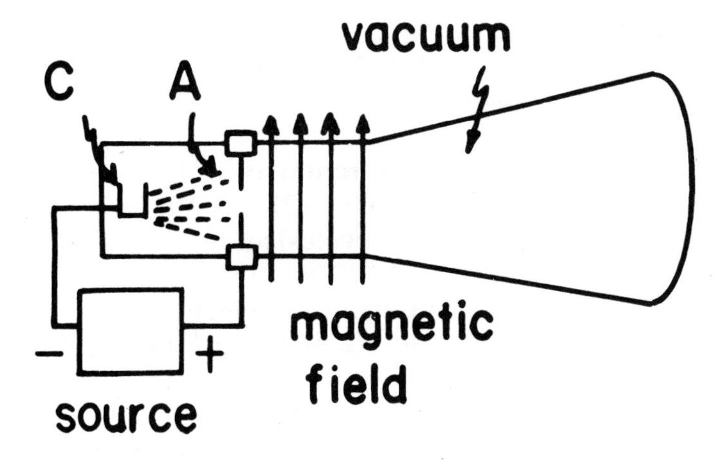


**101.** If the kinetic energy acquired by each electron is $6.4 \times 10^{-16}$ joule, then the potential difference between the cathode and anode is (1) $8.0 \times 10^{2}$ volts (2) $1.3 \times 10^{3}$ volts (3) $2.5 \times 10^{3}$ volts (4) $4.0 \times 10^{3}$ volts 101........

**102.** If $1.0 \times 10^{12}$ electrons pass through the magnetic field each second, the current through the field is (1) $1.6 \times 10^{-31}$ amp. (2) $1.6 \times 10^{-19}$ amp. (3) $2.6 \times 10^{-12}$ amp. (4) $1.6 \times 10^{-7}$ amp. 102.........

**Note that questions 103 through 105 have only three choices.**

**103.** As an electron moves through the magnetic field, its speed will (1) decrease (2) increase (3) remain the same 103.........

**104.** As an electron moves through the magnetic field, the magnitude of the magnetic force on the electron will (1) decrease (2) increase (3) remain the same 104.........

**105.** If the velocity of the electrons moving into the magnetic field were increased, the magnitude of the magnetic force on the electrons would (1) decrease (2) increase (3) remain the same 105.........

## GROUP 4—Atomic and Nuclear Physics

*Answer all fifteen questions, 106 through 120, in this group. Each question counts 1 credit. Record your answer to each of the fifteen questions.*

Base your answers to questions 106 through 108 on the information below.

Gamma radiation with a photon energy of $6.6 \times 10^{-12}$ joule strikes a surface and causes electrons to be ejected. The work function of the material is $3.3 \times 10^{-12}$ joule.

**106.** The frequency of the gamma radiation is approximately (1) $1 \times 10^{-46}$ cycle/sec. (2) $1 \times 10^{14}$ cycles/sec. (3) $1 \times 10^{22}$ cycles/sec. (4) $1 \times 10^{34}$ cycles/sec. 106.........

**107.** The photoelectric threshold frequency of the material is approximately (1) $2 \times 10^{21}$ cycles/sec. (2) $2 \times 10^{22}$ cycles/sec. (3) $5 \times 10^{21}$ cycles/sec. (4) $5 \times 10^{22}$ cycles/sec. 107.........

**108.** The maximum kinetic energy of the ejected electrons is approximately (1) $3.3 \times 10^{-12}$ joule (2) $9.9 \times 10^{-12}$ joule (3) 3.3 joules (4) 9.9 joules 108.........

**109.** In the nuclear equation ${}^{27}_{13}\text{Al} + {}^{4}_{2}\text{He} \rightarrow {}^{30}_{15}\text{P} + X$, $X$ represents (1) a neutron (2) a proton (3) an electron (4) an alpha particle 109.........

Base your answers to questions 110 and 111 on the information below.

${}^{210}_{83}\text{Bi}$ emits a beta particle.

**110.** What is the mass number of the new element? (1) 206 (2) 209 (3) 210 (4) 212 110.........

**111.** What is the atomic number of the new element? (1) 82 (2) 83 (3) 84 (4) 85 111.........

**112.** In a nuclear reactor, energy is released after an atom of uranium captures (1) a water molecule (2) a proton (3) an electron (4) a neutron 112.........

**113.** A decay product of ${}^{218}_{84}\text{Po}$ is (1) ${}^{214}_{85}X$ (2) ${}^{214}_{82}X$ (3) ${}^{218}_{82}X$ (4) ${}^{222}_{86}X$ 113.........

**114.** Which element exists in three isotopic forms in the Uranium Disintegration Series shown on the *Physics Reference Tables?* (1) U (2) Th (3) Bi (4) Pb 114.........

**115.** When an atom of U-238 changes to U-234, there is an emission of (1) one alpha particle, only (2) one beta particle, only (3) three beta particles (4) one alpha and two beta particles 115.........

**116.** The most stable element in the Uranium Disintegration Series is (1) ${}^{238}_{92}\text{U}$ (2) ${}^{234}_{92}\text{U}$ (3) ${}^{214}_{82}\text{Pb}$ (4) ${}^{206}_{82}\text{Pb}$ 116.........

**117.** The half-life of radium is 1,620 years. What fraction of a radium sample will remain after 3,240 years? (1) $1/16$ (2) $1/8$ (3) $1/4$ (4) $1/2$ 117.........

**118.** The combining of nuclei of light elements to form a heavier nucleus is called (1) alpha emission (2) beta emission (3) fusion (4) fission 118.........

**119.** The minimum photon energy level that would raise a hydrogen atom from the $n = 2$ to the $n = 3$ energy state is (1) 1.9 e.v. (2) 10.2 e.v. (3) 12.1 e.v. (4) 13.6 e.v. 119.........

**120.** When ultraviolet radiation is incident upon a surface, no photoelectrons are emitted. If a second beam causes photoelectrons to be ejected, it may consist of (1) radio waves (2) infrared rays (3) visible light rays (4) X-rays 120.........

## Examination June, 1973 Physics

**PART ONE** *A number of questions on this examination require the use of the Physics Reference Tables. Make sure that you have a copy for your use.*

DIRECTIONS (1-60): *In the space provided, write the* number *preceding the word or expression that, of those given, best completes the statement or answers the question.* [70]

1. Which is a vector quantity? (1) displacement (2) mass (3) speed (4) energy 1..........

2. A car is accelerated at 4.0 m./sec.$^2$ from rest. The car will reach a speed of 28 meters per second at the end of (1) 3.5 sec. (2) 7.0 sec. (3) 14 sec. (4) 24 sec. 2..........

3. An object is projected horizontally at 400 meters per second. If the object falls freely to the ground in 2.0 seconds, at what height did it begin its fall? (1) 4.9 meters (2) 9.8 meters (3) 19.6 meters (4) 200 meters 3..........

4. If the mass of an object were doubled, its acceleration due to gravity would be (1) halved (2) doubled (3) unchanged (4) quadrupled 4..........

5. Two concurrent forces of 40 newtons and $X$ newtons have a resultant of 100 newtons. Force $X$ could be (1) 20 newtons (2) 40 newtons (3) 80 newtons (4) 150 newtons 5..........

6. An elevator containing a man weighing 800 newtons is rising at a constant speed. The force exerted by the man on the floor of the elevator is (1) less than 80 nt. (2) between 80 and 800 nt. (3) 800 nt. (4) more than 800 nt. 6..........

7. A force of $F$ newtons gives an object with a mass of $M$ an acceleration of $A$. The same force $F$ will give a second object with a mass of $2M$ an acceleration of (1) $A/2$ (2) $2A$ (3) $A$ (4) $A/4$ 7..........

**8.** Two 1-kilogram masses are moving apart at a constant speed. Which graph best represents the force of gravitational attraction between the two masses during the time they are moving?

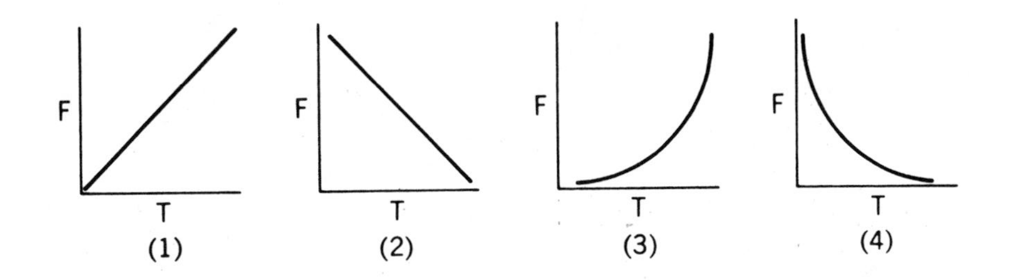


8..........

**9.** A 2-kilogram mass moving along a horizontal surface at 10 meters per second is acted upon by a 5-newton force of friction. The time required to bring the mass to rest is (1) 10 sec. (2) 2 sec. (3) 5 sec. (4) 4 sec. 9..........

**10.** If a 2-kilogram mass has a kinetic energy of 16 joules, it must have a speed of (1) 16 m./sec. (2) 2 m./sec. (3) 8 m./sec (4) 4 m./sec. 10..........

**11.** A force of 70 newtons must be exerted to keep a car moving with a constant speed of 10 meters per second .What is the rate at which energy must be supplied? (1) 1/7 watt (2) 7.0 watts (3) 700 watts (4) 7,000 watts 11..........

**12.** In order to keep an object weighing 20 newtons moving at constant speed along a horizontal surface, a force of 10 newtons is required. The force of friction between the surface and the object is (1) 0 nt. (2) 10 nt. (3) 20 nt. (4) 30 nt. 12..........

**13.** The internal energy of a solid is equal to the (1) absolute temperature of the solid (2) total kinetic energy of its molecules (3) total potential energy of its molecules (4) sum of the kinetic and potential energy of its molecules 13..........

**14.** What is the boiling point of ammonia in degrees Kelvin? (1) –306 degrees (2) –33 degrees (3) 240 degrees (4) 316 degrees 14..........

**15.** The ratio of a Kelvin degree to a Celsius degree is (1) 1:1 (2) 5;9 (3) 1:273 (4) 273:1 15..........

**16.** At a given location on the earth's surface, which graph best represents the relationship between an object's mass ($M$) and weight ($W$)?

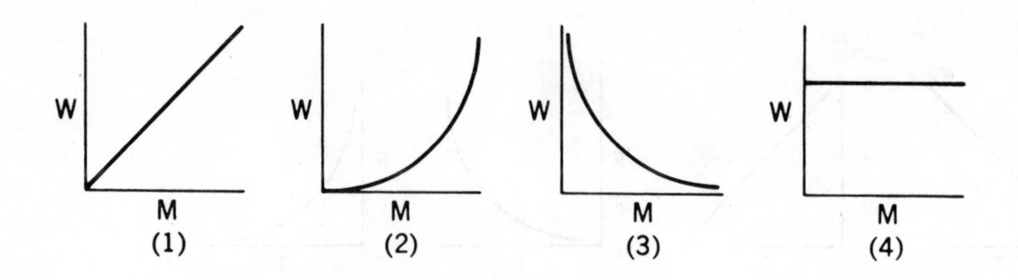


16..........

**17.** Which graph best represents the relationship between average molecular kinetic energy ($KE$) of an ideal gas and absolute temperature ($T$)?

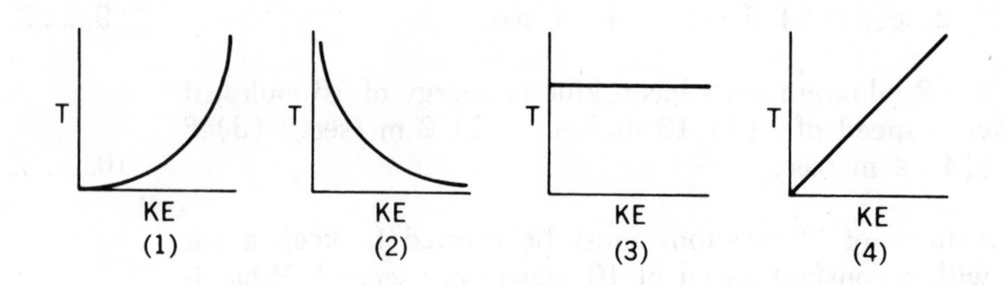


17..........

**18.** Which electromagnetic radiation has a wavelength shorter than that of visible light? (1) ultraviolet waves (2) infrared waves (3) radio waves (4) microwaves 18..........

**19.** An object is placed in front of a plane mirror as shown in the diagram.

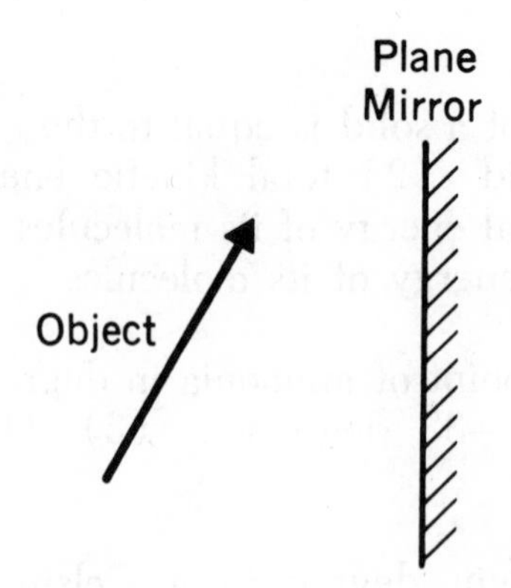

The image produced by the plane mirror is best represented by

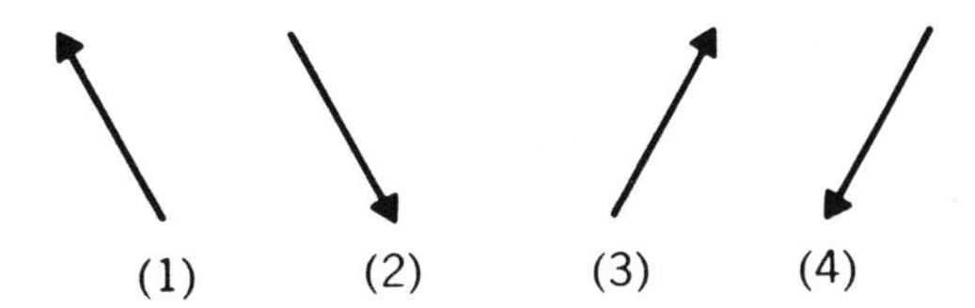


19..........

**20.** When a pulse traveling in a medium strikes the boundary of a different medium, the energy of the pulse will be (1) completely absorbed by the boundary (2) entirely transmitted into the new medium (3) entirely reflected back into the original medium (4) partly reflected back into the original medium and partly transmitted or absorbed into the new medium 20..........

**21.** Which wave requires a medium for transmission? (1) light (2) infrared (3) radio (4) sound 21..........

**22.** Light waves can be polarized because they (1) have high frequencies (2) have short wavelengths (3) are transverse (4) can be reflected 22..........

**23.** What is the frequency of a wave with a period of $5 \times 10^{-3}$ second? (1) $1 \times 10^2$ cycles/sec. (2) $2 \times 10^2$ cycles/sec. (3) $5 \times 10^2$ cycles/sec. (4) $1 \times 10^3$ cycles/sec. 23..........

**24.** Which wave has the greatest amplitude?

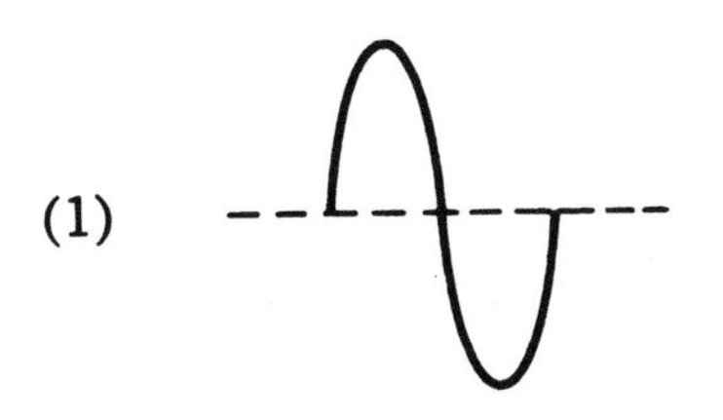


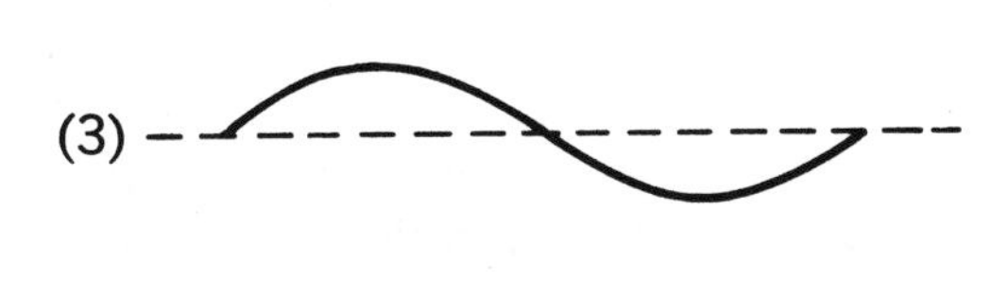


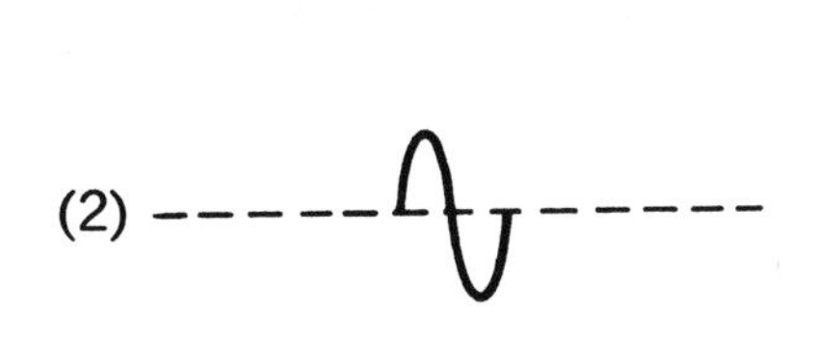


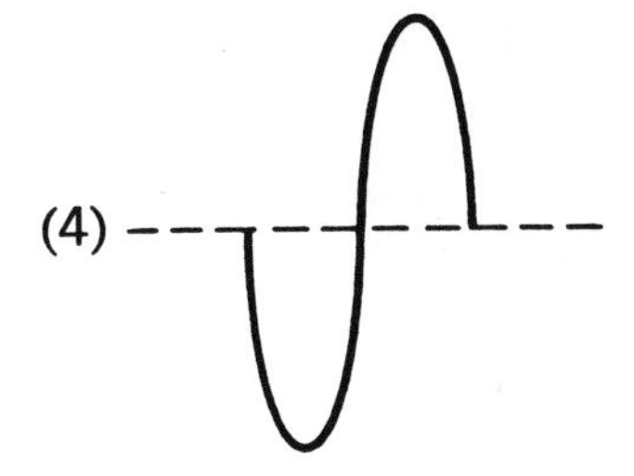


24..........

**25.** Two points on a periodic wave in a medium are said to be in phase if they (1) have the same amplitude, only (2) are moving in the same direction, only (3) have the same period (4) have the same amplitude and are moving in the same direction

25..........

**26.** What is the distance bewteen two consecutive points in phase on a wave called? (1) frequency (2) period (3) amplitude (4) wavelength

26..........

**27.** Waves are traveling with a speed of three meters per second toward *P* as shown in the diagram. If four crests pass *P* in one second, the wavelength is
(1) 1 m.
(2) 6 m.
(3) 3 m.
(4) 9 m.

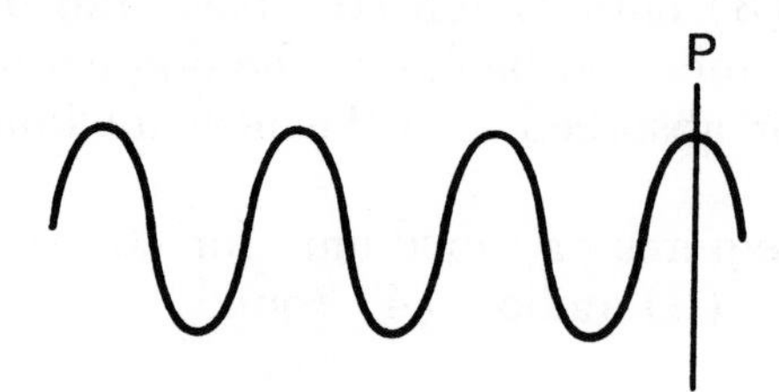


27..........

**28.** The diagram at the right shows a ray of light *R* incident on a water-air surface. Through which point is the ray most likenly to pass? (1) *A* (2) *B* (3) *C* (4) *D*

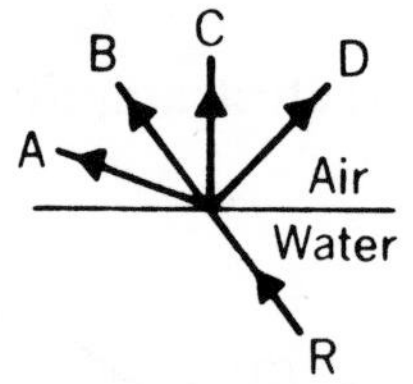


28..........

**29.** The spreading of a wave into the region behind an obstruction is (1) refraction (2) diffraction (3) dispersion (4) superposition

29..........

**30.** Two pulses in a stretched spring approach *P* as shown in the diagram.

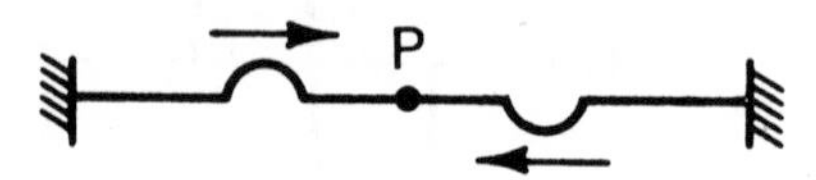

Which diagram best illustrates the appearance of the spring when each pulse meets at $P$?

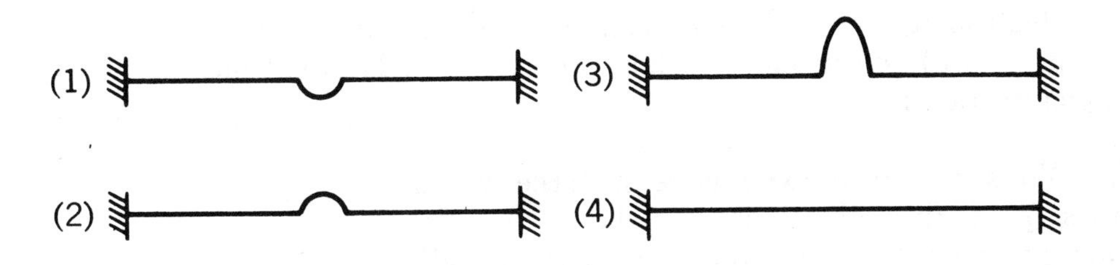


30..........

**31.** Which graph best represents the relationship between the resistance ($R$) of a solid conductor of constant cross section and its length ($L$)?

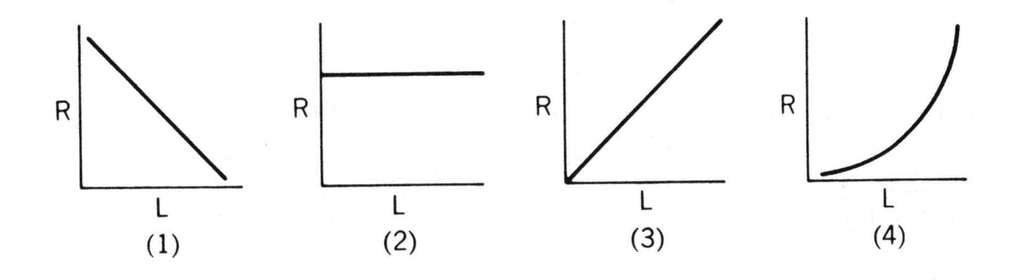


31..........

**32.** The work required to move 2 coulombs of charge through a potential difference of 5 volts is (1) 10 joules (2) 2 joules (3) 25 joules (4) 50 joules

32..........

**33.** Sphere *A* has a charge of +2 units and sphere *B*, which is identical to sphere *A*, has a charge of −4 units. If the two spheres are brought together and then separated, the charge on each sphere will be (1) −1 unit (2) −2 units (3) +1 unit (4) + 4 units

33........

**34.** An electrically neutral object can be attracted by a positively charged object because (1) like charges repel each other (2) the charges on a neutral body can be redistributed (3) the netural body becomes charged by contact (4) the net charge in a closed system varies

34..........

**35.** When an object is placed near a negatively charged electroscope, the leaves of the electroscope diverge farther. Which statement about the object is true? (1) It must be neutral.

(2) It must be positively charged. (3) It must be negatively charged. (4) It may be either positively or negatively charged. 35..........

**36.** Electromagnetic radiations can *not* be generated by accelerating (1) electrons (2) protons (3) neutrons (4) alpha particles 36..........

**37.** Wires $x$ and $y$ experience a force when electrons pass through them as shown in the diagram at the right. The force on wire $y$ will be toward (1) $A$ (2) $B$ (3) $C$ (4) $D$

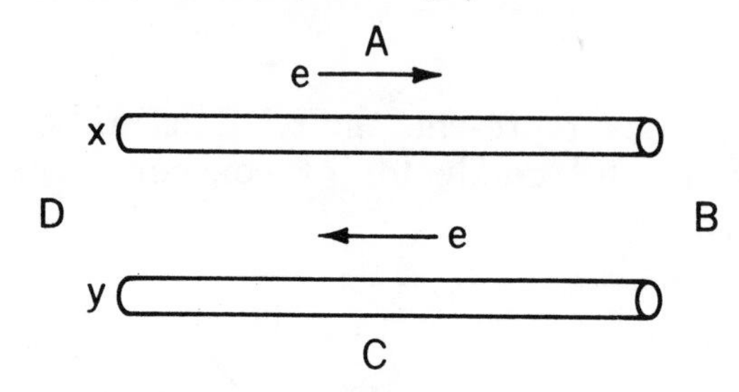


37..........

**38.** The magnetic lines of force near a long straight current-carrying wire are (1) straight lines parallel to the wire (2) straight lines perpendicular to the wire (3) circles in a plane perpendicular to the wire (4) circles in a plane parallel to the wire 38..........

**39.** Which diagram best illustrates the direction of the magnetic field between the unlike poles of two bar magnets? 39..........

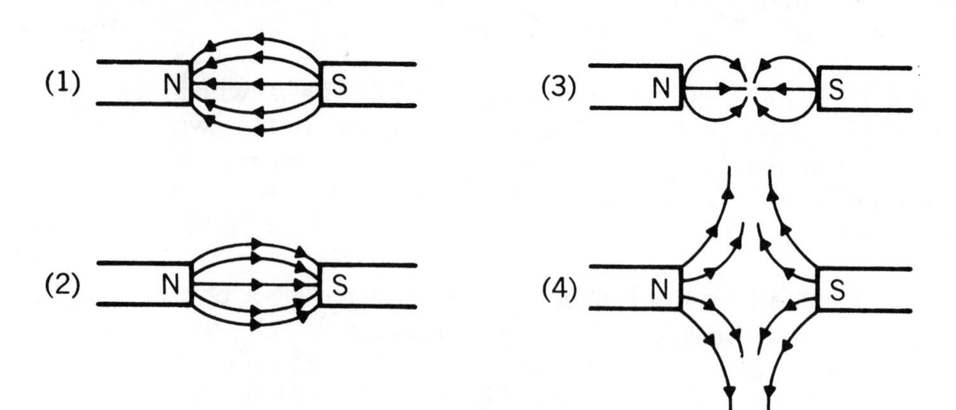


**40.** Magnetic fields are produced by (1) motion of electric charges (2) static electric charges (3) photon motion (4) gamma radiation 40..........

**41.** How long must a 100-watt light bulb be used in order to dissipate 1,000 joules of electrical energy? (1) 10 sec. (2) 100 sec. (3) 1,000 sec. (4) 100,000 sec. 41..........

**42.** Two identical resistors connected in series have a combined resistance of 8 ohms. When connected in parallel, the resistance of the combination will be (1) 8 ohms (2) 2 ohms (3) 16 ohms (4) 4 ohms 42..........

**43.** In the diagram shown, how many amperes is the reading of ammeter *A*?
(1) 5 amp.
(2) 2 amp.
(3) 3 amp.
(4) 7 amp.

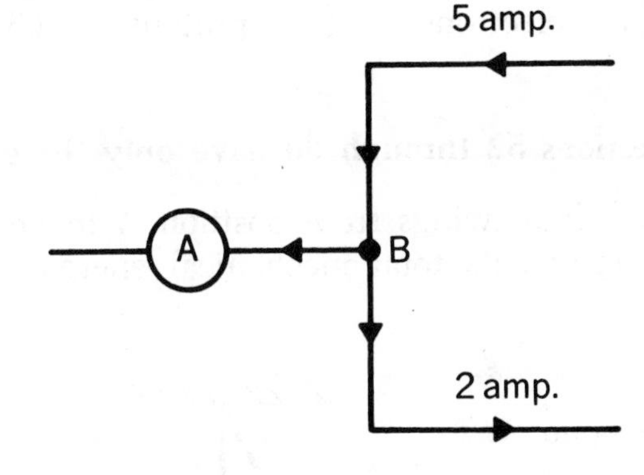


43..........

**44.** What is the current in a conductor if $6.25 \times 10^{18}$ electrons pass a given point each second? (1) 1 ampere (2) $1.6 \times 10^{-19}$ ampere (3) 2.6 amperes (4) $6.25 \times 10^{18}$ amperes 44..........

**45.** An electric current that exists in a liquid is caused by the movement of (1) electrons (2) positive ions, only (3) negative ions, only (4) positive and negative ions 45..........

**46.** Which atom is an isotope of $^{238}_{92}U$? (1) $^{238}_{91}X$ (2) $^{235}_{92}X$ (3) $^{238}_{93}X$ (4) $^{235}_{93}X$ 46..........

**47.** What is the minimum amount of energy required to excite a hydrogen atom in the ground state to the $n = 3$ energy level? (1) 14.5 electron-volts (2) 12.1 electron-volts (3) 11.9 electron-volts (4) 10.2 electron-volts 47..........

**48.** The nuclear binding energy of an atom is proportional to the (1) mass of the atom (2) mass defect of the atom

(3) nuclear protons of the atom (4) nuclear neutrons of the atom 48..........

**49.** An energy of 13.6 electron-volts is equivalent to (1) $1.60 \times 10^{-19}$ joule (2) $2.18 \times 10^{-18}$ joule (3) $6.25 \times 10^{-19}$ joule (4) $6.63 \times 10^{-18}$ joule 49..........

**50.** A model of the atom in which the electrons can exist only in specified orbits was suggested by (1) Bohr (2) Planck (3) Einstein (4) Rutherford 50..........

**51.** As excited hydrogen atoms return to the ground state, they emit (1) electrons (2) protons (3) photons (4) neutrons 51..........

**Note that questions 52 through 60 have only three choices.**

**52.** As a pendulum swings from position *A* to Position *B* as shown in the diagram, its total mechanical energy (neglecting friction)
(1) decreases
(2) increases
(3) remains the same

B
A

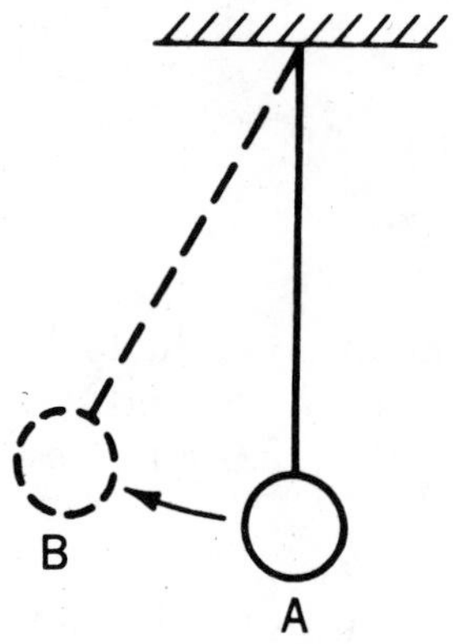

52..........

**53.** As a block is accelerated from rest along a horizontal surface, its gravitational potential energy (1) decreases (2) increases (3) remains the same 53..........

**54.** As water changes to ice without sublimation at zero degrees Celsius, its mass (1) decreases (2) increases (3) remains the same 54..........

**55.** As an object approaches a converging lens, the size of its real image (1) decreases (2) increases (3) remains the same 55..........

**56.** When a wave goes from a medium of higher refractive index to one of lower refractive index, its speed (1) decreases (2) increases (3) remains the same 56..........

**57.** As the resistance of a lamp operated at a constant voltage increases, the power used by the lamp (1) decreases (2) increases (3) remains the same 57..........

**58.** The current in a circuit is supplied by a generator. If the resistance in the circuit is increased, the force required to keep the generator turning at the same speed is (1) decreased (2) increased (3) the same 58..........

**59.** If the number of neutrons in an atom increases, the atomic number of the atom (1) decreases (2) increases (3) remains the same 59..........

**60.** Compared to the wavelength of a moving electron, the wavelength of a proton moving at the same speed is (1) shorter (2) longer (3) the same 60..........

**PART TWO** *This part consists of four groups, each group testing a major area in the course. Choose two of these four groups. Write your answers in the spaces provided.*

## GROUP 1 — Mechanics

*Answer all fifteen questions, 61 through 75 in this group. Each question counts 1 credit. For each of the fifteen questions, write the* number *of the word or expression that best completes that statement or answers that question.*

Base your answers to questions 61 through 65 on the graph below, which represents the motion of cars $A$ and $B$ on a straight track. Car $B$ passes car $A$ at the same instant that car $A$ starts from rest at $t = 0$ seconds.

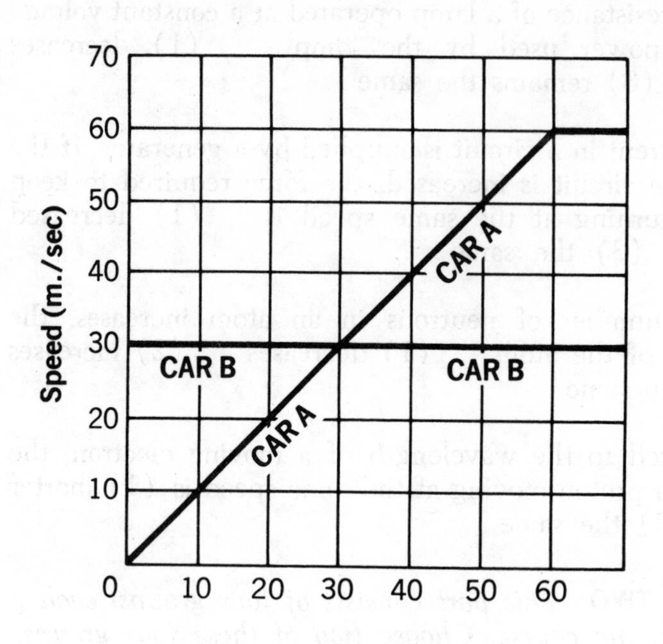


**61.** What is the acceleration of car $A$ during the interval between $t = 0$ and $t = 60$? (1) 1 m./sec.$^2$ (2) 10 m./sec.$^2$ (3) 20 m./sec.$^2$ (4) 30 m./sec.$^2$ 61.........

**62.** How far did car $A$ travel in the interval between $t = 0$ and $t = 60$? (1) 30 m. (2) 360 m. (3) 1,800 m. (4) 3,600 m. 62.........

**63.** How long after $t = 0$ did it take car $A$ to catch up to car $B$? (1) 10 sec. (2) 20 sec. (3) 30 sec. (4) 60 sec. 63.........

**64.** During the time intervals given below, which car traveled the greatest distance? (1) car $A$ from $t = 0$ to $t = 30$ (2) car $A$ from $t = 30$ to $t = 60$ (3) car $B$ from $t = 0$ to $t = 30$ (4) car $B$ from $t = 30$ to $t = 60$ 64.........

**65.** Which distance-time graph best represents the motion of car *B* during the time interval between $t = 0$ and $t = 60$?

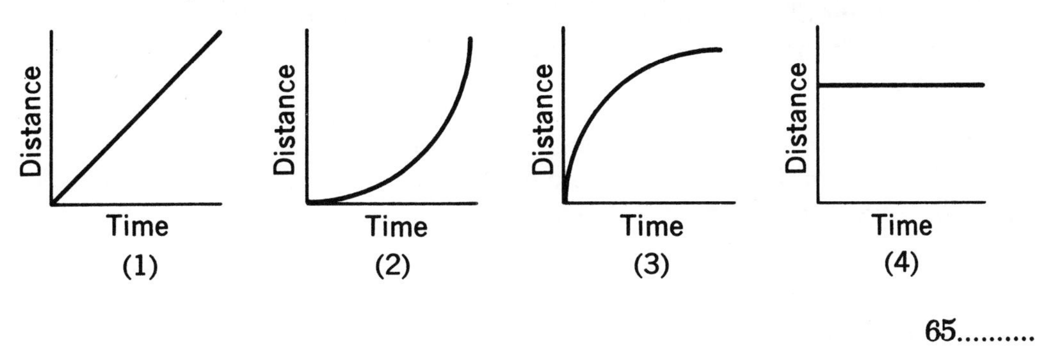


65..........

Base your answers to questions 66 through 69 on the diagram below which shows a cart held motionless by an external force *F* on a frictionless incline.

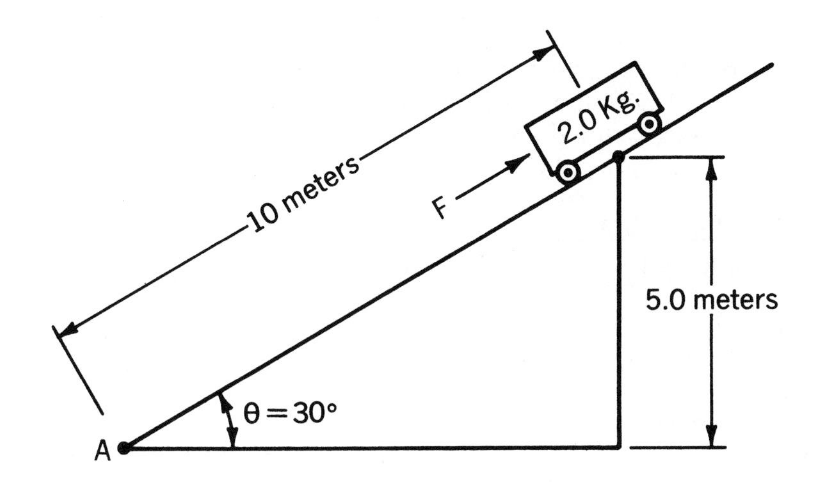


**66.** What is the magnitude of the external force *F* necessary to hold the cart motionless at point *C*? (1) 4.9 nt. (2) 2.0 nt. (3) 9.8 nt. (4) 19.6 nt.

66..........

**67.** If the gravitational potential energy of the cart at point *A* is zero, the gravitational potential energy of the cart at point *C* is (1) 4.9 joules (2) 10 joules (3) 49 joules (4) 98 joules

67..........

**Note that questions 68 and 69 have only three choices.**

**68.** If the cart were allowed to move from point *C* to point *A*, the gravitational potential energy of the cart would (1) decrease (2) increase (3) remain the same

68..........

**69.** Compared to the magnitude of the external force $F$ required to hold the cart at point $C$ on the frictionless surface, the magnitude of the external force $F$ required to hold the cart at point $C$, if friction were present, is (1) less (2) greater (3) the same 69..........

Base your answers to questions 70 through 73 on the diagram below that shows an object at $A$ that moves over a frictionless surface from $A$ to $E$. The object has a mass of $M$.

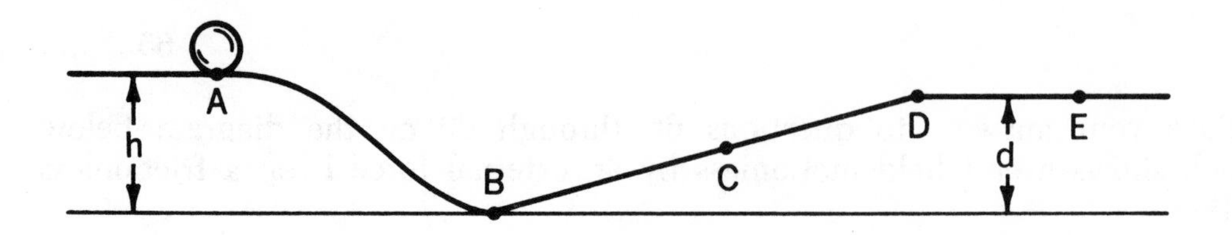


**70.** As the object moves from point $A$ to point $D$, the sum of its gravitational potential and kinetic energies (1) decreases, only (2) decreases and then increases (3) increases and then decreases (4) remains the same 70..........

**71.** The object will have a minimum gravitational potential energy at point (1) $A$ (2) $B$ (3) $C$ (4) $D$ 71..........

**72.** The object's kinetic energy at point $C$ is less than its kinetic energy at point (1) $A$ (2) $B$ (3) $D$ (4) $E$ 72..........

**73.** The object's kinetic energy at point $D$ is equal to (1) $Mgd$ (2) $Mg(d + h)$ (3) $Mgh$ (4) $Mg(h - d)$ 73..........

**74.** A mass of 10 kilograms is revolving at a linear speed of 5 meters per second in a circle with a radius of 10 meters. The centripetal force acting on the mass is (1) 5 nt. (2) 10 nt. (3) 20 nt. (4) 25 nt. 74..........

**75.** The pressure of a gas exerted on the walls of a balloon is produced by (1) the collisions of the gas molecules with the walls of the balloon (2) the repulsion between the gas molecules (3) collisions between the gas molecules (4) the expansion of the gas molecules 75..........

## GROUP 2 — Wave Phenomena

*Answer all fifteen questions, 76 through 90, in this group. Each question counts 1 credit. For each of the fifteen questions, write the* number *of the word or expression that best completes that statement or answers that question.*

Base your answers to questions 76 through 81 on the information and diagram below. The diagram represents two sound waves that are produced in air by two tuning forks. The frequency of wave *A* is 400 cycles per second.

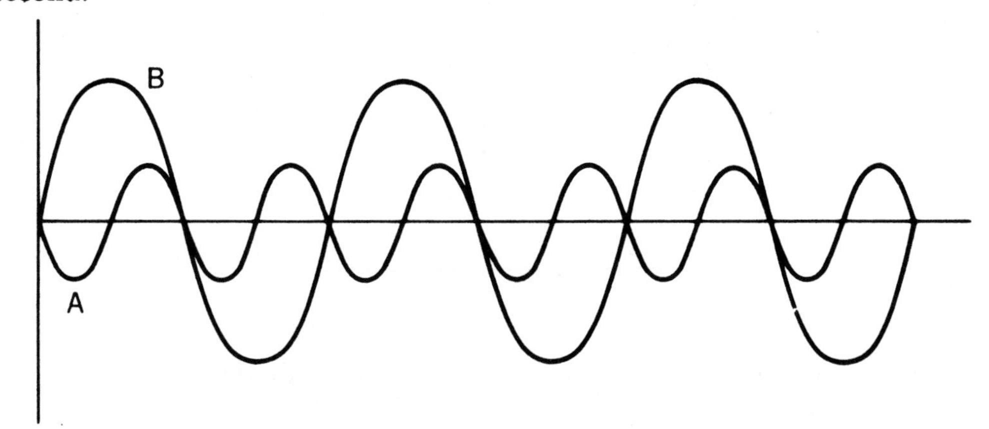


**76.** The period of wave *A* is (1) 400 sec. (2) 6 sec. (3) 1/4 sec. (4) 1/400 sec. 76..........

**77.** Under standard conditions of temperature and pressure, the wavelength in air of *A* is (1) 2.5 meters (2) 12 meters (3) 331/400 meters (4) 331 × 400 meters 77..........

**78.** The frequency of wave *B* is (1) 200 cycles/sec. (2) 400 cycles/sec. (3) 600 cycles/sec. (4) 800 cycles/sec. 78..........

**79.** Sound waves produced by tuning forks are (1) longitudinal (2) hyperbolic (3) torsional (4) elliptical 79..........

**Note that questions 80 and 81 have only three choices.**

**80.** Compared to the amplitude of Wave *B*, the amplitude of Wave *A* is (1) less (2) greater (3) the same 80..........

**81.** Compared to the speed of wave *A*, the speed of wave *B*.......... is (1) less (2) greater (3) the same 81..........

Base your answers to questions 82 through 85 on the diagram below which shows a ray of monochromatic yellow light incident on a surface of a rectangular prism at an angle of incidence of 45 degrees. The prism is made of glass and the angle of refraction is 30 degrees.

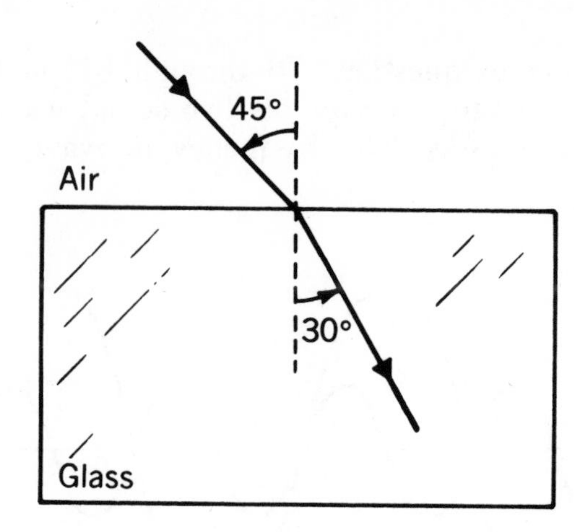


**82.** If the ray of yellow light continues through the glass back into air, the angle of refraction that the ray makes with the normal in air is (1) less than 45° (2) 45° (3) between 45° and 50° (4) greater than 50° 82..........

**83.** If the monochromatic yellow light is replaced by monochromatic red light, the angle of refraction in the glass will be (1) less than 30° (2) 30° (3) between 30° and 45° (4) greater than 45° 83..........

**84.** If the angle of incidence were increased to 50 degrees, the angle of refraction in the glass would be (1) less than 30° (2) 30° (3) between 30° and 40° (4) 90° 84..........

**85.** What is the index of refraction of air? (1) 1.0 (2) .21 (3) 1.1 (4) 1.3 85..........

**86.** The diffraction pattern produced by a double slit will show greatest separation of maxima when the color of the light source is (1) red (2) orange (3) blue (4) green 86..........

**87.** Total internal reflection occurs when substance $x$ is
(1) Canada balsam
(2) benzene
(3) glycerol
(4) water

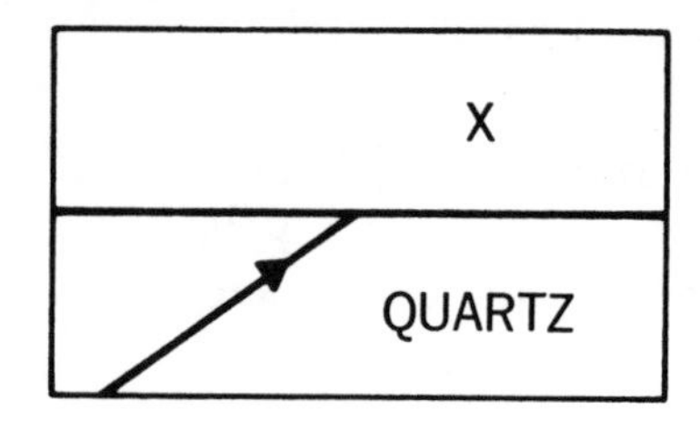


87..........

**88.** The apparent speed of light is *least* in (1) Canada balsam (2) lucite (3) diamond (4) alcohol 88..........

**89.** An object placed 0.07 meter from a converging lens produces a real image 0.42 meter from the lens. The focal length of the lens is (1) 0.03 meter (2) 0.06 meter (3) 0.07 meter (4) 2.4 meters 89..........

**Note that question 90 has only three choices.**

**90.** If the electromagnetic waves received on earth from a source in outer space appear to be increasing in frequency, the distance between the source and the earth is probably (1) decreasing (2) increasing (3) remaining the same 90..........

## GROUP 3 — Electricity

*Answer all fifteen questions, 91 through 105, in this group. Each question counts 1 credit. For each of the fifteen questions, write the* number *of the word or expression that best completes that statement or answers that question.*

Base your answers to questions 91 through 95 on the diagram below which shows a positive point charge placed at *A*.

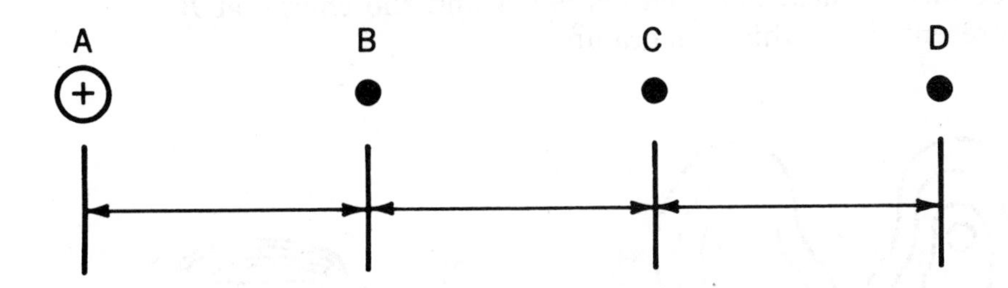


**91.** The electric field intensity at point *B* is *E*. At point *D* the field intensity will be equal to (1) $\frac{1}{9}E$ (2) $\frac{1}{3}E$ (3) $3E$ (4) $9E$ 91..........

**92** If a positive charge is placed at point *B*, the force exerted on this charge by charge *A* will be directed toward (1) the top of the page (2) the bottom of the page (3) *A* (4) *C* 92..........

**Note that question 93 has only three choices.**

**93.** If the charge is moved from point *B* to point *C*, the force between the two charges will (1) decrease (2) increase (3) remain the same 93..........

**94.** The electric field surrounding charge *A* is best represented by which diagram?

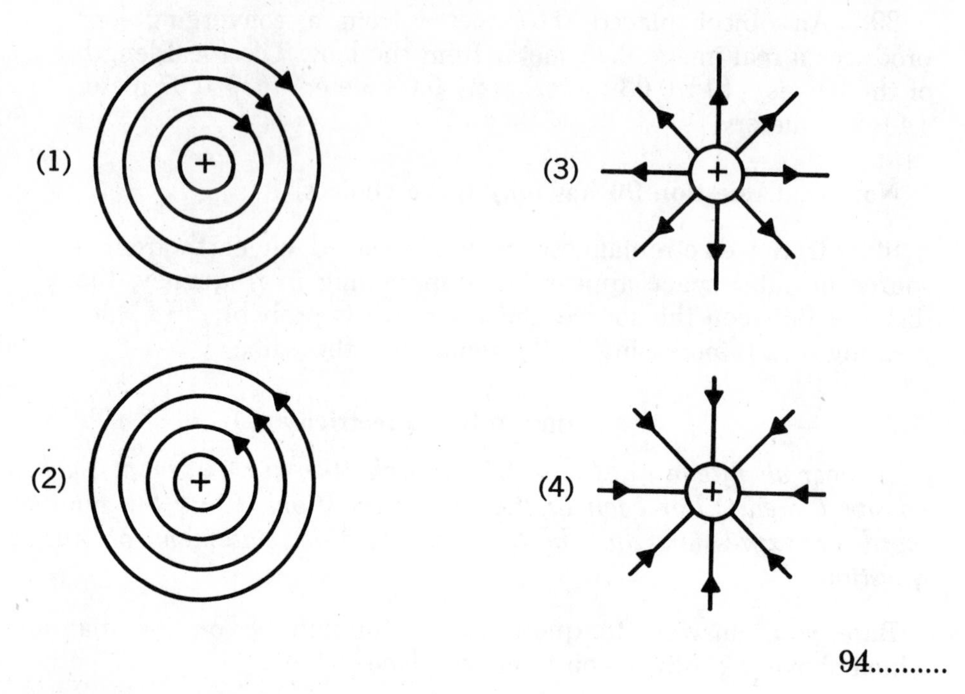


94..........

**95.** The electric field between charge *A* and the charge at *B* is best represented by which diagram?

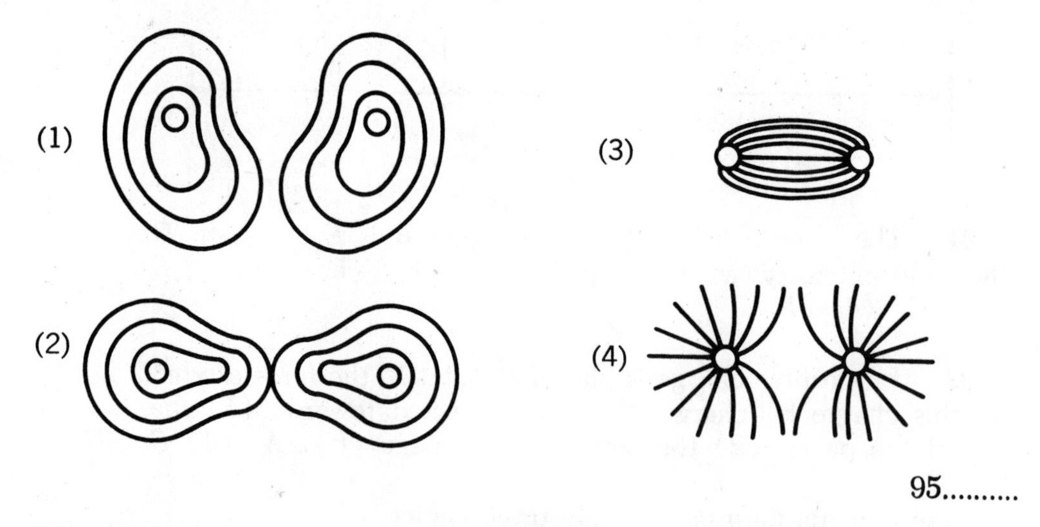


95..........

**96.** If the intensity of the electric field between two parallel charged plates is 5.0 newtons per coulomb, then the force exerted on a 2.0-coulomb charge placed between the plates is (1) 0.40 nt. (2) 2.5 nt. (3) 7.0 nt. (4) 10. nt.

96..........

Base your answers to questions 97 through 100 on the diagram below.

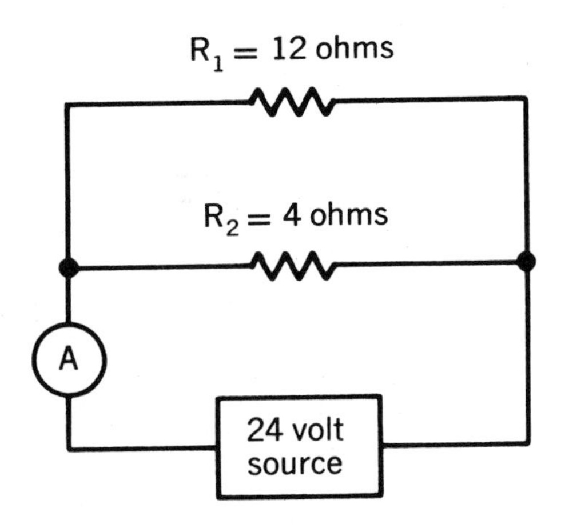


**97.** The current reading of ammeter *A* is (1) 6.0 amp. (2) 2.0 amp. (3) 8.0 amp. (4) 12 amp. 97..........

**98.** The potential difference across $R_2$ is (1) 24 volts (2) 18 volts (3) 6.0 volts (4) 4.0 volts 98..........

**99.** The combined resistance of resistors $R_1$ and $R_2$ is (1) 8.0 ohms (2) 2.0 ohms (3) 3.0 ohms (4) 16 ohms 99..........

**Note that question 100 has only three choices.**

**100.** If a third resistance is added in parallel to the 24-volt source, the combined resistance of the circuit will (1) decrease (2) increase (3) remain the same 100..........

Base your answers to questions 101 through 104 on the diagram below which shows a conductor that is moving toward point *A* through a uniform magnetic field with a speed of 2 meters per second. The direction of the magnetic field is downward and the conductor is moving perpendicularly to the direction of the magnetic field.

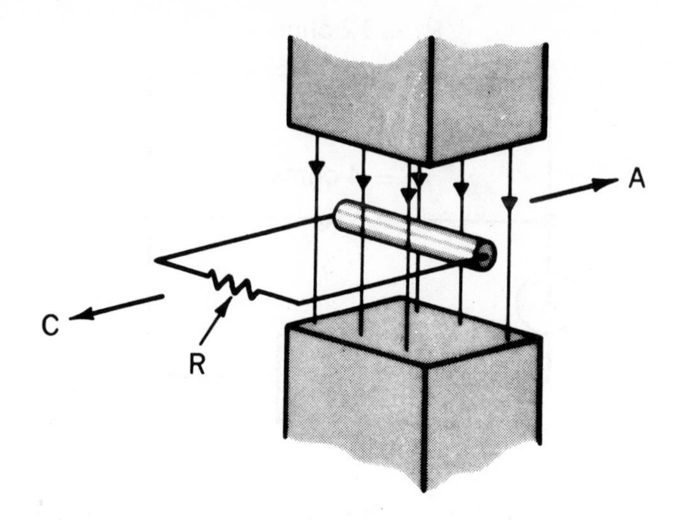


**Note that questions 101 through 103 have only three choices.**

**101.** If the speed of the conductor is increased, the induced electromotive force (1) decreases (2) increases (3) remains the same 101..........

**102.** If resistance $R$ is decreased, the force needed to move the conductor at the same speed (1) decreases (2) increases (3) remains the same 102..........

**103.** If the motion of the conductor is changed to 2 meters per second upwards, the magnitude of the induced electromotive force (1) decreases (2) increases (3) remains the same 103..........

**104.** If the motion of the conductor is changed to 2 meters per second toward point $C$, the induced electromotive force (1) will remain the same and its direction will be reversed (2) will decrease and its direction will remain the same (3) will increase and its direction will be reversed (4) will increase and its direction will remain the same 104..........

**Note that question 105 has only three choices.**

**105.** As the permeability of a substance in a magnetic field decreases, the flux density within the substance (1) decreases (2) increases (3) remains the same 105..........

### GROUP 4 — Atomic and Nuclear Physics

*Answer all fifteen questions, 106 through 120, in this group. Each question counts 1 credit. For each of the fifteen questions, write the* number *of the word or expression that best completes that statement or answers that question.*

Base your answers to questions 106 through 110 on Rutherford's experiments in which alpha particles were allowed to pass into a thin gold foil. All alpha particles had the same speed.

**106.** The paths of the scattered alpha particles were (1) hyperbolic (2) circular (3) parabolic (4) elliptical 106..........

**107.** Some of the alpha particles were deflected. The explanation for this phenomenon is that (1) electrons have a small mass (2) electrons have a small charge (3) the gold leaf was only a few atoms thick (4) the nuclear charge and mass are concentrated in a small volume 107..........

**108.** The alpha particles were scattered because of (1) gravitational forces (2) coulomb forces (3) magnetic forces (4) nuclear forces 108..........

**Note that questions 109 and 110 have only three choices.**

**109.** As the distance between the nuclei of the gold atoms and the paths of the alpha particles increases, the angle of scattering of the alpha particles (1) decreases (2) increases (3) remains the same 109..........

**110** If a foil were used whose nuclei had a greater atomic number, the angle of scattering of the alpha particles would (1) decrease (2) increase (3) remain the same 110..........

Base your answers to questions 111 through 116 on the information below:

When tellurium is bombarded with protons, the following reaction occurs:

$$^{1}_{1}H + ^{130}_{52}Te \rightarrow ^{130}_{53}I + y$$

The iodine produced is radioactive, has a half-life of 12.6 hours, and decays according to the reaction.

$$^{130}_{53}I \rightarrow ^{130}_{54}Xe + ^{0}_{-1}e$$

**111.** The number of neutrons in nucleus of $^{130}_{52}Te$ is (1) 52 (2) 78 (3) 130 (4) 182 111..........

**112.** Particle *y* is (1) a proton (2) a neutron (3) a positron (4) an electron 112..........

**113.** The first equation represents an example of (1) induced transmutation (2) natural radioactivity (3) nuclear fission (4) nuclear fusion 113..........

**114.** A $2.4 \times 10^{-5}$ kilogram sample of $^{130}_{53}I$ is left to decay. After 37.8 hours, the amount of iodine present will be (1) $8.0 \times 10^{-6}$ kg. (2) $2.0 \times 10^{-5}$ kg. (3) $3.0 \times 10^{-6}$ kg. (4) $4.0 \times 10^{-6}$ kg. 114..........

**115.** When it decays, $^{130}_{53}I$ emits (1) a proton (2) an alpha particle (3) a beta particle (4) a neutron 115..........

**116.** One atomic mass unit is defined as 1/12 of the mass of an isotope of the element (1) hydrogen (2) oxygen (3) uranium (4) carbon 116..........

Base your answers to questions 117 through 120 on the graph below that represents the maximum kinetic energy of the photoelectrons emitted by a metal surface upon exposure to a beam of light of varying frequency.

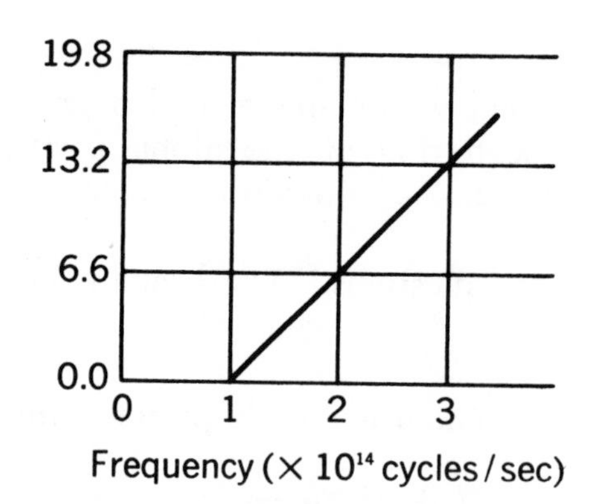


**117.** The work function for metal *A* is (1) 0.0 joule (2) $6.6 \times 10^{-20}$ joule (3) $6.6 \times 10^{-34}$ joule (4) $6.6 \times 10^{-48}$ joule 117..........

**118.** The slope of the line for metal *A* is (1) 6.6 joule-sec (2) $6.6 \times 10^{-6}$ joule-sec. (3) $6.6 \times 10^{-20}$ joule-sec. (4) $6.6 \times 10^{-34}$ joule-sec. 118..........

**Note that question 119 has only three choices.**

**119.** As the frequency of the light falling on metal *A* increases above the threshold frequency, the maximum kinetic energy of the photoelectrons (1) decreases (2) increases (3) remains the same

119..........

**120.** Metal *B* with a work function of $2.0 \times 10^{-19}$ joule is substituted for metal *A*. If photons with an energy of $6.0 \times 10^{-19}$ joule are incident on the surface of metal *B*, the kinetic energy of the emitted photoelectrons will be (1) $12.0 \times 10^{-19}$ joule (2) $2.0 \times 10^{-19}$ joule (3) $3.0 \times 10^{-19}$ joule (4) $4.0 \times 10^{-19}$ joule

120..........

# Summary of Equations Used in New York State Physics Course

## Unit 1—MECHANICS

$\bar{v} = \frac{s}{t}$ $\quad \bar{v} = \frac{v_f + v_i}{2}$

$a = \frac{\Delta v}{\Delta t}$

$s = \frac{1}{2} at^2$

$v^2 = 2as$

$F = ma$

$w = mg$

$F_c = \frac{mv^2}{r}$ $\quad a_c = \frac{v^2}{r}$

$F\Delta t = \Delta mv$ $\quad m_1v_1 = m_2v_2$

$W = Fs$

$E_p = mgh$

$E_k = \frac{1}{2} mv^2$

$P = \frac{W}{t} = \frac{Fs}{t} = Fv$

$\Delta E_p = -\Delta E_k$

$T_k = t_c + 273$

$\Delta Q = mc\Delta t$

$Q_e = Q_g$

$Q_f = mH_f$

$Q_v = mH_v$

$F = \frac{Gm_1m_2}{d^2}$

P = power

$T_k$ = Kelvin temperature

$t_c$ = Celsius temperature

Q = heat

$Q_e$ = heat lost

$Q_g$ = heat gained

c = specific heat

$H_f$ = heat of fusion

$H_v$ = heat of vaporization

$v_i$ = initial speed

$v_f$ = final speed

$\bar{v}$ = average speed

a = acceleration

s = distance

F = force

m = mass

g = acceleration of gravity

w = weight

$F_c$ = centripetal force

r = radius of curvature

t = time

W = work

$E_p$ = potential energy

$E_k$ = kinetic energy

G = Universal gravitational constant

d = distance

$a_c$ = centripetal acceleration

## Unit 2—WAVE PHENOMENA

$T = \frac{1}{f}$

$v = f\lambda$

$\frac{\sin \theta_1}{\sin \theta_2} = \frac{n_2}{n_1} = \frac{v_1}{v_2} = K$

$v = \frac{c}{n}$

$\sin \theta_c = \frac{1}{n}$

$\frac{1}{d_o} + \frac{1}{d_i} = \frac{1}{f}$

$\frac{s_o}{s_i} = \frac{d_o}{d_i}$

$\lambda = \frac{dx}{L}$ (first order)

T = period

f = frequency

v = speed

$\lambda$ = wavelength

$\theta_1$ = angle of incidence

$\theta_2$ = angle of refraction

n = index of refraction

$\theta_c$ = critical angle

$d_o$ = object to lens distance

$d_i$ = image to lens distance

$s_o$ = size of object

$s_i$ = size of image

f = focal length

d = distance between slits

x = distance from central maximum to line

L = distance from slit to screen

K = relative index of refraction

## Unit 3—ELECTRICITY

$F = \frac{kq_1q_2}{d^2}$

$E = \frac{F}{q}$

$V = \frac{W}{q}$

$E = \frac{V}{d}$

$V = IR$

$P = VI = I^2R = \frac{V^2}{R}$

$W = Pt$

$JQ = VIt = I^2Rt = Pt$

$F = qvB$

$V = Blv$

Series Circuits

$I_t = I_1 = I_2 = I_3 = \text{------}$

$V_t = V_1 + V_2 + V_3 + \text{------}$

$R_t = R_1 + R_2 + R_3 + \text{------}$

Parallel Circuits

$I_t = I_1 + I_2 + I_3 + \text{------}$

$V_t = V_1 = V_2 = V_3 = \text{------}$

$\frac{1}{R_t} = \frac{1}{R_1} + \frac{1}{R_2} + \frac{1}{R_3} + \text{------}$

F = force

q = charge

d = distance

E = electric field intensity

V = electric potential

I = current

R = resistance

P = power

t = time

W = energy

Q = heat

B = flux density

l = length of conductor

v = velocity

k = electrostatic constant

J = mechanical equivalent of heat

## Unit 4—ATOMIC AND NUCLEAR ENERGY

$E = hf$

$E_k = hf - W_o$

$E_{photon} = E_i - E_f = hf$

$m_f = \frac{1}{2^n} m_i$

$E = mc^2$

E = energy

h = Planck's constant

f = frequency

$W_o$ = work function

$E_k$ = kinetic energy of photoelectron

$E_i$ = initial energy level

$E_f$ = final energy level

$m_f$ = final mass

$m_i$ = initial mass

n = number of half lives

m = mass

c = speed of light

# APPENDIX I

## Physics Reference Tables

### List of Physical Constants

| | |
|---|---|
| Gravitational Constant (G) | $6.67 \times 10^{-11}$ newton-meters²/kg² |
| Acceleration of gravity (g) (near earth's surface) | 9.81 meters/second² |
| Speed of light (c) | $3.00 \times 10^{8}$ meters/second |
| Speed of sound at STP | $3.31 \times 10^{2}$ meters/second |
| Mechanical equivalent of heat | $J = 4.19 \times 10^{3}$ joules/kilocalorie<br>$\frac{1}{J} = 2.39 \times 10^{-4}$ kilocalories/joule |
| Mass energy relationship | 1 amu $= 9.31 \times 10^{2}$ Mev |
| Electrostatic constant | $k = 9.00 \times 10^{9}$ newton-meters²/coul² |
| Charge of the electron = 1 elementary charge | $1.60 \times 10^{-19}$ coulomb |
| One coulomb | $6.25 \times 10^{18}$ electrons<br>$6.25 \times 10^{18}$ elementary charges |
| Electron volt (ev) | $1.60 \times 10^{-19}$ joule |
| Planck's Constant (h) | $6.63 \times 10^{-34}$ joule-second |
| Rest mass of the electron ($m_e$) | $9.11 \times 10^{-31}$ kilogram |
| Rest mass of the proton ($m_p$) | $1.67 \times 10^{-27}$ kilogram |
| Rest mass of the neutron ($m_n$) | $1.67 \times 10^{-27}$ kilogram |

### Trigonometric Functions*

| | |
|---|---|
| sine 0° = | .000 |
| sine 30° = | .500 |
| sine 45° = | .707 |
| sine 60° = | .866 |
| sine 90° = | 1.000 |

### Wavelengths of Light in a Vacuum

| | |
|---|---|
| VIOLET | $< 4.5 \times 10^{-7}$ meters |
| BLUE | $4.5 - 5.0 \times 10^{-7}$ meters |
| GREEN | $5.0 - 5.7 \times 10^{-7}$ meters |
| YELLOW | $5.7 - 5.9 \times 10^{-7}$ meters |
| ORANGE | $5.9 - 6.1 \times 10^{-7}$ meters |
| RED | $> 6.1 \times 10^{-7}$ meters |

*A more complete Table of Trigonometric Functions appears on a later page.

## Heat Constants

| | Specific Heat (average) | Melting Point | Boiling Point | Heat of Fusion | Heat of Vaporization |
|---|---|---|---|---|---|
| | — | °C | °C | $\frac{kcal}{kg}$ | $\frac{kcal}{kg}$ |
| Alcohol, ethyl | 0.58 (liq) | −115 | 78 | 25 | 204 |
| Aluminum | 0.21 (sol) | 660 | 2057 | 77 | 2520 |
| Ammonia | 1.13 (liq) | −78 | −33 | 84 | 327 |
| Copper | 0.09 (sol) | 1083 | 2336 | 49 | 1150 |
| Ice | 0.50 (sol) | 0 | — | 80 | — |
| Iron | 0.11 (sol) | 1535 | 3000 | 7.9 | 1600 |
| Lead | 0.03 (sol) | 327 | 1620 | 5.9 | 207 |
| Mercury | 0.03 (liq) | −39 | 357 | 2.8 | 71 |
| Platinum | 0.03 (sol) | 1774 | 4300 | 27 | — |
| Silver | 0.06 (sol) | 961 | 1950 | 26 | 565 |
| Steam | 0.48 (gas) | — | — | — | — |
| Water | 1.00 (liq) | — | 100 | — | 540 |
| Tungsten | 0.04 (sol) | 3370 | 5900 | 43 | — |
| Zinc | 0.09 (sol) | 419 | 907 | 23 | 420 |

## Absolute Indices of Refraction

($\lambda = 5.9 \times 10^{-7}$ m.)

| | | | | | |
|---|---|---|---|---|---|
| Air | 1.00 | Carbon Tetrachloride | 1.46 | Glycerol | 1.47 |
| Alcohol | 1.36 | Diamond | 2.42 | Lucite | 1.50 |
| Benzene | 1.50 | Glass, Crown | 1.52 | Quartz, Fused | 1.46 |
| Canada Balsam | 1.53 | Glass, Flint | 1.61 | Water | 1.33 |

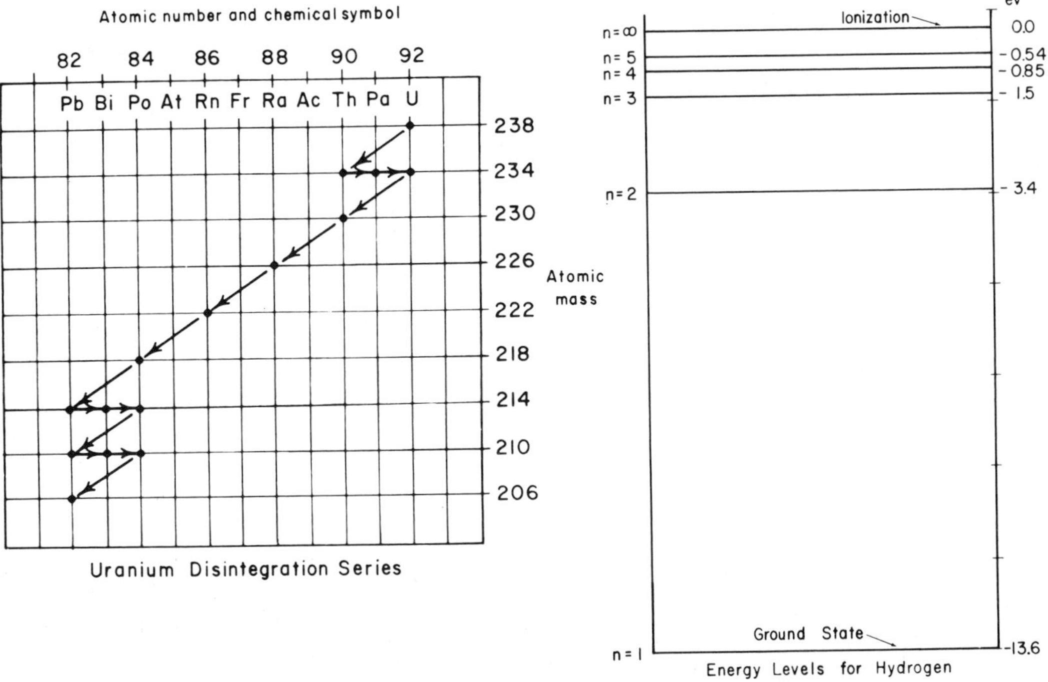

# APPENDIX II

## Density of Some Substances in Grams/$cm^3$

### Solids

| Substance | Density | Substance | Density |
|---|---|---|---|
| Aluminum | 2.7 | iron | 7.8 |
| Brass | 8.5 | lead | 11.3 |
| Copper | 8.9 | marble | 2.6 |
| cork | 0.24 | paraffin | 0.9 |
| glass | 2.6 | platinum | 21.5 |
| gold | 19.3 | silver | 10.5 |
| ice | 0.92 | steel | 7.7 |
| | | sulfur | 2.0 |

### Liquids

| Substance | Density | Substance | Density |
|---|---|---|---|
| alcohol | 0.79 | kerosene | 0.8 |
| carbon tetrachloride | 1.6 | mercury | 13.6 |
| gasoline | 0.7 | sulfuric acid | 1.84 |
| glycerine | 1.3 | water | 1.0 |

## International Atomic Masses

1. These are also known as atomic weights.
2. These are based on assigning the carbon-12 isotope an atomic mass of 12.
3. Most of the listed atomic masses are for naturally occurring mixtures of the isotopes.

# INTERNATIONAL ATOMIC MASSES

## Based on the Atomic Mass of C-12=12

| Name | Symbol | Atomic Number | Atomic Mass | Name | Symbol | Atomic Number | Atomic Mass |
|---|---|---|---|---|---|---|---|
| Actinium | Ac | 89 | [227] | Hydrogen | H | 1 | 1.01 |
| Aluminium | Al | 13 | 26.98 | Indium | In | 49 | 114.82 |
| Americium | Am | 95 | [243] | Iodine | I | 53 | 126.90 |
| Antimony | Sb | 51 | 121.75 | Iridium | Ir | 77 | 192.2 |
| Argon | Ar | 18 | 39.95 | Iron | Fe | 26 | 55.85 |
| Arsenic | As | 33 | 74.92 | Krypton | Kr | 36 | 83.80 |
| Astatine | At | 85 | [210] | Lanthanum | La | 57 | 138.91 |
| Barium | Ba | 56 | 137.34 | Lawrencium | Lw | 103 | [257] |
| Berkelium | Bk | 97 | [249] | Lead | Pb | 82 | 207.19 |
| Beryllium | Be | 4 | 9.01 | Lithium | Li | 3 | 6.94 |
| Bismuth | Bi | 83 | 208.98 | Lutetium | Lu | 71 | 174.97 |
| Boron | B | 5 | 10.81 | Magnesium | Mg | 12 | 24.31 |
| Bromine | Br | 35 | 79.90 | Manganese | Mn | 25 | 54.94 |
| Cadmium | Cd | 48 | 112.40 | Mendelevium | Md | 101 | [256] |
| Caesium | Cs | 55 | 132.91 | Mercury | Hg | 80 | 200.59 |
| Calcium | Ca | 20 | 40.08 | Molybdenum | Mo | 42 | 95.94 |
| Californium | Cf | 98 | [249] | Neodymium | Nd | 60 | 144.24 |
| Carbon | C | 6 | 12.01 | Neon | Ne | 10 | 20.18 |
| Cerium | Ce | 58 | 140.12 | Neptunium | Np | 93 | [237] |
| Chlorine | Cl | 17 | 35.45 | Nickel | Ni | 28 | 58.71 |
| Chromium | Cr | 24 | 52.00 | Niobium | Nb | 41 | 92.90 |
| Cobalt | Co | 27 | 58.93 | Nitrogen | N | 7 | 14.01 |
| Copper | Cu | 29 | 63.55 | Nobelium | No | 102 | [253] |
| Curium | Cm | 96 | [245] | Osmium | Os | 76 | 190.2 |
| Dysprosium | Dy | 66 | 162.50 | Oxygen | O | 8 | 16.00 |
| Einsteinium | Es | 99 | [245] | Palladium | Pd | 46 | 106.4 |
| Erbium | Er | 68 | 167.26 | Phosphorus | P | 15 | 30.97 |
| Europium | Eu | 63 | 151.96 | Platinum | Pt | 78 | 195.09 |
| Fermium | Fm | 100 | [253] | Plutonium | Pu | 94 | [242] |
| Fluorine | F | 9 | 19.00 | Polonium | Po | 84 | [210] |
| Francium | Fr | 87 | [223] | Potassium | K | 19 | 39.10 |
| Gadolinium | Gd | 64 | 157.25 | Praseodymium | Pr | 59 | 140.90 |
| Gallium | Ga | 31 | 69.72 | Promethium | Pm | 61 | [145] |
| Germanium | Ge | 32 | 72.59 | Protactinium | Pa | 91 | [231] |
| Gold | Au | 79 | 196.97 | Radium | Ra | 88 | [226] |
| Hafnium | Hf | 72 | 178.49 | Radon | Rn | 86 | [222] |
| Hahnium | Ha | 105 | ...... | Rhenium | Re | 75 | 186.2 |
| Helium | He | 2 | 4.00 | Rhodium | Rh | 45 | 102.91 |
| Holmium | Ho | 67 | 164.93 | Rubidium | Rb | 37 | 85.47 |

*A value given in brackets is the mass number of the most stable known isotope.

| Name | Symbol | Atomic Number | Atomic Mass | Name | Symbol | Atomic Number | Atomic Mass |
|---|---|---|---|---|---|---|---|
| Ruthenium | Ru | 44 | 101.07 | Thallium | Tl | 81 | 204.37 |
| Rutherfordium | Rf** | 104 | ......... | Thorium | Th | 90 | 232.03 |
| Samarium | Sm | 62 | 150.35 | Thulium | Tm | 69 | 168.93 |
| Scandium | Sc | 21 | 44.95 | Tin | Sn | 50 | 118.69 |
| Selenium | Se | 34 | 78.96 | Titanium | Ti | 22 | 47.90 |
| Silicon | Si | 14 | 28.09 | Tungsten | W | 74 | 183.85 |
| Silver | Ag | 47 | 107.87 | Uranium | U | 92 | 238.03 |
| Sodium | Na | 11 | 22.99 | Vanadium | V | 23 | 50.94 |
| Strontium | Sr | 38 | 87.62 | Xenon | Xe | 54 | 131.30 |
| Sulfur | S | 16 | 32.06 | Ytterbium | Yb | 70 | 173.04 |
| Tantalum | Ta | 73 | 180.95 | Yttrium | Y | 39 | 88.91 |
| Technetium | Tc | 43 | [99] | Zinc | Zn | 30 | 65.37 |
| Tellurium | Te | 52 | 127.60 | Zirconium | Zr | 40 | 91.22 |
| Terbium | Tb | 65 | 158.92 | | | | |

**Another suggested name is Khurchatovium.

1. These are also known as atomic weights.
2. These are based on assigning the carbon-12 isotope an atomic mass of 12.
3. Most of the listed atomic masses are for naturally occurring mixtures of the isotopes.

## Resistivity (at 20°C), ohm-m

| | | | |
|---|---|---|---|
| aluminum | $2.8 \times 10^{-8}$ | nichrome | $1.1 \times 10^{-6}$ |
| copper | $1.7 \times 10^{-8}$ | nickel | $6.8 \times 10^{-8}$ |
| iron | $1.0 \times 10^{-7}$ | silver | $1.6 \times 10^{-8}$ |
| manganin | $4.4 \times 10^{-7}$ | steel | $1.8 \times 10^{-7}$ |
| | | tungsten | $5.6 \times 10^{-8}$ |

# APPENDIX III

## Some Famous Physicists

Ampere, André (1775–1836) – related magnetism and electricity
Balmer, Johann (1825–1898) – found a mathematical relationship for visible lines in the hydrogen spectrum
Becquerel, Henri (1852–1908) – discovered radioactivity
Bohr, Niels (1885–1963) – developed the planetary model of the atom
Boyle, Robert (1627–1691) – discovered a gas law
Celsius, Anders (1701–1744) – devised a temperature scale
Charles, Jacques (1746–1823) – discovered a gas law
Compton, Arthur (1892–1962) – discovered some characteristics of X-rays
Copernicus, Nicholas (1473–1543) – proposed heliocentric solar system
Coulomb, Charles (1736–1806) – discovered laws of electric and magnetic force
Curie, Pierre (1850–1906), and Marie (1867–1934) – isolated radium
De Broglie, Louis (1892– ) – proposed wave nature of matter
Doppler, Christian (1803–1853) – effect of motion on apparent frequency of a wave
Einstein, Albert (1879–1955) – originated modern theory of relativity
Fahrenheit, Gabriel (1686–1736) – devised a temperature scale
Faraday, Michael (1791–1867) – discovered principles of electromagnetic induction
Feynman, Richard (1918– ) – developed modern quantum theory
Franklin, Benjamin (1706–1790) – demonstrated electrical nature of lightning
Fraunhofer, Joseph (1787–1826) – analyzed solar spectrum
Galilei, Galileo (1564–1642) – father of *modern* physics
Henry, Joseph (1791–1878) – independently discovered electromagnetic induction
Hertz, Heinrich (1857–1894) – discovered Maxwell's electromagnetic waves
Huygens, Christian (1629–1695) – advocated wave theory of light
Joule, James (1818–1889) – advocated principle of conservation of energy
Kelvin, Lord, Wm. Thomson (1824–1907) – proposed an absolute temperature scale
Kepler, Johannes (1571–1630) – discovered laws of planetary motion
Lenz, Heinrich (1804–1865) – discovered law for direction of induced current
Michelson, Albert (1852–1931) – measured the speed of light
Millikan, Robert (1868–1953) – measured the charge of an electron
Newton, Isaac (1642–1727) – stated the theory of universal gravitation
Oersted, Hans (1777–1851) – discovered magnetic field around electric current
Ohm, Georg (1787–1854) – defined resistance in a circuit
Planck, Max (1858–1947) – proposed the quantum theory
Roemer, Olaus (1644–1710) – measured the speed of light
Roentgen, Wilhelm (1845–1923) – discovered X-rays
Rutherford, Ernest (1871–1937) – proposed the nuclear atom
Schwinger, Julian (1918– ) – developed modern quantum theory
Snell, Willebrord (1591–1626) – discovered law of refraction
Thomson, Joseph J. (1856–1940) – measured the ratio of the charge of an electron to its mass
Townes, Charles (1915– ) – invented maser and theory of coherent atomic radiation
Volta, Alessandro (1745–1827) – invented electric cell
Watt, James (1736–1819) – invented steam engine
Young, Thomas (1773–1829) – demonstrated interference of light

# Values of the Trigonometric Functions

| Angle | Sin | Cos | Tan | Angle | Sin | Cos | Tan |
|---|---|---|---|---|---|---|---|
| 1° | .0175 | .9998 | .0175 | 46° | .7193 | .6947 | 1.0355 |
| 2° | .0349 | .9994 | .0349 | 47° | .7314 | .6820 | 1.0724 |
| 3° | .0523 | .9986 | .0524 | 48° | .7431 | .6691 | 1.1106 |
| 4° | .0698 | .9976 | .0699 | 49° | .7547 | .6561 | 1.1504 |
| 5° | .0872 | .9962 | .0875 | 50° | .7660 | .6428 | 1.1918 |
| 6° | .1045 | .9945 | .1051 | 51° | .7771 | .6293 | 1.2349 |
| 7° | .1219 | .9925 | .1228 | 52° | .7880 | .6157 | 1.2799 |
| 8° | .1392 | .9903 | .1405 | 53° | .7986 | .6018 | 1.3270 |
| 9° | .1564 | .9877 | .1584 | 54° | .8090 | .5878 | 1.3764 |
| 10° | .1736 | .9848 | .1763 | 55° | .8192 | .5736 | 1.4281 |
| 11° | .1908 | .9816 | .1944 | 56° | .8290 | .5592 | 1.4826 |
| 12° | .2079 | .9781 | .2126 | 57° | .8387 | .5446 | 1.5399 |
| 13° | .2250 | .9744 | .2309 | 58° | .8480 | .5299 | 1.6003 |
| 14° | .2419 | .9703 | .2493 | 59° | .8572 | .5150 | 1.6643 |
| 15° | .2588 | .9659 | .2679 | 60° | .8660 | .5000 | 1.7321 |
| 16° | .2756 | .9613 | .2867 | 61° | .8746 | .4848 | 1.8040 |
| 17° | .2924 | .9563 | .3057 | 62° | .8829 | .4695 | 1.8807 |
| 18° | .3090 | .9511 | .3249 | 63° | .8910 | .4540 | 1.9626 |
| 19° | .3256 | .9455 | .3443 | 64° | .8988 | .4384 | 2.0503 |
| 20° | .3420 | .9397 | .3640 | 65° | .9063 | .4226 | 2.1445 |
| 21° | .3584 | .9336 | .3839 | 66° | .9135 | .4067 | 2.2460 |
| 22° | .3746 | .9272 | .4040 | 67° | .9205 | .3907 | 2.3559 |
| 23° | .3907 | .9205 | .4245 | 68° | .9272 | .3746 | 2.4751 |
| 24° | .4067 | .9135 | .4452 | 69° | .9336 | .3584 | 2.6051 |
| 25° | .4226 | .9063 | .4663 | 70° | .9397 | .3420 | 2.7475 |
| 26° | .4384 | .8988 | .4877 | 71° | .9455 | .3256 | 2.9042 |
| 27° | .4540 | .8910 | .5095 | 72° | .9511 | .3090 | 3.0777 |
| 28° | .4695 | .8829 | .5317 | 73° | .9563 | .2924 | 3.2709 |
| 29° | .4848 | .8746 | .5543 | 74° | .9613 | .2756 | 3.4874 |
| 30° | .5000 | .8660 | .5774 | 75° | .9659 | .2588 | 3.7321 |
| 31° | .5150 | .8572 | .6009 | 76° | .9703 | .2419 | 4.0108 |
| 32° | .5299 | .8480 | .6249 | 77° | .9744 | .2250 | 4.3315 |
| 33° | .5446 | .8387 | .6494 | 78° | .9781 | .2079 | 4.7046 |
| 34° | .5592 | .8290 | .6745 | 79° | .9816 | .1908 | 5.1446 |
| 35° | .5736 | .8192 | .7002 | 80° | .9848 | .1736 | 5.6713 |
| 36° | .5878 | .8090 | .7265 | 81° | .9877 | .1564 | 6.3138 |
| 37° | .6018 | .7986 | .7536 | 82° | .9903 | .1392 | 7.1154 |
| 38° | .6157 | .7880 | .7813 | 83° | .9925 | .1219 | 8.1443 |
| 39° | .6293 | .7771 | .8098 | 84° | .9945 | .1045 | 9.5144 |
| 40° | .6428 | .7660 | .8391 | 85° | .9962 | .0872 | 11.4301 |
| 41° | .6561 | .7547 | .8693 | 86° | .9976 | .0698 | 14.3007 |
| 42° | .6691 | .7431 | .9004 | 87° | .9986 | .0523 | 19.0811 |
| 43° | .6820 | .7314 | .9325 | 88° | .9994 | .0349 | 28.6363 |
| 44° | .6947 | .7193 | .9657 | 89° | .9998 | .0175 | 57.2900 |
| 45° | .7071 | .7071 | 1.0000 | 90° | 1.0000 | .0000 | |

# Index

# GO METRIC!

**METRIC CONVERSION TABLES** includes the most frequently used conversions in an easy-to-use format. Comprehensive with separate tables for length and speed, area, volume, capacity, weight, temperature, and price. $3.95

**METRICS MADE EASY** is a fully illustrated concise guide that contains many useful conversion values and charts. With an explanation of metrics in a language that everyone can understand. $.75

BARRON'S, 113 Crossways Park Drive, Woodbury, New York 11797

Please ship the following quantities as indicated:

_____ METRIC CONVERSION TABLES, $3.95

_____ METRICS MADE EASY, $.75

Please enclose check or money order. Add 8% transportation charges. Add sales tax where required. Prices subject to change without notice.

NAME _______________

ADDRESS _______________

CITY _______________ STATE _____ ZIP _____